Chemical React
Equilibrium An

Chemical Reaction Equilibrium Analysis:
Theory and Algorithms

WILLIAM R. SMITH
Professor of Mathematics and Statistics
University of Guelph

RONALD W. MISSEN
Professor of Chemical Engineering
University of Toronto

A Wiley-Interscience Publication

JOHN WILEY & SONS
New York / Chichester / Brisbane / Toronto / Singapore

Copyright © 1982 by John Wiley & Sons, Inc.

All rights reserved. Published simultaneously in Canada.

Reproduction or translation of any part of this work beyond that permitted by Section 107 or 108 of the 1976 United States Copyright Act without the permission of the copyright owner is unlawful. Requests for permission or further information should be addressed to the Permissions Department, John Wiley & Sons, Inc.

Library of Congress Cataloging in Publication Data:

Smith, William R. (William Robert), 1943–
  Chemical reaction equilibrium analysis.

  "A Wiley-Interscience publication."
  Bibliography: p.
  Includes indexes.
    1. Chemical equilibrium.   2. Chemical equilibrium—Computer programs.   I. Missen, Ronald W. (Ronald William), 1928–   II. Title.

QD503.S58   541.3'92   82-6968
ISBN 0-471-09347-5   AACR2

Printed in the United States of America

10 9 8 7 6 5 4 3 2 1

To Elaine and Bobbie

Preface

We have written this book to describe and develop the fundamental basis for determining the composition at equilibrium of a chemical system and to provide computer algorithms for such a determination. Implicit in this purpose is the additional objective of exploring the usefulness of the equilibrium state as a model for predicting the behavior of chemically reacting systems, in full knowledge that the applicability of such a model is limited.

Our overall goal is thus to show both the usefulness and the limitations of chemical reaction equilibrium analysis and to provide for the requirements of those who wish to perform equilibrium computations for systems of any size or complexity. The first part of the goal recognizes that the model of the equilibrium state is not the appropriate model in many cases, and hence there is the need to discriminate between cases where it is useful and cases where it is not. The second part of the goal is not easily achieved, in view of the apparent abundance of approaches available and the present incomplete state of knowledge about nonideal systems. We attempt to provide a more unified approach to computing equilibrium compositions subject to these limitations.

The current state of affairs in this subject can be considered in terms of three aspects: the establishment of the physicochemical principles and structure underlying the problem; the accumulation and representation of requisite data; and the development of computational procedures for exploiting this structure in conjunction with appropriate numerical techniques. The first aspect was essentially completed by J. W. Gibbs just over a century ago, the second has been taking place since before that time and is still far from complete, and the third is a relatively recent occurrence that has been greatly facilitated by developments in high-speed digital computation over the past quarter century.

This book is intended to be used by students, teachers, and practitioners in fields in which chemical equilibrium is important—that is, in such fields as chemistry, chemical engineering, metallurgy, geochemistry, and physics. It is addressed to those who need to know the basis for equilibrium calculations and to have available computer algorithms, whether they are engaged in research and development work, engineering design and field work, or teaching. We have thus attempted to provide a book that can be used widely and flexibly—for formal senior undergraduate and graduate courses, for continuing education and professional development courses, and for self-study. It is assumed that the reader has some prior knowledge of chemical thermodynamics and associated physicochemical data. Some knowledge of such mathematical topics as multivariable calculus and linear algebra is also required, and an elementary understanding of computer programming is desirable.

The scope of the book is reflected in some of its main features. The treatment embraces both reaction and phase equilibrium—that is, physicochemical equilibrium in general. It also deals with both ideal and nonideal chemical systems. Types of systems addressed include those involving solutions of nonelectrolytes and electrolytes in addition to gases, and gases together with many single-species phases. Included in the appendixes are computer program listings for implementing algorithms discussed in the text. The problems at the ends of the chapters are an important part of the book. They are intended both to illustrate the discussion in the text itself and to extend it. The problems are self-contained in the sense that sufficient information is given for their solution, and in most cases references are given to literature sources for comparison of approaches and results. The answers to selected problems are given at the back of the book.

The book is divided into nine chapters. Chapter 1 provides an introduction to the general nature of the chemical equilibrium problem and its potentially useful applications; included is a brief historical sketch of the concept of chemical equilibrium and of attendant computational methods. Chapters 2 and 3 describe the two main criteria. The first is the closed-system constraint, which implies a careful examination of chemical stoichiometry. The second is the equilibrium condition, which involves criteria provided by chemical thermodyanamics. Two computational methods are illustrated in Chapter 4 for relatively simple systems so that the basic approaches can be seen. In Chapter 5 we present a brief survey of numerical methods for solving optimization problems and sets of nonlinear equations. Then in Chapter 6 we present and analyze algorithms for solving chemical equilibrium problems for ideal systems. In Chapter 7 we show how the methods of Chapter 6 can be extended to treat nonideal systems. We discuss the dependence of equilibrium composition on the data used in such computations in Chapter 8. In Chapter 9 we discuss some important practical matters concerning equilibrium computations, including special situations, and emphasizing potential difficulties.

The book can be used effectively in various ways. For example, those interested in the fundamental basis will follow the text and study the worked examples; those interested in solving equilibrium problems arising in their own work will concentrate on the examples and the problems provided. An undergraduate course can be based on the first four chapters and the use of the programs as "black boxes"; a graduate course could use the first four chapters for review and begin at Chapter 5.

We acknowledge our indebtedness to those who have contributed to this subject and on whose work we have drawn. Mr. J.-P. Schoch developed the BASIC computer programs and assisted in solving problems and examples. Dr. Henry Tan developed the HP-41C programs. Dr. T. W. Melnyk and Dr. O. H. Scalise assisted in developing the BNR and VCS FORTRAN programs. Mrs. Debora Foster was a research assistant in the development of much of the manuscript. Miss Claudette Gariepy did all the typing of a rather difficult final manuscript. Miss Diana Cowan, Miss Louisa Franceschi, Mrs. Donna Mokren, and Mrs. Jeanne O'Sullivan assisted in typing parts of drafts of the manuscript and in preparing copy for the appendixes and illustrations. Financial assistance for research in this field over the years has been received from the Natural Sciences and Engineering Research Council of Canada.

WILLIAM R. SMITH

Guelph, Ontario

RONALD W. MISSEN

Toronto, Ontario

May 1982

Contents

ONE   INTRODUCTION   1

1.1   The Nature of Chemical Equilibrium, 1
1.2   The Importance of Chemical Reaction Equilibrium Analysis, 2
1.3   The Problem of Computation, 2
1.4   The Constraints, 4
1.5   Applications of Chemical Equilibrium Analysis, 5
1.6   Historical Sketch, 9

TWO   THE CLOSED-SYSTEM CONSTRAINT AND CHEMICAL STOICHIOMETRY   14

2.1   The Approach, 14
2.2   The Closed-System Constraint, 15
    2.2.1   The Element-Abundance Equations, 15
    2.2.2   Some Terminology, 16
2.3   Chemical Stoichiometry, 18
    2.3.1   Introductory Concepts, 18
    2.3.2   General Treatment of Chemical Stoichiometry, 19
    2.3.3   The Stoichiometric Procedure/Algorithm, 23
    2.3.4   Illustration of the Treatment and Procedure, 25
    2.3.5   The Extent of Reaction, 26
2.4   Expressing Compositional Restrictions in Standard Form, 27
    2.4.1   Introduction, 27
    2.4.2   Reduction of a Given Stoichiometric Matrix to Standard Form, 28
    2.4.3   Stoichiometric Degrees of Freedom and Additional Stoichiometric Restrictions, 29
    2.4.4   Additional Stoichiometric Restrictions that Arise Explicitly, 31
    2.4.5   Additional Stoichiometric Restrictions that Arise Implicitly, 33

THREE CHEMICAL THERMODYNAMICS AND EQUILIBRIUM CONDITIONS   40

3.1 Thermodynamic Potential Functions and Criteria for Equilibrium, 40
3.2 Thermodynamic Description of a Chemical System, 42
3.3 Two Formulations of the Equilibrium Conditions, 44
3.4 The Stoichiometric Formulation, 45
3.5 The Nonstoichiometric Formulation, 46
3.6 Equivalence of the Two Formulations, 48
3.7 The Chemical Potential, 48
    3.7.1 Expressions for the Chemical Potential, 48
    3.7.2 Assigning Numerical Values to the Chemical Potential, 56
3.8 Implications of the Nonnegativity Constraint, 57
    3.8.1 Single-Species Phase, 58
    3.8.2 Ideal Solution, 59
    3.8.3 Nonideal Solution, 59
3.9 Existence and Uniqueness of Solutions, 60
3.10 The Stoichiometric Formulation in Terms of Equilibrium Constants, 62
3.11 Electrochemical Cells, 64
3.12 Ways in Which Standard Free-Energy Information Is Available, 65
    3.12.1 Use of Free Energy of Formation or Equilibrium Constants, 65
    3.12.2 Use of the Free-Energy Function, 66
    3.12.3 Use of Conventional Entropies, 67
    3.12.4 Use of Standard Electrode Potentials, 69
3.13 Sources of Data, 72

FOUR COMPUTATION OF CHEMICAL EQUILIBRIUM FOR RELATIVELY SIMPLE SYSTEMS   75

4.1 Relatively Simple Systems and Their Treatment, 75
4.2 Remarks on Choice of Formulation, 76
4.3 Stoichiometric Formulation for Relatively Simple Systems, 77
    4.3.1 System Involving One Stoichiometric Equation ($R = 1$), 77
    4.3.2 System Involving Two Stoichiometric Equations ($R = 2$), 77
    4.3.3 Stoichiometric Algorithm, 81

Contents xiii

4.4 Nonstoichiometric Formulation for Relatively Simple Systems, 84
    4.4.1 System Consisting of One Element ($M = 1$), 84
    4.4.2 Nonstoichiometric Algorithm, 88

FIVE SURVEY OF NUMERICAL METHODS   100

5.1 Two Classes of Numerical Problem, 101
5.2 Minimization Problems, 103
    5.2.1 Unconstrained Minimization Methods, 104
    5.2.2 Constrained Minimization Methods, 105
5.3 Nonlinear Equation Problems, 109
    5.3.1 Newton-Raphson Method, 110
    5.3.2 Parameter-Variation Methods, 111
5.4 Step-Size Parameter and Convergence Criteria, 113
    5.4.1 Computation of the Step-Size Parameter, 113
    5.4.2 Convergence Criteria, 115

SIX CHEMICAL EQUILIBRIUM ALGORITHMS FOR IDEAL SYSTEMS   117

6.1 Classifications of Algorithms, 118
6.2 Structure of Chapter, 119
6.3 Nonstoichiometric Algorithms, 120
    6.3.1 First-Order Algorithms, 120
    6.3.2 Second-Order Algorithms—the Brinkley-NASA-RAND (BNR) Algorithm, 122
    6.3.3 Other Nonstoichiometric Algorithms, 136
    6.3.4 Illustrative Example for the BNR Algorithm, 137
6.4 Stoichiometric Algorithms, 139
    6.4.1 Introduction, 139
    6.4.2 First-Order Algorithm, 140
    6.4.3 Second-Order Algorithm, 141
    6.4.4 Optimized Stoichiometry—the Villars-Cruise-Smith (VCS) Algorithm, 141
    6.4.5 Illustrative Example for the VCS Algorithm, 146
6.5 Composition Variables Other Than Mole Fraction, 147

SEVEN CHEMICAL EQUILIBRIUM ALGORITHMS FOR NONIDEAL SYSTEMS   153

7.1 The Transition from Ideality to Nonideality, 153
7.2 Further Discussion of Chemical Potentials in Nonideal Systems, 155

- 7.2.1 Use of Excess Free-Energy Expressions, 155
- 7.2.2 Use of Equations of State, 158
- 7.2.3 Use of Corresponding States Theory, 160
- 7.2.4 Electrolyte Solutions, 162
- 7.3 Algorithms for Nonideal Systems, 164
  - 7.3.1 Indirect Methods Based on Algorithms for Ideal Systems, 164
  - 7.3.2 Direct Methods, 167
  - 7.3.3 Intermediate Methods Based on Algorithms for Ideal Systems, 169
  - 7.3.4 Discussion, 169

EIGHT THE EFFECTS OF PROBLEM PARAMETER CHANGES ON CHEMICAL EQUILIBRIA (SENSITIVITY ANALYSIS) 173

- 8.1 Overall Effects of Parameter Changes, 174
  - 8.1.1 Effect of Temperature, 174
  - 8.1.2 Effect of Pressure, 177
  - 8.1.3 Effect of Inert Species, 178
- 8.2 Effects of Problem Parameter Changes on Optimization Problems and Nonlinear Equations, 178
  - 8.2.1 Qualitative Effects of Parameter Changes, 179
  - 8.2.2 Quantitative Effects of Parameter Changes and the Sensitivity Matrix, 180
- 8.3 Qualitative Effects of Parameter Changes on Chemical Equilibria, 181
- 8.4 Sensitivity Matrices for (T, P) Problems, 183
  - 8.4.1 Review of Literature, 183
  - 8.4.2 Stoichiometric Formulation, 184
  - 8.4.3 Nonstoichiometric Formulation, 192
  - 8.4.4 Second-Order Sensitivity Coefficients, 195
- 8.5 Effect of Errors in μ^* on Equilibrium Composition, 196
- 8.6 Calculation of Thermodynamic Derivatives for a System at Chemical Equilibrium, 198

NINE PRACTICAL CONSIDERATIONS AND SPECIAL TOPICS 201

- 9.1 Initial Estimates for Equilibrium Algorithms, 201
  - 9.1.1 Estimate of Mole Numbers, 201
  - 9.1.2 Estimate of Lagrange Multipliers, 203
- 9.2 Numerical Singularities and the Nonnegativity Constraint, 204
  - 9.2.1 Single-Species Phases, 204
  - 9.2.2 Multispecies Phases, 211

9.3	The Case of $C \equiv \text{Rank}(\mathbf{A}) \neq M$, 213	
9.4	Standard Free-Energy Data from Equilibrium Constants, 214	
	9.4.1 The Case $R = 1$, 214	
	9.4.2 The General Case, 215	
	9.4.3 Treatment of Possibly Incompatible Data, 217	
9.5	Species Present in Small Amounts, 217	
9.6	Restricted Equilibrium Problems, 218	
9.7	Treatment of Isomers, 219	
9.8	Treatment of Isotopes, 222	
	9.8.1 Calculation by Probability Theory, 223	
	9.8.2 Calculation by Equilibrium Algorithms, 224	
	9.8.3 Calculation from Equilibrium Constants, 224	
9.9	Thermodynamic Constraints Other Than (T, P), 226	
	9.9.1 Equilibrium Computation, 226	
	9.9.2 Sensitivity Analysis, 228	

APPENDIX A: COMPUTER PROGRAMS FOR GENERATING STOICHIOMETRIC EQUATIONS   **235**

A.1   HP-41C Program, 236
      A.1.1   Program Listing, 236
      A.1.2   *User's Guide*, 238
A.2   BASIC Program, 240
      A.2.1   Program Listing, 240
      A.2.2   Sample Input and Output, 242
A.3   FORTRAN Program with Sample Input and Output, 244

APPENDIX B: COMPUTER PROGRAMS FOR CALCULATING EQUILIBRIUM FOR RELATIVELY SIMPLE SYSTEMS   **247**

B.1   Stoichiometric Algorithms, 248
      B.1.1   HP-41C Program, 248
           B.1.1.1   Program Listing, 248
           B.1.1.2   *User's Guide*, 251
      B.1.2   BASIC Program, 256
           B.1.2.1   Program Listing, 256
           B.1.2.2   Sample Input and Output, 262
B.2   Nonstoichiometric Algorithms, 265
      B.2.1   HP-41C Program, 265
           B.2.1.1   Program Listing, 265
           B.2.1.2   *User's Guide*, 268
      B.2.2   BASIC Program, 270

APPENDIX C: FORTRAN COMPUTER PROGRAM FOR BNR ALGORITHM   277

C.1 Program Listing, 278
C.2 *User's Guide*, 292

APPENDIX D: FORTRAN COMPUTER PROGRAM FOR VCS ALGORITHM   295

D.1 Program Listing, 296
D.2 *User's Guide*, 324

LIST OF SYMBOLS   326

REFERENCES   334

ANSWERS TO SELECTED PROBLEMS   342

AUTHOR INDEX   347

SUBJECT INDEX   353

Chemical Reaction Equilibrium Analysis

CHAPTER ONE

Introduction

Chemical reaction equilibrium analysis has two principal concerns: to establish whether the equilibrium state of a chemical system is a useful model for a particular purpose; and to determine the equilibrium composition of the system under specified conditions, such as stipulated temperature and pressure. A related concern is the effect of changing conditions on this composition.

There are thus three questions regarding this type of analysis:

1 Is an equilibrium model appropriate?
2 What is the composition of the equilibrium state at specified conditions?
3 What is the effect of changing conditions?

The basis for obtaining answers to these questions is the subject of this book. We begin in this first chapter by considering chemical equilibrium in its various aspects, such as its nature, importance, relation to kinetics, scope in terms of constraints and applications, and historical development in outline.

1.1 THE NATURE OF CHEMICAL EQUILIBRIUM

We use the phrase "equilibrium state of a closed system" (i.e., a *locally stable* equilibrium state), and by this, as a preliminary statement, we mean a state that is

1 Independent of time.
2 Independent of the previous history of the system.
3 Resistant to fluctuations (within limits of magnitude but not of direction) of composition; that is, any such fluctuations are damped out.
4 Independent of position within the system; this excludes, for example, a system subject to a potential gradient resulting in a flux of mass or energy within it.

A stable equilibrium state is distinguished from an unsteady (kinetic) state by point 1, from a nearby, path-dependent (kinetic) state by point 2, from an unstable or a metastable equilibrium state by point 3, and from what is sometimes referred to as a *steady state* for such a system by point 4.

The chemical system of interest is by definition undergoing change in chemical composition in approaching an equilibrium state from some initial nonequilibrium state. This takes place by reaction or by mass transfer or by a combination of the two. The possibility of mass transfer implies that the system may consist of more than one phase. We thus include both reaction equilibrium and phase equilibrium in our use of the term "chemical equilibrium."

1.2 THE IMPORTANCE OF CHEMICAL REACTION EQUILIBRIUM ANALYSIS

The importance of chemical reaction equilibrium analysis derives from the circumstances in which the equilibrium state is a useful model for describing the state of an actual system. The following circumstances include the usual ones for which this is the case:

1. The equilibrium model may be useful when rates of change (reaction and mass transfer) are relatively rapid. This tends to be the case when temperature is relatively high, as in a rocket engine, or when catalytic activity is relatively high, as in the case of a sulfur dioxide converter. The inferences of analytical chemistry involving ionic species are also normally based on this model.
2. The equilibrium model may be useful in a positive sense as a reference state to which rate considerations are applied, as in the cases of maximum conversion in a chemical reactor, electromotive force (emf) of a chemical cell, and stagewise operations in separation processes.
3. The equilibrium model may be useful in a negative sense, such as in predicting too low a conversion or yield or in avoiding equilibrium with respect to certain undesired species.
4. The equilibrium model may be useful as a guide in choosing process conditions, including the evaluation of a catalyst, particularly in conjunction with the effects of changing conditions.

The usefulness of the equilibrium model will normally diminish in favor of a kinetic model whenever rates of change are relatively slow. We may thus be led astray by a particular equilibrium model, if this is the case in the actual system.

1.3 THE PROBLEM OF COMPUTATION

Once we have decided that the equilibrium model is worth investigating, we must determine the equilibrium state (its composition) under particular conditions and perhaps also its dependence on these conditions. This involves an application of the second law of thermodynamics since this law provides the requisite criteria. To exploit these criteria, we must construct appropriate forms of the potential functions involved and must have appropriate information (equations of state or their equivalent and energy and "free"-energy data), to specify the parameters of the problem.

The problem then becomes essentially a mathematical one. As such, it can be viewed either as one of optimization (e.g., the minimization of the Gibbs function subject to certain constraints, including the mass-balance constraints) or as one involving the solution of a set of simultaneous nonlinear algebraic equations (which arise from the necessary mathematical conditions for this minimization). These viewpoints are essentially equivalent, and we stress this in later chapters. A great deal of misconception and fruitless argument has resulted in the literature from overly enthusiastic proponents of one viewpoint to the exclusion of the other.

Since the mass-balance constraints are linear in the mole numbers of the species present and the potential function is nonlinear in these variables, any general method of solution is necessarily an iterative one. For small systems, solutions may be effected by means of *hand calculation methods* or programmable hand calculators. For large systems, however (e.g., 100 species involving 10 elements), we must of necessity resort to the use of numerical methods in conjunction with digital computers. Throughout most of this book we are interested mainly in general computational algorithms that can handle large systems, but we also consider algorithms that can be used for smaller systems. The latter approach can exploit the growing use of small computers in addition to programmable hand calculators.

To understand the many computation methods that have appeared in the literature, some mathematics results must be used. Our point of view in this matter is that of an investigator using mathematical sophistication where necessary, rather than that of a mathematician exploring the use of mathematics in a chemical setting. We mention briefly here some of the concepts used in later chapters.

We need to use ideas from the fields of optimization and numerical analysis. For example, among other approaches, we consider the method of Lagrange multipliers for constrained optimization and the Newton-Raphson method for solving nonlinear algebraic equations. Since the molecular formulas of chemical species can be associated with vectors and the mass-balance constraints are linear in the mole numbers, we also need to use some concepts from linear algebra. Among the most important notions are vector-matrix manipulation (e.g., vector-matrix multiplication), the solution of systems of linear equations, the concept of linear dependence and independence of vectors, and the concept of the rank of a matrix.

An understanding of the mathematics underlying the computational techniques is necessary for appreciation of the differences among them. It is also necessary for those who wish to formulate their own computation algorithms.

However, for those whose primary goal is to use a working computer program for performing equilibrium calculations, we give in the appendixes program listings and user's instructions for algorithms discussed in the text.

1.4 THE CONSTRAINTS

We consider constraints in three categories:

1 Overriding constraints that exclude certain types of problem for the purpose of this book.
2 Mandatory constraints that are independent of the nature of a particular problem.
3 Specific constraints that serve to define the parameters of a particular problem; these parameters are normally taken to be temperature, pressure, elemental abundances, molecular formulas, and standard chemical potentials.

Category 1 includes the following: (a) no surface or tensile effects (we are dealing with bulk systems in which the only external mechanical stress is that of a uniform fluid pressure); (b) no field effects (gravitational, electrical, etc.); (c) no macroscopic kinetic energy effects; and (d) no nuclear effects. Category 2 includes the following: (a) the closed-system or mass-balance constraint, which in this context means the conservation of elemental species; this includes the charge-balance constraint for ionic systems; and (b) the nonnegativity constraint—the number of moles of species i, n_i, is either positive or zero for all i; that is,

$$n_i \geq 0. \qquad (1.4\text{-}1)$$

In category 3 there are various possibilities, but they would typically arise from the following: (a) specification of temperature T and pressure P by means of thermodynamic conditions or constraints; and (b) specification of the remaining parameters; normally these are given numerically at the outset, but there are situations in which they may also be specified by means of auxiliary conditions or constraints.

At this point we elaborate on category 3a but defer further consideration of category 3b to Chapter 8. Because of the importance of T and P as independent variables, which implies the Gibbs function G as the proper potential, we work entirely in terms of this set of variables—recasting, if necessary, other problems in terms of this standard (T, P) problem. Some of the important types of problem are:

1 Parameters (T, P) specified; T and P are fixed at given values.

2 Parameters (T, V) specified; T is fixed at a given value, and P is specified indirectly by an equation of state that links volume V to T, P, and the equilibrium composition. This specification is appropriate for a chemical system in a rigid container; the procedure involves the simultaneous solution of the equilibrium condition in terms of T, P, and the equation of state. The approach described is alternative to the use of T and V as independent variables, which implies the Helmholtz function A as the proper potential.

3 Parameters $(\Delta H, P)$ specified; P is fixed, and T is specified indirectly by an energy balance that links enthalpy change (ΔH) to T, P, and the equilibrium composition. This specification is appropriate, for example, when the inlet T to a reactor is specified and the energy balance is determined by heat transfer considerations between the inlet and the point at which equilibrium is postulated; a special case is that for adiabatic change ($\Delta H = 0$ under certain circumstances); the procedure involves the simultaneous solution of the equilibrium condition and the energy balance.

4 Parameters $(T, \Delta P)$ specified; T is fixed, and P is specified indirectly by a momentum balance that links pressure drop ΔP to T, P, and the equilibrium composition; this specification is appropriate, for example, when the inlet P to a reactor is specified, and the momentum balance is determined by pressure-drop considerations between the inlet and the point at which equilibrium is postulated. The procedure involves simultaneous solution of the equilibrium condition and the momentum balance.

5 Parameters $(\Delta H, \Delta P)$ specified; neither T nor P is fixed; both are specified indirectly by an energy balance and a momentum balance. This specification is appropriate when both inlet T and P are specified; that is, this situation is a combination of cases 3 and 4, and the procedure involves simultaneous solution of the equilibrium condition, the energy balance, and the momentum balance.

6 Parameters (S, P) specified; P is fixed, and T is specified indirectly by an equation that links entropy S to T, P, and the equilibrium composition; this specification is appropriate, for example, when a process can be considered adiabatic and reversible (i.e., isentropic), as in flow through a perfect nozzle.

These examples of the specification of thermodynamic constraints are summarized in Table 1.1.

1.5 APPLICATIONS OF CHEMICAL EQUILIBRIUM ANALYSIS

The following examples illustrate situations in which consideration of equilibrium is an important part of the overall analysis. Other situations are

Table 1.1 Some Examples of Specification of Thermodynamic Constraints

	Given				
Case	T	P	Auxiliary Variable(s)	Auxiliary Condition(s)[a]	Special Case
1	Yes	Yes	—	—	—
2	Yes	—	V	$V = V(T, P, \mathbf{n})$[b]	—
3	—	Yes	ΔH	$\Delta H = \Delta H(T, P, \mathbf{n})$	$\Delta H = 0$ (adiabatic reaction)
4	Yes	—	ΔP	$\Delta P = \Delta P(T, P, \mathbf{n})$	$\Delta P = 0$ (becomes case 1)
5	—	—	$\Delta H, \Delta P$	$\Delta H = \Delta H(T, P, \mathbf{n})$ $\Delta P = \Delta P(T, P, \mathbf{n})$	$\left.\begin{array}{l}\Delta H = 0\\ \Delta P = 0\end{array}\right\}$ (becomes case 3)
6	—	Yes	S	$\Delta S = \Delta S(T, P, \mathbf{n})$	$\Delta S = 0$ (isentropic change; e.g., ideal flow through nozzle)

[a] For simultaneous solution with equilibrium condition.
[b] Notation means that V is a function of T, P, and the mole numbers $\mathbf{n}$ of the species present at equilibrium.

examined in more detail throughout the remaining chapters, both as worked examples and as problems at the ends of the chapters.

1 Chemical Kinetics The equilibrium concept imposes a restriction on the form of a rate law, whenever we consider a net rate of reaction to be the difference between forward and reverse reaction rates (Blum and Luus, 1964; Van Rysselberghe, 1967; Swartz, 1969; Denbigh, 1981, pp. 444–449). In reaction mechanisms, equilibrium is frequently postulated for a relatively rapid reaction step and its reverse, to eliminate the unknown concentration of a transient intermediate species whose concentration is not involved in the observed rate law. This is the steady-state hypothesis. In the transition-state theory of reaction rates, equilibrium between the activated complex and reactants is also an important postulate.

2 Inorganic Chemistry The formation of a complex ion in a solvent usually involves the association of a central cation with anions or neutral molecules as ligands. A measure of the stability of such a complex is given by a formation (equilibrium) constant. One such constant is associated with each stepwise addition of a ligand to the central cation, up to the maximum coordination number. A knowledge of the stepwise constants allows calculation of the distribution of the various species at equilibrium (Harvey and Porter, 1963, pp. 346–355).

3 Organic Chemistry The existence of isomers of organic substances in general has implications for yield and conversion at equilibrium (Missen, 1963; Smith and Missen, 1974).

4 *Analytical Chemistry* Equilibrium analysis plays a large role in quantitative chemical analysis (Laitenen, 1960, p. 2), because the reactions involved are usually sufficiently rapid for a state of equilibrium to be attained. If the concentration of a solution is sufficiently high, nonideality must be taken into account.

5 *Chemical Processes* Classical examples of the importance of equilibrium analysis in chemical processes include ammonia synthesis and the conversion of SO_2 to SO_3 in a sulfuric acid plant. In such cases the analysis provides information about maximum possible conversions, as functions of the problem parameters, and, in conjunction with kinetic and thermochemical data, information for development of reactor design criteria. Methanol synthesis from hydrogen and carbon oxides has been extensively discussed (Bakemeier, et al., 1970; Strelzoff, 1970).

6 *Energy Conversion* An electrochemical cell may be regarded as a chemical reactor or as an energy-conversion device. A fuel cell is an example of the latter. The overall reaction in a fuel cell (e.g., a hydrogen-oxygen fuel cell) is equivalent to a combustion reaction. The advantage of using a fuel cell, compared with combustion in a heat engine in the normal sense, is that the efficiency of energy conversion is not limited by the Carnot relation. This is essentially because the former involves conversion of chemical free energy directly into electrical energy rather than through chemical enthalpic energy. Another aspect of equilibrium analysis in this case is that the emf of the cell is determined in terms of electrode potentials associated with electrode processes in the usual way through the Nernst equation.

7 *Rocket Propellants* In rocket propulsion, equilibrium analysis, together with thermochemical analysis, plays a role in evaluating the performance of the propellant system, fuel + oxidizer (Verhoek, 1969).

8 *Solutions of Nonelectrolytes* Interpretation of the behavior of equilibrium properties of nonideal solutions of nonelectrolytes ranges between the extremes of chemical interactions (characterized by equilibrium constants) and physical interactions (characterized by intermolecular potential energy functions). This is reflected in various theories of solutions (Prausnitz, 1969, Chapters 4 and 7) and their predictions concerning deviation from ideality in terms of excess functions (Missen, 1969). Chemical effects in solution may be classed as association (involving one species) or solvation (involving more than one species). Hydrogen bonding is a chemical interaction that can be involved in either class. The ideal continuous-association model for solutions of an alcohol such as methanol with a relatively inert, nonpolar substance such as carbon tetrachloride pictures self-association of alcohol molecules through hydrogen bonding in polymerlike fashion, together with equilibrium among the monomers, dimers, trimers, and so on of alcohol, and ideal-solution behavior of these species and the second component. Thus all the deviations from ideality of the actual solution are attributed to chemical equilibrium effects.

This is probably an oversimplification, and other interactions usually must be built into the model to achieve even a satisfactory accounting of the observed behavior. An example of hydrogen bonding in solvation occurs in complex formation between acetone and chloroform. This can also be described by means of an equilibrium constant, and such complexes can frequently be detected by experimentally determining phase behavior and constructing equilibrium phase diagrams, wherein they are revealed as maxima in freezing-point curves.

9 *Environmental Chemistry* Considerable attention has been focused on equilibrium aspects of environmental problems. For example, the Division of Water, Air and Waste Chemistry of the American Chemical Society sponsored a symposium on equilibrium concepts in natural water systems in 1966. Included was a discussion by Kramer (1967) of equilibrium models for the Great Lakes in which various chemical and phase equilibria were taken into account. The main reason why equilibrium models may be valid is that the time available ("residence" time) for mixing of runoff water bringing in dissolved material is many years. The models evidently assume the lakes to be well-stirred tanks. Kramer concluded that, to a first approximation, the Great Lakes fit a model involving equilibrium of calcite, dolomite, apatite, kaolinite, gibbsite, feldspars, water and air with $p_{CO_2} = 3.5 \times 10^{-4}$ atm partial pressure at 5°C and 1 atm. Other models with different constraints were also investigated.

Air pollution by nitrogen oxides has received much attention since it is largely caused by emissions from internal combustion engines, in which the ubiquitous automobile plays a large role (in creating smog). Whenever air is used in combustion, there is the possibility of forming nitric oxide (NO). Since the free energy of formation of NO is a relatively large positive value at low temperature, the equilibrium concentration is extremely small when NO is formed at ambient temperature. But combustion occurs at high temperature, and the endothermic nature of the formation reaction for NO means that the equilibrium concentration increases with increase in temperature. This increase seems to be particularly rapid above 1800 K (Ermenc, 1970). Studies on a particular furnace showed that the actual concentration of NO, as governed by the rate of reaction, approached the equilibrium value in the range 2000 to 2100 K but became significantly less than the equilibrium value as the (wall) temperature increased above 2200 K. Air pollution by nitric oxide, which converts to nitrogen dioxide and thence to organic irritants on photochemical reaction with hydrocarbons in the atmosphere, can be reduced if air is replaced by oxygen for combustion or if combustion is made more efficient so as to reduce the amount of excess air (Ermenc, 1970).

10 *Chemical Economics* A rather unusual application of equilibrium data has been made by Sommerfeld and Lenk (1970) in correlating the selling price of a number of "bulk" chemicals, such as calcium carbonate and ethylene, with the standard free energy of formation (ΔG_f°) and annual sales volume.

Such a correlation is apparently useful in profitability analysis and determination of optimal plant sizes.

1.6 HISTORICAL SKETCH

In this brief sketch we trace the development of the concept of chemical equilibrium and associated computational techniques to provide perspective for the main treatment to follow. A thorough treatment of the former, in particular, would help to clarify some points of confusion that tend to carry over to this day and would show the importance of the role of chemical equilibrium in industrial and technological developments (cf. Le Chatelier, 1888, cited by Lewis and Randall, 1961, p. 2). At least one brief essay has been presented (Lindauer, 1962), and the subject is implicitly treated to varying extents in books on the history of chemistry, with a considerable discussion given by Partington (1964, Chapters 18–21).

The concept of chemical equilibrium is rooted in both chemical and physical notions. In a chemical sense it has developed from ideas on chemical change and affinity, at one stage contemporaneously with ideas on rate and mechanism of reaction. In a physical sense it has developed from the concept of balanced forces in static equilibrium in mechanical systems. The former may have resulted in some unnecessary confusion between the equilibrium state and events that precede it, as reflected, for example, in having kinetic as well as thermodynamic derivations of equilibrium relations (van't Hoff, 1898, p. 104). The mechanical analogy is also imperfect; one aspect of this is that the chemist is usually at pains to point out the "dynamic," as opposed to "static," nature of chemical equilibrium in terms of opposing rate processes. For example, van't Hoff, although using divisions of statics and dynamics in his treatment of physical chemistry, included chemical equilibrium in *dynamics* because he apparently preferred to associate it with chemical change, affinity, and rate of reaction. Nernst (1916), on the other hand, regarded chemical equilibrium in terms of chemical *statics*, but he included equilibrium and kinetics, together with energy transformations, in a "Doctrine of Affinity." To understand the historical development of the concept of equilibrium in chemistry, one must be able to distinguish among these various ideas and terms.

The concept of affinity is an ancient one in chemistry (Walden, 1954), and in its qualitative origins referred to the tendency or relative tendency for reaction to occur. Two significant contributions to its qualitative development were made by Bergman in 1775 and Berthollet in 1801 (Lindauer, 1962). Bergman's concept of elective affinity asserted that affinity depended solely on the nature of the reacting substances; although this is erroneous, we see a vestige of it in the familiar electromotive series. Berthollet emphasized the importance of the quantity (i.e., concentration) of reacting species, which later gave rise to the idea of *mass action*. Affinity was given quantitative expression by van't Hoff in the late 1800s in terms of the maximum work obtainable from a chemical process. This is equivalent to interpreting the Gibbs free energy of reaction ($\Delta G_{T,P}$) as the "driving force" of a reaction, and it is in this form that the word "affinity" is now used in connection with reaction tendency. The culmination of this idea is due to De Donder (1936, Chapter 4). For a reaction at equilibrium, the affinity is zero.

Perhaps the first important milestone for the quantitative development of equilibrium, as distinct from affinity, was the work of Berthelot and St. Gilles (1862), which showed that the same equilibrium composition was obtained in the system of acetic acid, ethanol, water, and ethyl acetate from the two directions of esterification and hydrolysis. This led Guldberg and Waage (1864) to a quantitative equilibrium equation, through the so-called *law of mass action*, by balancing "chemical forces" at equilibrium. Although the term "mass action" is firmly embedded in the chemical literature, it has given rise to considerable confusion in the teaching of both kinetics and thermodynamics (Mysels, 1956; Guggenheim, 1956). This is not to detract from the contribution of Guldberg and Waage, although the proper significance of this has been the subject of some dispute (Guggenheim, 1956; Lund, 1965). Their derivation of the equilibrium constant expression, together with kinetic (rate) expressions being used at that time, presumably led to the kinetic derivation of the equilibrium constant, as given, for example, by van't Hoff (1898, p. 9).

The parameters affecting equilibrium received attention by van't Hoff, who proposed the concept of *mobile equilibrium*, which was supported by evidence that the elevation of temperature favors endothermic reactions. Le Chatelier (1884) subsequently focused greater attention on the forces affecting equilibrium and, as a result, postulated—without proof but with examples—a "law" (Le Chatelier's principle) regarding how a system in stable chemical equilibrium adjusts to external factors tending to change its temperature, pressure, or molar density or to change the concentration of one or more of the species involved. This is to the effect that the system undergoes changes, which if they occurred by themselves, would result in a change in temperature, pressure, and other parameters opposite to that resulting from the external effect. In this Le Chatelier was following a statement by van't Hoff concerning temperature effects only. It has been argued (de Heer, 1957) that Le Chatelier's statement is vague and ambiguous and thus difficult to use operationally and perhaps open to exception (Mellon, 1979). More precise formulations have been made (de Heer, 1957; Callen, 1960; Prigogine and Defay, 1954, Chapter 17), and the theorem is then sometimes referred to as the *Le Chatelier-Braun principle*, in recognition of the later contribution of F. Braun (1887, 1888).

The thermodynamic formulation of the equilibrium constant expression was also obtained by van't Hoff with the use of reversible cycles. A more elegant approach was provided by Gibbs (1876) with his introduction of the concept of the chemical potential. The influence of van't Hoff, with his cycles, seems to have been greater in the first part of this century than that of Gibbs, with his chemical potential. However, in relatively recent times, Gibbs has been "rediscovered," and the methods used in this book in developing the principles are essentially his, bearing little resemblance to those of van't Hoff. With these contributions and the later ones by Nernst (third law) and Lewis (activity and fugacity), we may consider the principles underlying the concepts of chemical equilibrium to be firmly established. Of course, there still remains the difficult problem of being able to express and predict the way in which the chemical potentials of substances behave in nonideal mixtures. The solution to this problem, however, lies in statistical mechanics, the molecular basis for thermodynamics (Prausnitz, 1969; Reed and Gubbins, 1973), and not in classical thermodynamics.

Although the theoretical principles were established by the early part of this century, the practical matter of computing equilibrium became more difficult as more complex systems were considered in connection with developments in chemical processing, in military and space programs, and in methods of chemical analysis. The difficulty arises because of the juxtaposition of the logarithmic nature of the chemical potential–activity relation and the linear nature of the mass-balance constraints. This difficulty is present even for the simplest case, that of a mixture of ideal gases.

Algorithms for the computation of chemical equilibrium prior to the early 1950s were oriented primarily toward hand calculation methods and have been reviewed by Kobe and Leland (1954). They revolved around the equilibrium-constant expressions written for a set of stoichiometric equations in terms of concentrations. This involves the solution of a set of nonlinear algebraic equations, usually equal in number to the difference between the number of species and the number of elements involved. Before the advent of electronic computers, the number of such equations that the investigator could readily solve was limited to a small number that could be treated by hand.

The power of hand computation methods was extended somewhat in the early 1940s coincidentally with the development of rockets in World War II, the design of which required the ability to calculate equilibrium compositions of systems containing larger numbers of species. A paper by Damköhler and Edse (1943) outlines a method that systematically reduces the number of equations for a hydrogen-carbon-oxygen system to two equations in two unknowns, which are then solved graphically. Such reduction in the number of working equations is usually possible for any problem, but the actual reduction procedure depends greatly on the individual problem at hand and the ingenuity of the investigator.

The foundations for general-purpose algorithms for the computation of chemical equilibrium were laid by Brinkley (1947). He outlined a systematic procedure that he intended for use with desk calculators, although it also lends itself to implementation on digital computers, the development of which was just beginning at that time. By focusing attention on a set of "component species," the number of which is usually equal to the number of elements M

involved, Brinkley formulated an algorithm wherein the number of working equations is essentially equal to M, thus permitting calculations for systems with large numbers of species that were not possible before. It is important historically to note that Brinkley's algorithm was first derived by considering equilibrium-constant expressions.

The rapid development of electronic computers from about 1950 and the interest in rocket propellants stimulated by the space program in the United States in the late 1950s led to a great revival of interest in equilibrium computations. By viewing the problem of equilibrium computation as a nonlinear optimization problem, White et al. (1958) developed an algorithm that solved the problem by "minimizing the free energy directly." In effect, this meant that they did not use stoichiometric equations or reactions. The popular impact of this and other "direct minimization methods" has been so great that they are today among the most popular methods used for calculating equilibrium compositions with a digital computer (see, e.g., Mehrotra et al., 1979; Eriksson, 1979). By the late 1950s controversy had developed between approaches characterized as "classical equilibrium-constant methods" and "direct free-energy minimization methods" (Zeleznik and Gordon, 1968). This polarization of views was unnecessary, since Zeleznik and Gordon (1960) showed that they were computationally equivalent. This had also been pointed out by Brinkley (1960) in conjunction with his method, and it should not be surprising in view of the fact that the working equations due to Brinkley (1947) and White et al. (1958) are remarkably similar.

Since the late 1950s there has been a revival of interest in implementing with computers some of the techniques that employed stoichiometric equations and extent-of-reaction variables (Villars, 1959; Naphtali, 1959; Cruise, 1964; Smith and Missen, 1968; Bos and Meershoek, 1972). These stoichiometric methods have been shown (Smith, 1976, 1978) to have considerable computational advantages in the treatment of a certain class of equilibrium problems and hence form an important part of the development described in this book. This class is characterized by multiphase equilibrium problems involving the presence of many single-species phases (usually solids). These typically arise in geochemical (e.g., Nordstrom et al., 1979; Giggenbach, 1980) and metallurgical (e.g., Samuels, 1971; Mehrotra et al., 1979) situations.

These stoichiometric algorithms can also be motivated by free-energy minimization considerations. The only difference is that the independent variables with which the minimization is effected are the set of extent-of-reaction variables, whereas with the so-called direct methods, the mass-balance constraints are usually handled computationally by the use of Lagrange multipliers; this might be called an "indirect" way of incorporating constraints.

There has also been a limited revival of interest in the type of approach proposed by Damköhler and Edse (1943) for use on computers (Erickson et al., 1966). For a particular problem, such an approach can be very efficient. However, the amount of prior algebraic manipulation required for each problem makes it of limited utility. In this book we largely ignore such methods.

All computational methods referred to thus far were developed originally for use in ideal systems. (For the rocket-propellant problems, the high temperatures involved make the assumption of ideal-gas behavior not a grossly unreasonable one.) Most methods usually consider a single ideal-gas phase and any number of pure condensed phases. Although the treatment of nonideal systems is often mentioned briefly by many authors in connection with their computation methods, the nonideal problem has not been considered much in its own right before. This is one of the new areas for work in this subject, and in Chapter 7 we discuss some general ideas that are useful in such a connection.

In summary, two events have served most to stimulate the growth of the equilibrium computation methods: (1) the introduction of the digital computer, which permitted the routine handling of large problems that could scarcely be contemplated by hand calculation methods; and (2) the development of rocket technology. The first stimulus by this means occurred in Germany in the 1940s, and the second occurred with the development of the space program in the United States in the 1950s. Rocket technology posed chemical equilibrium computational problems that demanded immediate answers. Since this type of problem (large numbers of species) had rarely been considered in the past, it stimulated the development of new algorithms.

Our task in this book is mainly an expository one. We attempt to bring the main computational methods that have appeared in the literature over the last 20 years together in one place and compare them. We also want to point out that the subject is not by any means a closed one, especially regarding calculations for nonideal systems. Reviews of the subject, which include two previous monographs, have been given by Kobe and Leland (1954), Zeleznik and Gordon (1968), Van Zeggeren and Storey (1970), Klein (1971), Holub and Vonka (1976), Seider et al. (1980), and Smith (1980a, 1980b).

CHAPTER TWO

The Closed-System Constraint and Chemical Stoichiometry

Here we develop the basis for the constraint on the computation of chemical equilibrium that is due to the requirement for conservation of elements in a closed system undergoing chemical change, a special form of the law of conservation of mass. This constraint is intimately bound up with what is usually called *chemical stoichiometry*,* whether it is expressed directly in terms of conservation equations or indirectly in terms of chemical equations. The specific purpose of this chapter is to develop chemical stoichiometry for a closed system in a form suitable for incorporation in an equilibrium-computation algorithm. Useful tools for this are provided by linear algebra since the conservation equations are themselves linear algebraic equations, and in what follows we make use of vector-matrix notation. The end result is a stoichiometric-coefficient algorithm, computer programs for which are contained in Appendix A.

2.1 THE APPROACH

We first define a closed system and then develop a method (Smith and Missen, 1979) for treating chemical stoichiometry that involves generating, *a priori*, an appropriate set of chemical equations. Since another approach involves starting with a set of such equations, we discuss the implications of this and finally consider some special stoichiometric situations. In this sense the treatment is more general than, and goes beyond, the specific purpose indicated previously.

---

*The word "stoichiometry," of Greek origin, literally concerns measurement (-metry) of an element (stoichion); in chemical stoichiometry the element is a chemical element.

2.2 THE CLOSED-SYSTEM CONSTRAINT

2.2.1 The Element-Abundance Equations

A closed system has a fixed mass; that is, it does not exchange matter with its surroundings, although it may exchange energy. It may consist of one or more than one phase and may undergo reaction and mass transfer internally. Its importance in equilibrium computations is that the equilibrium conditions of thermodynamics (Chapter 3) apply primarily to such a system.

In the laboratory and in chemical processing, the concept of a closed system obviously applies to a batch system. Perhaps less obviously, it also applies to a fluid system undergoing "plug" flow in which there is no mixing or dispersion in the direction of flow and in which all elements of fluid have the same residence time in a particular vessel or conduit (Levenspiel, 1972, p. 97). In such a case each portion of fluid, of arbitrary size, acts as a batch system in moving through the vessel. This description is most suitably applied to a fluid flowing at a relatively high velocity in a conduit of uniform cross section.

Operationally, any description of a closed system is an expression of the law of conservation of mass. A closed system can be defined by a set of element-abundance equations expressing the conservation of the chemical elements making up the species of the system. There is one equation for each element, as follows:

$$\sum_{i=1}^{N} a_{ki} n_i = b_k; \qquad k = 1, 2, \ldots, M, \qquad (2.2\text{-}1)$$

where a_{ki} is the subscript to the kth element in the molecular formula of species i; n_i is the number of moles of i (in some basis amount of system); b_k is the fixed number of moles of the kth element in the system; M is the number of elements; and N is the number of species. Alternatively, equations (2.2-1) may be written so as to express the change from one compositional state to another:

$$\sum_{i=1}^{N} a_{ki} \delta n_i = 0; \qquad k = 1, 2, \ldots, M, \qquad (2.2\text{-}2)$$

where δn_i is the change in the number of moles of the ith species between two compositional states of the system.

In vector-matrix form, the element-abundance equations 2.2-1 and 2.2-2 are, respectively,

$$\mathbf{An} = \mathbf{b}, \qquad (2.2\text{-}3)$$

and

$$\mathbf{A}\,\delta\mathbf{n} = \mathbf{0}, \qquad (2.2\text{-}4)$$

where, as described in more detail in the following paragraphs, **A** is the formula matrix, **n** is the species-abundance vector,* and **b** is the element-abundance vector. Again, it is the fact that **b** is fixed that characterizes a closed system. Any one of equations 2.2-1 to 2.2-4 expresses the closed-system constraint.

Example 2.1 Write equations 2.2-1 and 2.2-3 for a reaction involving the species NH_3, O_2, NO, NO_2, and H_2O. Assume that the initial state of the system consists of NH_3 and O_2 in the molar ratio $4:7$.

Solution The $NH_3 : O_2$ molar ratio establishes a basis amount of system such that $b_N = 4$, $b_H = 12$, and $b_O = 14$. The three element-abundance equations 2.2-1 for the three elements in the order nitrogen, hydrogen, and oxygen are then:

$$1n_{NH_3} + 0n_{O_2} + 1n_{NO} + 1n_{NO_2} + 0n_{H_2O} = b_N = 4,$$

$$3n_{NH_3} + 0n_{O_2} + 0n_{NO} + 0n_{NO_2} + 2n_{H_2O} = b_H = 12,$$

$$0n_{NH_3} + 2n_{O_2} + 1n_{NO} + 2n_{NO_2} + 1n_{H_2O} = b_O = 14.$$

Equation 2.2-3 for this system is

$$\begin{pmatrix} 1 & 0 & 1 & 1 & 0 \\ 3 & 0 & 0 & 0 & 2 \\ 0 & 2 & 1 & 2 & 1 \end{pmatrix} \begin{pmatrix} n_{NH_3} \\ n_{O_2} \\ n_{NO} \\ n_{NO_2} \\ n_{H_2O} \end{pmatrix} = \begin{pmatrix} 4 \\ 12 \\ 14 \end{pmatrix},$$

where the matrix on the left is **A**, which is made up of the coefficients on the left in equations 2.2-1 and the two vectors are **n** and **b**, respectively.

The maximum number of linearly independent element-abundance equations, which is the same as the maximum number of linearly independent rows (or columns) in the matrix **A**, is given by the rank of **A** (Noble, 1969, p. 128).

2.2.2 Some Terminology

To provide a concise summary of unambiguous terminology, we define a number of terms in this section, mostly relating to a closed system, even

---

*All vectors are column vectors, and superscript T, used in Section 2.2.2 and later, denotes the transpose of a vector.

though some of them have already been introduced. These are as follows:

chemical species: a chemical entity distinguishable from other such entities by

1. Its molecular formula; or, failing that, by
2. Its molecular structure (e.g., to distinguish isomeric forms with the same molecular formula); or failing that, by
3. The phase in which it occurs [e.g., $H_2O(\ell)$ is a species distinct from $H_2O(g)$].

chemical substance: a chemical entity distinguishable by properties 1 or 2 (above), but not by 3; thus $H_2O(\ell)$ and $H_2O(g)$ are the same substance, water.

chemical system: a collection of chemical species and elements denoted by an ordered set of species and an ordered set of the elements contained therein as follows:

$$\{(A_1, A_2, \ldots, A_i, \ldots, A_N), \quad (E_1, E_2, \ldots, E_k, \ldots, E_M)\},$$

where A_i is the molecular formula, together with structural and phase designations, if necessary, of species i and E_k is element k; the order is immaterial but once decided, remains fixed. The list of elements includes (1) each isotope involved in isotopic exchange, (2) the protonic charge p, if ionic species are involved, and (3) a designation such as $X_1, X_2, \ldots$, for each inert substance in the species list, where an inert substance is one that is not involved in the system in the sense of physicochemical change.

formula vector (Brinkley, 1946) $\mathbf{a}_i$: the vector of subscripts (usually integers) to the elements in the molecular formula of a species; for instance, for $C_6H_5NO_2$, $\mathbf{a} = (6, 5, 1, 2)^T$.

formula matrix A: the $M \times N$ matrix in which column i is $\mathbf{a}_i$; $\mathbf{A} = (\mathbf{a}_1, \mathbf{a}_2, \ldots, \mathbf{a}_i, \ldots, \mathbf{a}_N)$; $\mathbf{A}$ is the coefficient matrix in the element-abundance equations 2.2-1.

species-abundance vector n: the vector of nonnegative real numbers representing the numbers of moles of the species in a basis amount of the chemical system; $\mathbf{n} = (n_1, n_2, \ldots, n_i, \ldots, n_N)^T$; $n_i \geq 0$; $\mathbf{n}$ also denotes the composition or compositional state of a system.

element-abundance vector b: the vector of (usually nonnegative) real numbers representing the number of moles of elements in a basis amount of the chemical system: $\mathbf{b} = (b_1, b_2, \ldots, b_k, \ldots, b_M)^T$; $\mathbf{b}$ is often specified by the relative amounts of reactants for the system.

closed chemical system: one for which all possible $\mathbf{n}$ satisfy the element-abundance equations 2.2-3 for some given $\mathbf{b}$.

species-abundance-change vector, $\delta \mathbf{n} = \mathbf{n}^{(2)} - \mathbf{n}^{(1)}$: the changes in mole numbers between compositional states (1) and (2) of the closed chemical system; it must satisfy equation 2.2-4.

feasibility or infeasibility (of a closed system): whether a given $\mathbf{b}$ is compatible with the species list and the preceding definitions of $\mathbf{A}$ and $\mathbf{n}$; for example, for the system $\{(NO_2, N_2O_4), (N, O)\}$; $\mathbf{b} = (b_N, b_O)^T$, $\mathbf{b} = (1, 2)^T$ is feasible, but $\mathbf{b} = (2, 2)^T$ is infeasible; a necessary condition for feasibility is that the rank of the augmented matrix $(\mathbf{A}, \mathbf{b})$, obtained from the system of linear equations $\mathbf{An} = \mathbf{b}$ be equal to the rank of $\mathbf{A}$; this is not a sufficient condition because the algebraic theorem on which it is based allows for the possibility of solutions involving negative values for some or all of the n_i; a sufficient condition for infeasibility is that the ranks be unequal; we assume throughout that all systems are feasible.

2.3 CHEMICAL STOICHIOMETRY

2.3.1 Introductory Concepts

In a closed chemical system we are interested in the various compositional states that can arise, subsequent to an initial state, as a result of chemical change within the system. The determination of any of these states is subject to the element-abundance equations. These algebraic equations may alternatively be cast in the form of chemical equations, which is what we usually think of when we speak about chemical stoichiometry. Whether the equations are algebraic or chemical, one of the purposes of chemical stoichiometry is to determine the appropriate number of them, that is, the maximum number that are linearly independent. This number is different for the two types of equation, as described subsequently. For the algebraic equations, it is usually M, but it may be less than this.

The conservation equations usually do not, of course, provide all the information required to determine the composition $\mathbf{n}$. This is most easily seen in terms of equation 2.2-3. The difference between the number of variables N used to describe the composition and the maximum number of linearly independent equations relating $\{n_i\}$ is called the *number of stoichiometric degrees of freedom* F_s. This is then the number of additional relations among the variables required to determine any compositional state. If the state is an equilibrium state, the additional relations arise from thermodynamic conditions, as described in Chapter 3; otherwise, they may arise from kinetic rate laws or from analytical determinations.

Thus far the only linear equations relating $\{n_i\}$ that we have considered are the element-abundance equations 2.2-1. The difference between N and the maximum number of linearly independent element-abundance equations is in general denoted by the symbol R. Throughout this section F_s and R are

Chemical Stoichiometry

numerically equal because only equations 2.2-1 are involved as linear equations relating to $\{n_i\}$. They are not equal in general, however, and this is discussed in Section 2.4.

Chemical stoichiometry enables us to determine the values of F_s and R for a given system (i.e., one for which $\mathbf{A}$ is known) and to write a permissible set of chemical equations. Before describing a method for doing this, however, we describe the genesis of chemical stoichiometry and chemical equations from the conservation equations.

2.3.2 General Treatment of Chemical Stoichiometry*

The general solution of equation 2.2-1 or 2.2-3, a set of M linear equations in N unknowns, is

$$\mathbf{n} = \mathbf{n}^\circ + \sum_{j=1}^{R} \boldsymbol{\nu}_j \xi_j, \tag{2.3 1}$$

where $\mathbf{n}^\circ$ is any particular solution (e.g., an initial composition), $(\boldsymbol{\nu}_1, \boldsymbol{\nu}_2, \ldots, \boldsymbol{\nu}_R)$ is any set of R linearly independent solutions of the homogeneous equation corresponding to equation 2.2-3 (i.e. equation 2.2-4), and the quantities ξ_j are a set of real parameters. Each $\boldsymbol{\nu}_j$ is called a *stoichiometric vector*, defined in general as follows:

stoichiometric vector $\boldsymbol{\nu}$: any nonzero vector of N real numbers satisfying the equation $\mathbf{A}\boldsymbol{\nu} = \mathbf{0}$. Hence

$$\mathbf{A}\boldsymbol{\nu}_j = \mathbf{0}; \quad (\boldsymbol{\nu}_j \neq \mathbf{0}); \quad j = 1, 2, \ldots, R, \tag{2.3-2}$$

which may also be written as

$$\sum_{i=1}^{N} a_{ki} \nu_{ij} = 0; \quad k = 1, 2, \ldots, M; \quad j = 1, 2, \ldots, R, \tag{2.3-3}$$

and $\nu_{ij} \neq 0$ for at least one i for every j. The quantity R is the maximum number of linearly independent solutions of equations 2.3-2 and is given by

$$R = N - C, \tag{2.3-4}$$

where

$$C = \text{rank } (\mathbf{A}). \tag{2.3-5}$$

Usually, but not always, $C = M$.

An alternative way of regarding the parameters $\{\xi_j\}$ and the quantities $\{\nu_{ij}\}$ may be obtained from further examination of equation 2.3-1, which may be

---

*An elementary treatment has been described by Smith and Missen (1979) and has been illustrated for a simple system.

written as

$$n_i = n_i^\circ + \sum_{j=1}^{R} \nu_{ij}\xi_j; \qquad i = 1, 2, \ldots, N. \tag{2.3-1a}$$

For fixed $\mathbf{n}^\circ$, we have

$$\left(\frac{\partial n_i}{\partial \xi_j}\right)_{\xi_{k \neq j}} = \nu_{ij}; \qquad i = 1, 2, \ldots, N; \qquad j = 1, 2, \ldots, R, \tag{2.3-6}$$

where the notation $\xi_{k \neq j}$ means all ξ's other than the jth, and ν_{ij} is called the *stoichiometric coefficient* of the ith species in the jth stoichiometric vector. Thus ν_{ij} is the rate of change of the mole number of the ith species n_i with respect to the reaction parameter ξ_j.

Further significance of ξ_j is discussed in Section 2.3.5. Here we note that equation 2.3-1 may be regarded as essentially a linear transformation from the N independent variables $\mathbf{n}$ to the R independent variables $\boldsymbol{\xi}$. The variables $\mathbf{n}$ are constrained by the element-abundance equations 2.2-3, whereas the variables $\boldsymbol{\xi}$ are not so constrained, since for any $\{\xi_j\}$, premultiplication of equation 2.3-1 by $\mathbf{A}$ gives

$$\mathbf{An} = \mathbf{An}^\circ + \sum_{j=1}^{R} \xi_j \mathbf{A}\boldsymbol{\nu}_j.$$

The first term on the right is $\mathbf{b}$, and the second term vanishes because of the definition of the stoichiometric vectors (see equation 2.3-2).

The chemical significance of equation 2.3-1 is that *any* compositional state of the system $\mathbf{n}$ can be written in terms of any *particular* state $\mathbf{n}^\circ$ and a linear combination of a set of R linearly independent vectors $\boldsymbol{\nu}_j$ satisfying equation 2.2-4.

Equation 2.3-2 leads naturally to the concept of *chemical equations*. What we call a "chemical equation" is simply a chemical shorthand way of writing equation 2.3-2 or 2.3-3, in which the columns of $\mathbf{A}$ are replaced by the corresponding molecular formulas of the species.

Equations 2.3-3 may be written in terms of the columns of $\mathbf{A}$ as

$$\sum_{i=1}^{N} \mathbf{a}_i \nu_{ij} = \mathbf{0}; \qquad j = 1, 2, \ldots, R. \tag{2.3-7}$$

A set of chemical equations results from equations 2.3-7 when we replace the formula vectors $\mathbf{a}_i$ by their species names A_i and the vector $\mathbf{0}$ by 0:

$$\sum_{i=1}^{N} A_i \nu_{ij} = 0; \qquad j = 1, 2, \ldots, R. \tag{2.3-8}$$

Chemical Stoichiometry

Such equations are a chemical shorthand way of writing the vector equations 2.3-7 (or equation 2.3-2).

To be able to use these concepts in actual situations, we must be able to determine the quantities R and a set of R linearly independent stoichiometric vectors $\{\nu_j\}$. We discuss a systematic numerical determination of these quantities in the next section but first use an example to illustrate the definitions.

Example 2.2 Consider the system $\{(NH_3, O_2, NO, NO_2, H_2O), (N, H, O)\}$ in Example 2.1, in which the formula matrix $\mathbf{A}$ is given. The vector $\nu_1 = (0, -\frac{1}{2}, -1, 1, 0)^T$ is a stoichiometric vector since it satisfies $\mathbf{A}\nu = \mathbf{0}$; that is,

$$\begin{pmatrix} 1 & 0 & 1 & 1 & 0 \\ 3 & 0 & 0 & 0 & 2 \\ 0 & 2 & 1 & 2 & 1 \end{pmatrix} \begin{pmatrix} 0 \\ -\frac{1}{2} \\ -1 \\ 1 \\ 0 \end{pmatrix} = \begin{pmatrix} 0 \\ 0 \\ 0 \end{pmatrix}.$$

Another stoichiometric vector for this system is $\nu_2 = (-\frac{2}{3}, -\frac{5}{6}, \frac{2}{3}, 0, 1)^T$. These two vectors are linearly independent because of the values of the last two entries of each vector. The rank of $\mathbf{A}$ is $C = 3$, and hence the maximum number of linearly independent vectors is $R = 5 - 3 = 2$.

Any composition of the system can be written from equation 2.3-1 as

$$\mathbf{n} = \mathbf{n}^\circ + \left(0, -\tfrac{1}{2}, -1, 1, 0\right)^T \xi_1 + \left(-\tfrac{2}{3}, -\tfrac{5}{6}, \tfrac{2}{3}, 0, 1\right)^T \xi_2.$$

Equations 2.3-7 for this system are

$$0\begin{pmatrix}1\\3\\0\end{pmatrix} - \tfrac{1}{2}\begin{pmatrix}0\\0\\2\end{pmatrix} - 1\begin{pmatrix}1\\0\\1\end{pmatrix} + 1\begin{pmatrix}1\\0\\2\end{pmatrix} + 0\begin{pmatrix}0\\2\\1\end{pmatrix} = \begin{pmatrix}0\\0\\0\end{pmatrix},$$

and

$$-\tfrac{2}{3}\begin{pmatrix}1\\3\\0\end{pmatrix} - \tfrac{5}{6}\begin{pmatrix}0\\0\\2\end{pmatrix} + \tfrac{2}{3}\begin{pmatrix}1\\0\\1\end{pmatrix} + 0\begin{pmatrix}1\\0\\2\end{pmatrix} + 1\begin{pmatrix}0\\2\\1\end{pmatrix} = \begin{pmatrix}0\\0\\0\end{pmatrix}.$$

Replacing the formula vectors by the names of the respective species A_i and $\mathbf{0}$ by 0, we have

$$0NH_3 - \tfrac{1}{2}O_2 - 1NO + 1NO_2 + 0H_2O = 0,$$

$$-\tfrac{2}{3}NH_3 - \tfrac{5}{6}O_2 + \tfrac{2}{3}NO + 0NO_2 + 1H_2O = 0.$$

Conventionally, species names with negative stoichiometric coefficients are written on the left side of a chemical equation and those with positive coefficients on the right side, so that negative numbers do not appear. Thus clearing of fractions and zero quantities and rearranging in accordance with this convention, we have

$$2NO + O_2 = 2NO_2$$

and

$$4NH_3 + 5O_2 = 4NO + 6H_2O.$$

A linearly independent set of R stoichiometric vectors $\{\nu_j\}$ is called a *complete* set of stoichiometric vectors for the system with formula matrix **A**. This is an appropriate name since, from equations 2.3-1, we can determine *any* possible solution **n** of the element-abundance equations by specifying, by some means other than chemical stoichiometry itself, an appropriate set of R ξ_j values (relative to a suitable $\mathbf{n}°$), along with the matrix. A concise way of writing any set of stoichiometric vectors is by defining a matrix **N** whose columns are the vectors ν_j; that is,

$$\mathbf{N} = (\nu_1, \nu_2, \ldots, \nu_q). \tag{2.3-9}$$

When $q = R$ and all ν_j are linearly independent, **N** is "complete," and hence we define the following:

complete stoichiometric matrix N: an $N \times R$ matrix whose R columns are linearly independent stoichiometric vectors, with the additional specification that $R = N - \text{rank}(\mathbf{A})$ (equations 2.3-4 and -5); this condition implies that rank $(\mathbf{N}) = R$. This enables us to write equations 2.3-2 as the single matrix equation

$$\mathbf{AN} = \mathbf{0}. \tag{2.3-10}$$

Analogous to the idea of a complete set of stoichiometric vectors, we define the following:

complete set of chemical equations: the set of equations 2.3-8, where the ν_{ij} form a complete stoichiometric matrix **N**, as defined previously. We emphasize that such a set of equations is not unique since any one equation can be replaced by a linear combination of any of the equations. It is generated solely from the list of species presumed (or demonstrated) to be present, that is, from **A**, and neither requires nor implies any knowledge of reactions presumed to be taking place, or of reaction mechanisms.

If we define

$$C = \text{rank}(\mathbf{A}), \tag{2.3-5}$$

as previously, the significance of C is as follows: given R n_i values, we can solve equations 2.2-3 for C n_i values, provided that the formula vectors of those C n_i values are linearly independent. This is equivalent to partitioning the species into two groups, components (numbering C) and noncomponents (numbering R). The components may be regarded as chemical "building blocks" for forming the noncomponents in chemical equations, one equation being required for each noncomponent. This leads to the following definition:

component: one of a set of C species of the chemical system, whose set of formula vectors $\{\mathbf{a}_1, \mathbf{a}_2, \ldots, \mathbf{a}_C\}$ satisfies rank $(\mathbf{a}_1, \mathbf{a}_2, \ldots, \mathbf{a}_C) = C$ [where $C =$ rank $(\mathbf{A})$].

Example 2.3 For the system described in Examples 2.1 and 2.2, a complete stoichiometric matrix is

$$\mathbf{N} = \begin{pmatrix} 0 & -\frac{2}{3} \\ -\frac{1}{2} & -\frac{5}{6} \\ -1 & \frac{2}{3} \\ 1 & 0 \\ 0 & 1 \end{pmatrix}.$$

Equation 2.3-10 becomes

$$\begin{pmatrix} 1 & 0 & 1 & 1 & 0 \\ 3 & 0 & 0 & 0 & 2 \\ 0 & 2 & 1 & 2 & 1 \end{pmatrix} \begin{pmatrix} 0 & -\frac{2}{3} \\ -\frac{1}{2} & -\frac{5}{6} \\ -1 & \frac{2}{3} \\ 1 & 0 \\ 0 & 1 \end{pmatrix} = \begin{pmatrix} 0 & 0 \\ 0 & 0 \\ 0 & 0 \end{pmatrix}.$$

This matrix equation is equivalent to the two vector equations in Example 2.2. Hence a complete set of chemical equations for this system is given by the two chemical equations written there.

Since $C = 3$ for this system, a set of components is given by $\{A_1, A_2, A_3\}$, where $\{\mathbf{a}_1, \mathbf{a}_2, \mathbf{a}_3\}$ are linearly independent. The nine possible sets of components are $\{NH_3, O_2, NO\}$, $\{NH_3, O_2, NO_2\}$, $\{NH_3, O_2, H_2O\}$, $\{NH_3, NO, NO_2\}$, $\{NH_3, NO, H_2O\}$, $\{NH_3, NO_2, H_2O\}$, $\{O_2, NO, H_2O\}$, $\{O_2, NO_2, H_2O\}$, and $\{NO, NO_2, H_2O\}$.

2.3.3 The Stoichiometric Procedure/Algorithm

The procedure simultaneously determines rank $(\mathbf{A})$ and a complete set of chemical equations. HP-41C, BASIC, and FORTRAN computer programs implementing it are given in Appendix A, and we describe the "hand calculation" procedure here. This procedure can also be used for balancing oxidation-reduction equations in inorganic and analytical chemistry, as an alternative to other methods, such as the half-reaction method that uses oxidation numbers (Mahan, 1975, pp. 257–265), and ion-electron and valence-electron methods (Engelder, 1942, pp. 122–127), which require additional concepts.

The procedure is similar to that used in the solution of linear algebraic equations by Gauss-Jordan reduction (Noble, 1969, pp. 65–66). It involves the reduction of the formula matrix **A** to unit matrix form (Noble, 1969, pp. 131–132) by elementary row operations (Noble, 1969, p. 78). The unit matrix form is represented by

$$\mathbf{A}^* = \begin{pmatrix} \mathbf{I}_C & \mathbf{Z} \\ \mathbf{0} & \mathbf{0} \end{pmatrix}, \qquad (2.3\text{-}11)$$

where $\mathbf{I}_C$ is a $(C \times C)$ identity matrix and $\mathbf{Z}$ is a $(C \times R)$ matrix, at least one of whose elements is nonzero; C is the rank of $\mathbf{A}^*$ as well as the rank of $\mathbf{A}$. In many cases the **0** submatrices are absent.

A complete stoichiometric matrix is formed from $\mathbf{A}^*$ by appending the $R \times R$ identity matrix below $-\mathbf{Z}$; thus

$$\mathbf{N} = \begin{pmatrix} -\mathbf{Z} \\ \mathbf{I}_R \end{pmatrix}, \qquad (2.3\text{-}12)$$

(Schneider and Reklaitis, 1975; Schubert and Hofmann, 1975, 1976). A complete stoichiometric matrix expressed in the form of equation 2.3-12, that is, one that contains the $R \times R$ identity matrix, is said to be in *canonical* form. Here **N** is a complete stoichiometric matrix for **A** since our Gauss-Jordan procedure essentially constructs the columns of **Z** in $\mathbf{A}^*$ to satisfy

$$\mathbf{A}_C \mathbf{Z} = \mathbf{A}_R, \qquad (2.3\text{-}13)$$

where the columns of $\mathbf{A}_C$ refer to a set of component species and the columns of $\mathbf{A}_R$ refer to the remaining species. Thus we have

$$\mathbf{Z} = \mathbf{A}_C^{-1} \mathbf{A}_R, \qquad (2.3\text{-}14)$$

and hence

$$\mathbf{AN} = (\mathbf{A}_C, \mathbf{A}_R) \begin{pmatrix} -\mathbf{A}_C^{-1} \mathbf{A}_R \\ \mathbf{I}_R \end{pmatrix} = -\mathbf{A}_R + \mathbf{A}_R = \mathbf{0}. \qquad (2.3\text{-}15)$$

In addition to row operations, column interchanges may be required to obtain the unit matrix form, depending on the way in which the species have been arbitrarily ordered at the outset as columns of **A**.

The steps of the procedure are as follows (Smith and Missen, 1979):

1 Write the formula matrix **A** for the given system, with each column identified above it by the chemical species represented.

Chemical Stoichiometry

2 Form a unit matrix as large as possible in the upper-left portion of **A** by elementary row operations, and column interchange if necessary; if columns are interchanged, the designation of the species (above the column) must be interchanged also. The final result is a matrix **A***, as in equation 2.3-11.

3 At the end of these steps, the following are established:

(a) The rank of the matrix **A**, which is C, the number of components, is the number of 1's on the principal diagonal of **A***;

(b) A set of components is given by the C species indicated above the columns of the unit matrix;

(c) The maximum number of linearly independent stoichiometric equations is given by $R = N - C$; and

(d) A complete stoichiometric matrix **N** in canonical form is obtained from the submatrix **Z** in equation 2.3-11, according to equation 2.3-12; each equation in a permissible set of chemical equations is obtained from a column of **N** by first writing equation 2.3-8 and then rearranging, using the convention described in Example 2.2.

2.3.4 Illustration of the Treatment and Procedure

The procedure described in Section 2.3.3 can be used for a chemically reacting system that involves inert species, charged species, and mass transfer between phases. The first two of these have been illustrated previously (Smith and Missen, 1979), and we illustrate the third here.

Example 2.4 Consider the esterification of ethyl alcohol (C_2H_6O) with acetic acid ($C_2H_4O_2$) to form water and ethyl acetate ($C_4H_8O_2$) in a liquid (ℓ)-vapor(g) contact, which allows for the presence of acetic acid dimer in the vapor phase (Sanderson and Chien, 1973). The system is represented by

$$\{(H_2O(\ell), C_2H_6O(\ell), C_2H_4O_2(\ell), C_4H_8O_2(\ell), H_2O(g), C_2H_6O(g),$$

$$C_2H_4O_2(g), C_4H_8O_2(g), (C_2H_4O_2)_2(g)), (C, H, O)\}.$$

For this system, we use the procedure described in Section 2.3.3 to determine the number of components C, a set of components, the number of chemical equations $R(= F_s)$, and a permissible set of chemical equations.

Following the steps outlined previously, we have, with $N = 9$ and $M = 3$,

1

$$\begin{array}{c} \quad (1)\ (2)\ (3)\ (4)\ (5)\ (6)\ (7)\ (8)\ (9) \\ \mathbf{A} = \begin{pmatrix} 0 & 2 & 2 & 4 & 0 & 2 & 2 & 4 & 4 \\ 2 & 6 & 4 & 8 & 2 & 6 & 4 & 8 & 8 \\ 1 & 1 & 2 & 2 & 1 & 1 & 2 & 2 & 4 \end{pmatrix} \end{array}$$

Here the numbers at the tops of the columns correspond to the species in the order given, and the rows are in the order of the elements given.

2. The matrix **A** can be put in the following form by means of elementary row operations and column interchanges.

$$\begin{array}{c} (2)\ (1)\ (3)\ \ (4)\ (5)\ (6)\ (7)\ \ \ (8)\ (9) \\ \mathbf{A}^* = \begin{pmatrix} 1 & 0 & 0 & 1 & 0 & 1 & 0 & 1 & 0 \\ 0 & 1 & 0 & -1 & 1 & 0 & 0 & -1 & 0 \\ 0 & 0 & 1 & 1 & 0 & 0 & 1 & 1 & 2 \end{pmatrix} \end{array}$$

3. (a) Rank (**A**) = C = 3;
   (b) a set of components is $\{H_2O(\ell)(1), C_2H_6O(\ell)(2), C_2H_4O_2(\ell)(3)\}$;
   (c) $R = N - C = 9 - 3 = 6$;
   (d) reordering the list of species according to the designations above $\mathbf{A}^*$, we have the following complete stoichiometric matrix:

$$\mathbf{N} = \begin{bmatrix} -1 & 0 & -1 & 0 & -1 & 0 \\ 1 & -1 & 0 & 0 & 1 & 0 \\ -1 & 0 & 0 & -1 & -1 & -2 \\ 1 & 0 & 0 & 0 & 0 & 0 \\ 0 & 1 & 0 & 0 & 0 & 0 \\ 0 & 0 & 1 & 0 & 0 & 0 \\ 0 & 0 & 0 & 1 & 0 & 0 \\ 0 & 0 & 0 & 0 & 1 & 0 \\ 0 & 0 & 0 & 0 & 0 & 1 \end{bmatrix}.$$

This corresponds to the following set of six chemical equations:

$$C_2H_6O(\ell) + C_2H_4O_2(\ell) = H_2O(\ell) + C_4H_8O_2(\ell),$$

$$H_2O(\ell) = H_2O(g),$$

$$C_2H_6O(\ell) = C_2H_6O(g),$$

$$C_2H_4O_2(\ell) = C_2H_4O_2(g),$$

$$C_2H_6O(\ell) + C_2H_4O_2(\ell) = H_2O(\ell) + C_4H_8O_2(g),$$

$$2C_2H_4O_2(\ell) = (C_2H_4O_2)_2(g).$$

2.3.5 The Extent of Reaction

The quantity ξ introduced in equation 2.3-1 is the extent-of-reaction parameter originally introduced by De Donder (1936, p. 2; Prigogine and Defay, 1954, p.

10) to measure the "degree of advancement of a reaction." The quantities ξ_j (i.e., for R chemical equations) were introduced previously as a set of real parameters in establishing the concept of chemical equations from the element-abundance equations. If we accept the existence of chemical equations *ab initio*, then equations 2.3-1 and 2.3-6 define a set of quantities ξ_j, one such quantity for each chemical equation written. The extent of reaction is a useful variable for equilibrium computations. From equation 2.3-1, it is an *extensive* quantity.

2.4 EXPRESSING COMPOSITIONAL RESTRICTIONS IN STANDARD FORM

2.4.1 Introduction

We discussed in Section 2.3 how a complete stoichiometric matrix **N** and a corresponding complete set of chemical equations can be obtained when the formula matrix **A** of a system is given. These procedures are essentially those used in the computer program VCS in Appendix D, which calculates equilibrium compositions. Whenever **A** is given at the outset, we advocate the formation of an **N** matrix in this way. This guarantees that

$$\text{rank}(\mathbf{N}) = N - \text{rank}(\mathbf{A}), \tag{2.4-1}$$

in which case we say that the compositional restrictions are expressed in *standard* form. However, if an **N** matrix is given at the outset, there is no guarantee that is the case. The purpose of this section is to show how the formula matrix **A** can be modified (if necessary) so that equation 2.4-1 is satisfied. An important situation in which an **N** matrix *is* specified at the outset is in problems involving only mass transfer of substances between phases.

The key feature of a complete stoichiometric matrix, corresponding to a given formula matrix **A**, is that its rank R is given by equation 2.3-4, where R is the proper number of stoichiometric equations needed to describe all possible compositions of the system. Normally, we do *not* advocate forming an **N** matrix for a set of chemical equations written or suggested *ab initio* by some means because it is not necessarily assured that such a matrix has the correct rank R. Typical situations giving rise to a stoichiometric matrix whose rank is incorrect are: (1) there may be too many equations in such a set, in the sense that they are not all linearly independent; and (2) even if the equations written *are* linearly independent, they do not necessarily represent the maximum possible number of linearly independent equations.

Occasionally, however, we may wish to consider a specific **N** matrix at the outset. For example, an **N** matrix may be suggested by a kinetic mechanism. Such a mechanism must be examined to ensure that rank $(\mathbf{N}) = R$. Even if rank $(\mathbf{N})$ is correct, kinetic schemes must pass other stoichiometric tests (Ridler et al., 1977; Oliver, 1980), which are related to the nonnegativity constraints on the mole numbers. Here we discuss how such **N** matrices may be utilized. Indeed, some authors approach chemical stoichiometry in this way (e.g., Aris and Mah, 1963).

We discuss such problems here partly because it is worthwhile to view this approach in terms of the formulation in Sections 2.2 and 2.3 and partly because certain types of **N** matrix have special properties, which are explored in Section 2.4.3. They also specify some special kinds of chemical equilibrium problem, which we treat in Chapter 9.

2.4.2 Reduction of a Given Stoichiometric Matrix to Standard Form

For the formation of hydrogen bromide from hydrogen and bromine, the relevant system in kinetic terms is $\{(Br_2, H_2, HBr, H, Br), (H, Br)\}$. The accepted chain-reaction mechanism (e.g., Moore, 1972, p. 398) is

$$Br_2 \rightarrow 2Br$$

$$Br + H_2 \rightarrow HBr + H$$

$$H + Br_2 \rightarrow HBr + Br$$

$$H + HBr \rightarrow H_2 + Br$$

$$2Br \rightarrow Br_2$$

A stoichiometric matrix **N** is constructed from the stoichiometric coefficients in the kinetic scheme, with each column corresponding to a given reaction, made up by the coefficients of the species in that reaction. With the equations and species ordered as indicated previously, a stoichiometric matrix is

$$\mathbf{N} = \begin{pmatrix} -1 & 0 & -1 & 0 & 1 \\ 0 & -1 & 0 & 1 & 0 \\ 0 & 1 & 1 & -1 & 0 \\ 0 & 1 & -1 & -1 & 0 \\ 2 & -1 & 1 & 1 & -2 \end{pmatrix}.$$

Since we can establish from the formula matrix of the system, according to the method described in Section 2.3.3, that there are at most three linearly independent chemical equations for this system, there must be two columns too many for **N**.

We can determine a linearly independent set of chemical equations by performing elementary column operations on **N**. This is equivalent to performing elementary row operations on $\mathbf{N}^T$ in the same way as for **A** in the computer

Expressing Compositional Restrictions in Standard Form

programs listed in Appendix A. We can thus use the rows of **N** as input to this program. Since the matrix in this case is small, we illustrate the procedure by hand. This results eventually in the matrix

$$(\mathbf{N}^*)^T = \begin{pmatrix} 1 & 0 & 0 & 0 & -2 \\ 0 & 1 & 0 & -2 & 0 \\ 0 & 0 & 1 & -1 & -1 \\ 0 & 0 & 0 & 0 & 0 \\ 0 & 0 & 0 & 0 & 0 \end{pmatrix}.$$

The rank of $(\mathbf{N}^*)^T$ is 3, and hence a complete stoichiometric matrix is given by the nonzero rows of $(\mathbf{N}^*)^T$,

$$\begin{pmatrix} 1 & 0 & 0 \\ 0 & 1 & 0 \\ 0 & 0 & 1 \\ 0 & -2 & -1 \\ -2 & 0 & -1 \end{pmatrix}.$$

A linearly independent set of chemical equations is hence given by

$$2\text{Br} = \text{Br}_2,$$

$$2\text{H} = \text{H}_2,$$

$$\text{H} + \text{Br} = \text{HBr}.$$

Usually, but not always, the number of linearly independent chemical equations that results coincides with

$$\text{rank}(\mathbf{N}) = R. \tag{2.4-2}$$

Equation 2.4-2 essentially means that the given **N** can be reduced to a complete stoichiometric matrix. The original **N** matrix is often in error in the sense that it has *too many* columns (as in the preceding example). We see in what follows, however, that occasionally *fewer* than the maximum number of chemical equations may occur.

2.4.3 Stoichiometric Degrees of Freedom and Additional Stoichiometric Restrictions

We have seen the way in which the specification of the formula matrix **A** for a chemical system consisting of N species restricts the allowable compositions **n** to those satisfying the element-abundance constraints of equations 2.2-1 or 2.2-3. The number of linearly independent constraints posed by these restrictions is given by $C = \text{rank}(\mathbf{A})$. Thus a total of $(N - C)$ mole numbers of appropriate species of the composition vector must be specified for the remaining mole numbers to be determined. The number ($N - C$) is essentially the number of degrees of freedom that are imposed by the element-abundance constraints. We have denoted this number by R, which is then defined by

$$R = N - C. \tag{2.3-4}$$

We also introduced in Section 2.3.1 the quantity F_s to denote the stoichiometric degrees of freedom. When we specify *a priori* a stoichiometric matrix **N** for a system, the number of stoichiometric degrees of freedom F_s is defined as

$$F_s = \text{rank}(\mathbf{N}). \tag{2.4-3}$$

When a formula matrix **A** is specified and the only compositional restrictions are the element-abundance constraints, we have seen that

$$F_s = R = N - \text{rank}(\mathbf{A}). \tag{2.4-4}$$

In general,

$$F_s \leq R, \tag{2.4-5}$$

since for the matrix equation

$$\mathbf{AN} = \mathbf{0}, \tag{2.3-10}$$

we must have (Noble, 1969, p. 142)

$$\text{rank } \mathbf{N} \leq N - \text{rank}(\mathbf{A}), \tag{2.4-6}$$

where N is the number of columns of **A** and the number of rows of **N**. For convenience, following equation 2.4-5, we define the quantity r as the difference between R and F_s*:

$$\begin{aligned} r &= R - F_s \\ &= N - \text{rank}(\mathbf{A}) - \text{rank}(\mathbf{N}), \end{aligned} \tag{2.4-7}$$

and we call r the number of special stoichiometric restrictions.

The purpose of this section is to show how the r additional stoichiometric restrictions can be combined with the usual element-abundance constraints to

---

*In the previous treatment (Smith and Missen, 1979) this distinction was not drawn, and hence the treatment was restricted to cases in which $r = 0$. The distinction is necessary in going beyond "pure" stoichiometry ($r \neq 0$). It has, in effect, been emphasized by Björnbom (1975, 1977, 1981) from another point of view.

Expressing Compositional Restrictions in Standard Form

form a modified matrix $\mathbf{A}'$, and modified element-abundance vector $\mathbf{b}'$ so that

$$F_s = N - \text{rank}(\mathbf{A}'). \tag{2.4-8}$$

The importance of this is that we can treat the *combined* set of constraints in the same manner as before by using the modified formula matrix $\mathbf{A}'$. Since the right side of equation 2.4-8 is R for the modified system, we have $r = 0$; that is,

$$r = N - \text{rank}(\mathbf{A}') - \text{rank}(\mathbf{N}) = 0. \tag{2.4-9}$$

Thus the compositional restrictions are expressed in standard form.

We call a pair of matrices $\mathbf{A}'$ and $\mathbf{N}$ satisfying equation 2.4-9 "compatible" matrices. In the remaining parts of this section we show how to obtain compatible matrices for two types of problem in which we effectively have $r > 0$. The two cases refer to situations in which the additional stoichiometric restrictions arise both explicitly and implicitly.

2.4.4 Additional Stoichiometric Restrictions that Arise Explicitly

Suppose that it is known experimentally that the amounts of two species p and q are always equal. This can be written as

$$n_p - n_q = 0. \tag{2.4-10}$$

In this case $r = 1$, and from equation 2.4-7, we obtain

$$F_s = R - 1 = N - C - 1.$$

We wish to be able to account for this restriction by means of a modified *complete* stoichiometric matrix $\mathbf{N}'$ that is compatible with a formula matrix $\mathbf{A}'$.

Example 2.5 Consider the system

$$\{(C_6H_5CH_3, H_2, C_6H_6, CH_4), (C, H)\}$$

discussed by Björnbom (1975) and by Schneider and Reklaitis (1975). Suppose that it is known experimentally that if the initial state is toluene, the resulting benzene and methane occur in equimolar amounts.

To see how this additional constraint is incorporated, we first write the usual element-abundance constraints and the compositional restriction of equation 2.4-10. Thus we have

$$\mathbf{An} = \begin{pmatrix} 7 & 0 & 6 & 1 \\ 8 & 2 & 6 & 4 \end{pmatrix} \begin{pmatrix} n_1 \\ n_2 \\ n_3 \\ n_4 \end{pmatrix} = \begin{pmatrix} b_1 \\ b_2 \end{pmatrix}, \tag{2.4-11}$$

and

$$(0 \quad 0 \quad 1 \quad -1) \begin{pmatrix} n_1 \\ n_2 \\ n_3 \\ n_4 \end{pmatrix} = 0. \qquad (2.4\text{-}12)$$

Here rank $(\mathbf{A}) = 2$, $N = 4$, $R = 2$, $r = 1$, and $F_s = 4 - 2 - 1 = 1$.

Let us now see how this example can be presented in standard form. Equations 2.4-11 and 2.4-12 are equivalent to the single set of equations

$$\mathbf{A'n} = \mathbf{b'}, \qquad (2.4\text{-}13)$$

where the matrix $\mathbf{A'}$ and the vector $\mathbf{b'}$ are

$$\mathbf{A'} = \begin{pmatrix} 7 & 0 & 6 & 1 \\ 8 & 2 & 6 & 4 \\ 0 & 0 & 1 & -1 \end{pmatrix},$$

and

$$\mathbf{b'} = \begin{pmatrix} b_1 \\ b_2 \\ 0 \end{pmatrix}.$$

Equation 2.4-13 now incorporates *all* the compositional constraints on the system. We treat $\mathbf{A'}$ as a modified system formula matrix and obtain a complete stoichiometric matrix for it. This yields

$$\mathbf{N'} = \begin{pmatrix} -1 \\ -1 \\ 1 \\ 1 \end{pmatrix}.$$

We have combined the element-abundance constraints and the additional stoichiometric restriction in equation 2.4-13 (cf. $\mathbf{An} = \mathbf{b}$). From this, rank $(\mathbf{A'}) = 3$, and $F_s = 4 - 3 = 1 = $ rank $(\mathbf{N'}) = N - $ rank $(\mathbf{A'})$.

In general, we may incorporate *any* additional constraints that are of the form

$$\mathbf{Dn} = \mathbf{d}, \qquad (2.4\text{-}14)$$

where

$$\text{rank } (\mathbf{D}) = r. \qquad (2.4\text{-}15)$$

Expressing Compositional Restrictions in Standard Form

We do this by forming the modified formula matrix $\mathbf{A}'$ and element-abundance vector $\mathbf{b}'$ by means of

$$\mathbf{A}' = \begin{pmatrix} \mathbf{A} \\ \mathbf{D} \end{pmatrix} \tag{2.4-16a}$$

and

$$\mathbf{b}' = \begin{pmatrix} \mathbf{b} \\ \mathbf{d} \end{pmatrix} \tag{2.4-16b}$$

All the constraints are now expressed in the single equation

$$\mathbf{A}'\mathbf{n} = \mathbf{b}'. \tag{2.4-13}$$

We form a complete stoichiometric matrix $\mathbf{N}'$ from $\mathbf{A}'$ in the usual way. Then the problem is formulated in standard form:

$$F_s = N - \text{rank}\,(\mathbf{A}') = \text{rank}\,(\mathbf{N}'). \tag{2.4-17}$$

It follows that every possible composition $\mathbf{n}$ of the system is given by the general solution of equation 2.4-13, which is

$$\mathbf{n} = \mathbf{n}^\circ + \sum_{j=1}^{F_s} \mathbf{\nu}_j \xi_j. \tag{2.4-18}$$

2.4.5 Additional Stoichiometric Restrictions that Arise Implicitly

We consider here the general situation in which an $\mathbf{N}$ matrix is specified *a priori* as determining the allowable compositions a system may attain, starting from some particular given composition $\mathbf{n}^\circ$. We assume that all the columns of $\mathbf{N}$ are linearly independent. If this is not the case, we use the methods described in Section 2.4.2 to achieve this. For a given stoichiometric matrix $\mathbf{N}$, we show how a "fictitious" formula matrix $\mathbf{A}$ can be found, thus enabling us to treat the problem in standard form. That is, we want to have

$$F_s = N - \text{rank}\,(\mathbf{A}) = \text{rank}\,(\mathbf{N}). \tag{2.4-19}$$

The solution of the problem is relatively straightforward, and we can perhaps appreciate this best by considering an example of an *unrestricted* stoichiometric system.

Example 2.6 Consider the system $\{(CH_4, O_2, CO_2, H_2O, H_2), (C, H, O)\}$. A complete stoichiometric matrix is

$$\mathbf{N}^T = \begin{pmatrix} -\frac{1}{2} & -1 & \frac{1}{2} & 1 & 0 \\ -\frac{1}{2} & -\frac{1}{2} & \frac{1}{2} & 0 & 1 \end{pmatrix}.$$

Now form the 3×5 matrix with the first three columns being the 3×3 identity matrix and the last two columns being the negative of the first three elements of the rows of $\mathbf{N}^T$. This yields the matrix

$$\mathbf{A}^* = \begin{pmatrix} 1 & 0 & 0 & \frac{1}{2} & \frac{1}{2} \\ 0 & 1 & 0 & 1 & \frac{1}{2} \\ 0 & 0 & 1 & -\frac{1}{2} & -\frac{1}{2} \end{pmatrix}.$$

It is readily verified that this $\mathbf{A}^*$ is compatible with $\mathbf{N}$ given previously since

$$\mathbf{A}^*\mathbf{N} = \mathbf{0}, \qquad (2.4\text{-}20)$$

and

$$N - \text{rank}(\mathbf{A}^*) = 2 = \text{rank}(\mathbf{N}).$$

Thus, for the matrix $\mathbf{A}^*$, $\mathbf{N}$ is a complete stoichiometric matrix. The system formula vectors are given by

$$\mathbf{a}^*(CH_4) = (1, 0, 0)^T$$

$$\mathbf{a}^*(O_2) = (0, 1, 0)^T$$

$$\mathbf{a}^*(CO_2) = (0, 0, 1)^T$$

$$\mathbf{a}^*(H_2O) = (\tfrac{1}{2}, 1, -\tfrac{1}{2})^T$$

$$\mathbf{a}^*(H_2) = (\tfrac{1}{2}, \tfrac{1}{2}, -\tfrac{1}{2})^T$$

The element-abundance equations corresponding to this formula matrix are

$$\sum_{i=1}^{N} a_{ji}^* n_i = \sum_{i=1}^{N} a_{ji}^* n_i^o = b_j^*, \qquad (2.4\text{-}21)$$

where $\mathbf{n}^o$ is any allowable composition of the system (e.g., the initial composition).

Expressing Compositional Restrictions in Standard Form

The fact that we have produced a rather strange looking formula matrix in this example is not really very strange at all if we examine the situation more carefully.* Normally, we are given $\mathbf{A}$ and then determine a complete stoichiometric matrix $\mathbf{N}$ satisfying

$$\mathbf{AN} = \mathbf{0}. \tag{2.3-10}$$

We saw in the previous discussion of this problem that the matrix $\mathbf{N}$ is not unique but is only subject to the requirement that it have $N - \text{rank}(\mathbf{A})$ linearly independent columns. If we now consider the transpose of equation 2.3-10, we have

$$\mathbf{N}^T \mathbf{A}^T = \mathbf{0}. \tag{2.4-22}$$

Now we consider the case when $\mathbf{N}^T$ is given, and we want to determine a "complete" matrix $\mathbf{A}^*$ that satisfies equation 2.4-22. Just as $\mathbf{N}$ in equation 2.3-10 is not unique, $\mathbf{A}^T$ in equation 2.4-22 is also not unique, but is an $N \times C$ matrix $(\mathbf{A}^*)^T$ that satisfies

$$C = \text{rank}(\mathbf{A}^*) = N - \text{rank}(\mathbf{N}). \tag{2.4-23}$$

The main point we wish to emphasize by means of the preceding example is that we can either start from a given formula matrix $\mathbf{A}$ and obtain a complete (but nonunique) stoichiometric matrix $\mathbf{N}$ or turn the situation around, starting from a given stoichiometric matrix $\mathbf{N}$ and obtain a (nonunique) compatible formula matrix $\mathbf{A}^*$. The actual recipe for constructing $\mathbf{A}^*$ when $\mathbf{N}$ is in a specific form is straightforward and can be performed by inspection.

The *general* prescription is as follows. We start with a matrix $\mathbf{N}$ in the form

$$\mathbf{N} = \begin{pmatrix} \mathbf{I}_{F_s} \\ \mathbf{N}_1 \end{pmatrix}, \tag{2.4-24}$$

where $\mathbf{I}_{F_s}$ is the identity matrix of order F_s and $\mathbf{N}_1$ is an $(N - F_s) \times F_s$ matrix. An arbitrary matrix $\mathbf{N}$ can be put in this form by the method discussed in Section 2.4.2. Then a compatible matrix $\mathbf{A}^*$ is given by

$$\mathbf{A}^* = (-\mathbf{N}_1, \mathbf{I}_{N-F_s}), \tag{2.4-25}$$

where $\mathbf{A}^*$ is compatible with $\mathbf{N}$ since it has $(N - F_s)$ linearly independent columns and satisfies

$$\mathbf{A}^* \mathbf{N} = (-\mathbf{N}_1, \mathbf{I}_{N-F_s}) \begin{pmatrix} \mathbf{I}_{F_s} \\ \mathbf{N}_1 \end{pmatrix} = -\mathbf{N}_1 + \mathbf{N}_1 = \mathbf{0}.$$

*cf. Björnbom (1981).

The element-abundance vector $\mathbf{b}^*$, corresponding to $\mathbf{A}^*$ is given by

$$\mathbf{A}^*\mathbf{n}^\circ = \mathbf{b}^*, \tag{2.4-26}$$

where $\mathbf{n}^\circ$ is any allowable composition of the system, such as the starting composition. Thus the element-abundance constraints are

$$\mathbf{A}^*\mathbf{n} = \mathbf{b}^*.$$

Smith (1976) has also treated Example 2.5 by this implicit approach.

Example 2.7 Consider the system $\{(C_2H_6(\ell), C_3H_6(\ell), C_3H_8(\ell), C_2H_6(g), C_3H_6(g), C_3H_8(g)), (C, H)\}$, in which only mass transfer of the substances between the two phases is allowed. Find $\mathbf{A}^*$ and $\mathbf{b}^*$ so that the problem may be treated in standard form.

Solution The $\mathbf{N}$ matrix is generated at the outset from the chemical equations

$$C_2H_6(g) = C_2H_6(\ell),$$

$$C_3H_6(g) = C_3H_6(\ell),$$

and

$$C_3H_8(g) = C_3H_8(\ell),$$

as

$$\mathbf{N} = \begin{pmatrix} 1 & 0 & 0 \\ 0 & 1 & 0 \\ 0 & 0 & 1 \\ -1 & 0 & 0 \\ 0 & -1 & 0 \\ 0 & 0 & -1 \end{pmatrix}.$$

This is in the same form as equation 2.4-24, with $F_s = 3$. A compatible formula matrix $\mathbf{A}^*$, from equation 2.4-25, is

$$\mathbf{A}^* = \begin{pmatrix} 1 & 0 & 0 & 1 & 0 & 0 \\ 0 & 1 & 0 & 0 & 1 & 0 \\ 0 & 0 & 1 & 0 & 0 & 1 \end{pmatrix}.$$

From equation 2.4-26,

$$\mathbf{b}^* = \mathbf{A}^*\mathbf{n}^\circ = \begin{pmatrix} n_1^\circ + n_4^\circ \\ n_2^\circ + n_5^\circ \\ n_3^\circ + n_6^\circ \end{pmatrix}.$$

PROBLEMS

2.1 (a) Write equation 2.2-3 in full for the system $\{(H_3PO_4, H_2PO_4^-, HPO_4^{2-}, PO_4^{3-}, H^+, OH^-, H_2O), (H, O, P, p)\}$, if the system results from dissolving 2 moles of H_3PO_4 in 1 mole of H_2O.

(b) From the result in part a, write equations 2.2-1 for the system, that is, by multiplying out equation 2.2-3.

2.2 Balance each of the following by Gauss-Jordan reduction, and in so doing show that only one chemical equation is required in each case:

(a) $Na_2O_2 + CrCl_3 + NaOH \neq Na_2CrO_4 + NaCl + H_2O$

(b) $K_2Cr_2O_7 + H_2SO_4 + H_2SO_3 \neq Cr_2(SO_4)_3 + H_2O + K_2SO_4$

(c) $KClO_3 + NaNO_2 \neq KCl + NaNO_3$

(d) $KMnO_4 + H_2O + Na_2SnO_2 \neq MnO_2 + KOH + Na_2SO_4$

2.3 For each of the following systems, determine the number C and a permissible set of components and the maximum number R and a permissible set of independent chemical equations:

(a) $\{(CO, CO_2, H, H_2, H_2O, O, O_2, OH, N_2, NO), (C, H, O, N)\}$

(b) $\{(CH_4, C_2H_2, C_2H_4, C_2H_6, C_6H_6, H_2, H_2O), (C, H, H_2O)\}$

(c) $\{(CH_4, CH_3D, CH_2D_2, CHD_3, CD_4), (C, H, D)\}$ (Apse and Missen, 1967)

(d) $\{(C(gr), CO(g), CO_2(g), Zn(g), Zn(\ell), ZnO(s)), (C, O, Zn)\}$

(e) $\{(Fe(C_2O_4)^+, Fe(C_2O_4)_2^-, Fe(C_2O_4)_3^{3-}, Fe^{3+}, SO_4^{2-}, H\dot{S}O_4^-, H^+, HC_2O_4^-, H_2C_2O_4, C_2O_4^{2-}), (C, Fe, H, O, S, p)\}$ (Swinnerton and Miller, 1959)

(f) $\{(H_2O, H_2O_2, H^+, K^+, MnO_4^-, Mn^{2+}, O_2, SO_4^{2-}), (H, K, Mn, O, S, p)\}$

(g) $\{(C_6H_6(\ell), C_6H_6(g), C_7H_8(\ell), C_7H_8(g), o\text{-}C_8H_{10}(\ell), o\text{-}C_8H_{10}(g), m\text{-}C_8H_{10}(\ell), m\text{-}C_8H_{10}(g), p\text{-}C_8H_{10}(\ell), p\text{-}C_8H_{10}(g)), (C, H)\}$

(h) $\{(O_2(g), H_2O(g), CH_4(g), CO(g), CO_2(g), H_2(g), N_2(g), CHO(g), CH_2O(g), OH(g), Fe(s), FeO(s), Fe_3O_4(s), C(gr), CaO(s), CaCO_3(s)), (O, H, C, Fe, Ca, N_2)\}$ (Madeley and Toguri, 1973b)

2.4 Ethylene can be made by the dehydrogenation of ethane. Methane is a possible by-product, and it is undesirable for the system to approach equilibrium with respect to all these species at the outlet of the reactor, as the following figures show. For a feed that contains 0.4 mole of steam (inert) per mole of C_2H_6, and for an outlet temperature of 1100 K and pressure of 1.6 atm, it can be calculated that if equilibrium obtained at the outlet, there would be 0.515 mole of ethylene per mole of ethane *in the feed* and 0.950 mole of methane. Calculate the mole fraction of each species in the outlet mixture on a steam-free basis.

2.5 One method proposed for making synthesis gas ($CO + H_2$) for methanol manufacture is the partial oxidation of natural gas (say, methane); CO_2 and water are also products. Calculate the complete composition of the resulting gas, if equilibrium is reached at 1310 K and 10 atm and if equilibrium calculations for a feed gas containing 0.522 mole of O_2 per mole of CH_4 indicate that oxygen is virtually used up, that there are 0.0524 mole of CH_4 remaining per mole of CH_4 in the feed, and that 1.820 moles of H_2 are formed.

2.6 In connection with a process for making synthesis gas (CO and H_2) for the Fischer-Tropsch process by the partial oxidation of methane (CH_4), with H_2O and CO_2 as other products, the following chemical equations could conceivably be written:

$$CH_4 + 2O_2 = CO_2 + 2H_2O,$$

$$2CH_4 + O_2 = 2CO + 4H_2,$$

$$2CO + O_2 = 2CO_2,$$

$$2H_2 + O_2 = 2H_2O,$$

$$H_2 + CO_2 = H_2O + CO,$$

and

$$CH_4 + CO_2 = 2CO + 2H_2.$$

How many of these equations are linearly independent?

2.7 For the system $\{(O_2, SO_2, SO_3, N_2), (O, S, N_2)\}$, with $\mathbf{b} = (42, 10, 158)^T$, corresponding to 100 moles of feed of composition 10 mole % SO_2, 11% O_2, and 79% N_2, if the fraction of SO_2 in the feed converted to SO_3 at equilibrium at 883 K and 1 atm is 0.697, calculate the value of the extent of reaction ξ.

2.8 From the results of Problem 2.3b, relate the number of moles of each species to the extents of reaction (ξ), assuming that the system initially consists only of $n^\circ_{C_2H_6}$ moles of C_2H_6 and $n^\circ_{H_2O}$ moles of H_2O.

2.9 Consider the system $\{(o\text{-}C_7H_8O, m\text{-}C_7H_8O, o\text{-}C_7H_8SO_4, m\text{-}C_7H_8SO_4, H_2O, H_2SO_4), (O, S, H, C)\}$ (Whitwell and Dartt, 1973), where o- and m- denote *ortho-* and *meta-* isomers, respectively.

(a) Determine C and a complete stoichiometric matrix for the system.

(b) Suppose that the composition is restricted such that $n_4/n_3 = n^\circ_2/n^\circ_1$, where the superscript denotes a specified initial amount and the subscripts represent species as ordered previously. Incorporate this restriction by obtaining a modified formula matrix **A′** and element-abundance vector **b′** such that **A′n** = **b′**.

(c) Determine a complete stoichiometric matrix for **A′**.

2.10 For the kinetic scheme (Oliver, 1980)

$$A_1 \rightarrow A_2 + A_4$$

$$A_3 \rightarrow A_1 + A_4$$

$$A_3 \rightarrow A_4,$$

where each A_i denotes a chemical species,

(a) construct a stoichiometric matrix.
(b) Reduce this matrix to the form of equation 2.4-24.
(c) Find a compatible formula matrix for the stoichiometric matrix of part b.
(d) Is this formula matrix unusual (Oliver, 1980)?

CHAPTER THREE

Chemical Thermodynamics and Equilibrium Conditions

In Chapter 2 we dealt with the stoichiometric description of a chemical system that is valid regardless of whether the system is at equilibrium. Here we deal with the thermodynamic description of a chemical system and the conditions for equilibrium in a closed system provided by chemical thermodynamics. The treatment is necessarily synoptic. Full developments and accounts are given, for example, by Prigogine and Defay (1954), Lewis and Randall (1961), and Denbigh (1981).

We first review conditions for equilibrium in terms of potential functions and the thermodynamic description of a chemical system, introducing the chemical potential. We then formulate the equilibrium conditions in terms of the chemical potential in two ways, corresponding to the two ways of incorporating the closed-system constraint discussed in Chapter 2. After showing the equivalence of these two formulations, we develop the expressions for the chemical potential that are necessary for their use. We conclude the chapter by commenting on the nonnegativity constraint and on the existence and uniqueness of solutions, introducing equilibrium constants, discussing reactions in electrochemical cells, and describing the ways by which the requisite information for the chemical potential is obtained.

3.1 THERMODYNAMIC POTENTIAL FUNCTIONS AND CRITERIA FOR EQUILIBRIUM

The second law of thermodynamics provides several potential functions governing the direction of natural or spontaneous processes. The particular potential function appropriate to a given situation is governed by the choice of thermodynamic variables, which are regarded as independent variables. Specification of the values of these variables defines the *state* of the system. Thus these functions are referred to as *state functions*, which implies that any change

Thermodynamic Potential Functions and Criteria for Equilibrium

in the function between two states of the system is independent of the "path" of the change.

Among the most important potential functions are the entropy function, the Helmholtz function, and the Gibbs function. For each such function, there is a statement of the second law of thermodynamics that includes both the criterion for a natural process to occur and for its ultimate equilibrium state; the statement must also incorporate any relevant constraints.

Thus, for the entropy function S, the statement is

$$dS_{ad} \geq 0, \qquad (3.1\text{-}1)$$

where subscript ad refers to an adiabatic system; for the Helmholtz function A,

$$dA_{T,V} \leq 0; \qquad (3.1\text{-}2)$$

and for the Gibbs function G,

$$dG_{T,P} \leq 0. \qquad (3.1\text{-}3)$$

In each case the symbol d refers to an infinitesimal change, and the inequality refers to a spontaneous process and the equality to equilibrium; for relation 3.1-2, there is no work interaction of any kind between the system and its environment, and for relation 3.1-3, there is no work involved other than that related to volume change (PV work). At equilibrium, depending on the appropriate constraint(s), entropy is at a (local) maximum, the Helmholtz function is at a minimum, and the Gibbs function is at a minimum.

Of these three potential functions, the most important, because of the constraints, temperature and pressure, is the Gibbs function. The development in this book is based almost entirely on this function, but the results can be recast into equivalent forms when appropriate to a particular situation.

The Helmholtz function and the Gibbs function are both sometimes referred to as *free-energy functions*. The Helmholtz function is also sometimes referred to as the *work* function and the Gibbs function as the *free-enthalpy* function. We do not use these last terms, but because of common usage, we frequently refer to the Gibbs function as *free energy*.

In fact, both A and G, as well as other potential functions, have interpretations as work quantities:

$$dA_T \leq -\delta w,$$

$$dA_{T,V} \leq -\delta w',$$

$$dG_{T,P} \leq -\delta w', \qquad (3.1\text{-}4)$$

where w is work of any kind, w' is work other than work of volume change, and the symbol δ denotes a path-dependent quantity. In each case, the inequality refers to a thermodynamically irreversible change in state and the equality (the maximum work obtainable) refers to a reversible change. We have occasion to use relation 3.1-4 in the case of an electrochemical cell in Section 3.11.

3.2 THERMODYNAMIC DESCRIPTION OF A CHEMICAL SYSTEM

A homogeneous (single-phase) chemical system, open or closed, is defined thermodynamically by one of the following natural sets of state function and independent variables:

$$U = U(S, V, \mathbf{n}), \qquad (3.2\text{-}1)$$

$$H = H(S, P, \mathbf{n}), \qquad (3.2\text{-}2)$$

$$A = A(T, V, \mathbf{n}), \qquad (3.2\text{-}3)$$

or

$$G = G(T, P, \mathbf{n}), \qquad (3.2\text{-}4)$$

where U is internal energy and H is enthalpy. Equation 3.2-4, for example, states that G is a (single-valued) function of T, P, and the (N) mole numbers $\mathbf{n}$. Each of these state functions is also homogeneous (in the mathematical sense; cf. physicochemical sense discussed previously) of degree 1 in each mole number n_i. Each of these equations gives rise to a corresponding equation for the (complete) differential of the function involved:

$$dU = T\,dS - P\,dV + \sum_{i=1}^{N} \mu_i\,dn_i, \qquad (3.2\text{-}5)$$

$$dH = T\,dS + V\,dP + \sum_{i=1}^{N} \mu_i\,dn_i, \qquad (3.2\text{-}6)$$

$$dA = -S\,dT - P\,dV + \sum_{i=1}^{N} \mu_i\,dn_i, \qquad (3.2\text{-}7)$$

and

$$dG = -S\,dT + V\,dP + \sum_{i=1}^{N} \mu_i\,dn_i, \qquad (3.2\text{-}8)$$

Thermodynamic Description of a Chemical System

where the chemical potential for the species i, μ_i, is defined by any of

$$\mu_i = \left(\frac{\partial U}{\partial n_i}\right)_{S,V,n_{j \neq i}} = \left(\frac{\partial H}{\partial n_i}\right)_{S,P,n_{j \neq i}}$$

$$= \left(\frac{\partial A}{\partial n_i}\right)_{T,V,n_{j \neq i}} = \left(\frac{\partial G}{\partial n_i}\right)_{T,P,n_{j \neq i}}. \quad (3.2\text{-}9)$$

Because of the homogeneity property of these functions, μ_i depends only on the intensive state of the system, such as defined by T, P, and composition.

Since the most important of these four functions is the Gibbs function G, we continue to use this function exclusively, with the understanding that corresponding descriptions can be written in terms of U, H, or A as required.

From equation 3.2-8 and the definition of G, the temperature and pressure derivatives for G and μ_i, in their most useful forms, are as follows:

$$\left[\frac{\partial(G/T)}{\partial T}\right]_{P,\mathbf{n}} = \frac{-H}{T^2}, \quad (3.2\text{-}10)$$

$$\left(\frac{\partial G}{\partial P}\right)_{T,\mathbf{n}} = V, \quad (3.2\text{-}11)$$

$$\left[\frac{\partial(\mu_i/T)}{\partial T}\right]_{P,\mathbf{n}} = \frac{-\bar{h}_i}{T^2}, \quad (3.2\text{-}12)$$

and

$$\left(\frac{\partial \mu_i}{\partial P}\right)_{T,\mathbf{n}} = \bar{v}_i, \quad (3.2\text{-}13)$$

where the subscript $\mathbf{n}$ means that all mole numbers are constant and $\bar{h}_i$ and $\bar{v}_i$ are the partial molar enthalpy and partial molar volume, respectively, of species i in the system:

$$\bar{h}_i = \left(\frac{\partial H}{\partial n_i}\right)_{T,P,n_{j \neq i}}, \quad (3.2\text{-}14)$$

$$\bar{v}_i = \left(\frac{\partial V}{\partial n_i}\right)_{T,P,n_{j \neq i}}. \quad (3.2\text{-}15)$$

The additivity equation for the total Gibbs function of the system is obtained by integration of equation 3.2-8 at fixed T, P, and composition:

$$G(T,P,\mathbf{n}) = \sum_{i=1}^{N} n_i \mu_i. \quad (3.2\text{-}16)$$

Differentiation of this equation and comparison of the result with equation 3.2-8 leads to the Gibbs-Duhem equation for the (homogeneous) system:

$$S\,dT - V\,dP + \sum_{i=1}^{N} n_i\,d\mu_i = 0. \qquad (3.2\text{-}17)$$

This result can also be obtained by applying Euler's theorem to the Gibbs function as a homogeneous function of degree 1 in the mole numbers.

The equations to this point may be applied to a homogeneous system or to each phase in a heterogeneous system. For a closed, heterogeneous (multiphase) system, we note that, as a consequence of the definition of a chemical species in Section 2.2.2, the chemical potential and partial molar quantities of a species in a given phase are determined by the variables that define the state of that phase only, and the implications for the equations in this section for G, μ_i, and so on are then as follows:

1. Equations 3.2-4, 3.2-8, 3.2-10, 3.2-11, and 3.2-16 apply to the system as a whole or to each phase, provided that the extensive quantities G, H, S, and V relate to the whole system or to the phase under consideration.
2. Equations 3.2-9, 3.2-12, 3.2-13, 3.2-14, and 3.2-15 apply to each species (and hence to a particular phase).
3. In particular, equation 3.2-17 applies to each phase; that is, there is a Gibbs-Duhem equation for each phase.

3.3 TWO FORMULATIONS OF THE EQUILIBRIUM CONDITIONS

For either a single-phase or multiphase system to be at equilibrium, G is at a (global) minimum subject to the closed-system constraint and the nonnegativity constraint at the given thermodynamic conditions (fixed T and P). This is essentially the statement of relation 3.1-3. Here and in Sections 3.4 to 3.7 we assume that $n_i > 0$ (mathematically this means that the nonnegativity constraints are "non-binding"); that is, we ignore the possibility in the nonnegativity constraint that $n_i = 0$ and return to the implications of this latter possibility in Section 3.8. At equilibrium, we thus deal with

$$dG_{T,P} = 0, \qquad (3.3\text{-}1)$$

although this by itself is a necessary but not a sufficient condition.

Our problem is essentially to express G as a function of the n_i and to seek those values of the n_i that make G a minimum subject to the constraints. We assume that we are given values for the element-abundance vector **b**, temperature T, pressure P, and the appropriate "free-energy" data.

The Stoichiometric Formulation

We describe two formulations of the minimization problem (cf. Smith, 1980a), referred to here as:

1. the stoichiometric formulation, in which the closed-system constraint is treated by means of stoichiometric equations so as to result in an essentially unconstrained minimization problem, and
2. the nonstoichiometric formulation, in which stoichiometric equations are not used but, instead, the closed-system constraint is treated by means of Lagrange multipliers. These two formulations are described in turn in Sections 3.4 and 3.5, following which their equivalence is shown.

3.4 THE STOICHIOMETRIC FORMULATION

From Chapter 2 the mole numbers **n** are related to the extents of reaction $\boldsymbol{\xi}$ of the R stoichiometric equations, which are the independent variables, by

$$\mathbf{n} = \mathbf{n}^\circ + \sum_{j=1}^{R} \boldsymbol{\nu}_j \xi_j. \tag{2.3-1}$$

Hence we may write (cf. equation 3.2-4)

$$G = G(T, P, \boldsymbol{\xi}), \tag{3.4-1}$$

and the problem is one of minimizing G, for fixed T and P, in terms of the R ξ_j's. Since these last are independent quantities, the first-order necessary conditions for a minimum in G are

$$\left(\frac{\partial G}{\partial \boldsymbol{\xi}}\right)_{T,P} = \mathbf{0}, \tag{3.4-2}$$

or

$$\left(\frac{\partial G}{\partial \xi_j}\right)_{T,P,\xi_{k \neq j}} = 0; \quad j = 1, 2, \ldots, R. \tag{3.4-3}$$

There are $R = N - C$ equations in the set 3.4-3. Since

$$\left(\frac{\partial G}{\partial \xi_j}\right)_{T,P,\xi_{k \neq j}} = \sum_{i=1}^{N} \left(\frac{\partial G}{\partial n_i}\right)_{T,P,n_{k \neq i}} \left(\frac{\partial n_i}{\partial \xi_j}\right)_{\xi_{k \neq j}}; \quad j = 1, 2, \ldots, R,$$

$$\tag{3.4-4}$$

$$\left(\frac{\partial G}{\partial n_i}\right)_{T,P,n_{k\neq i}} = \mu_i, \tag{3.2-9}$$

$$\left(\frac{\partial n_i}{\partial \xi_j}\right)_{\xi_{k\neq j}} = \nu_{ij}, \tag{2.3-6}$$

then, on combining equations 3.4-3, 3.2-9, and 2.3-6, we have

$$\sum_{i=1}^{N} \nu_{ij}\mu_i = 0; \quad j = 1, 2, \ldots, R. \tag{3.4-5}$$

The quantity on the left side of this equation is denoted by ΔG_j, and its negative has been called *the affinity* by De Donder (1936, Chapter 4). Equations 3.4-5 are R conditions for equilibrium in the system and are readily recognized as the "classical" forms of the equilibrium conditions (Denbigh, 1981, p. 173). When appropriate expressions for the μ_i are introduced into the equations in terms of free-energy data and the mole numbers, the solution of these equations provides the composition of the system at equilibrium.

Example 3.1 For the system described in Example 2.2, for which $R = 2$, the equations 3.4-5 corresponding to the two stoichiometric equations

$$-\tfrac{1}{2}O_2 - NO + NO_2 = 0 \quad \text{and} \quad -\tfrac{2}{3}NH_3 - \tfrac{5}{6}O_2 + \tfrac{2}{3}NO + H_2O = 0$$

are

$$-\tfrac{1}{2}\mu_{O_2} - \mu_{NO} + \mu_{NO_2} = 0 \quad \text{and} \quad -\tfrac{2}{3}\mu_{NH_3} - \tfrac{5}{6}\mu_{O_2} + \tfrac{2}{3}\mu_{NO} + \mu_{H_2O} = 0,$$

respectively.

3.5 THE NONSTOICHIOMETRIC FORMULATION

The problem is formulated as one of minimizing G, for fixed T and P, in terms of the N mole numbers, subject to the M element-abundance constraints. That is, from equation 3.2-16,

$$\min G(\mathbf{n}) = \sum_{i=1}^{N} n_i \mu_i, \tag{3.5-1}$$

subject to

$$\sum_{i=1}^{N} a_{ki} n_i = b_k; \quad k = 1, 2, \ldots, M. \tag{2.2-1}$$

We assume, for convenience, that $M = \text{rank}(\mathbf{A}) = C$.

The Nonstoichiometric Formulation

This is a simple form of constrained optimization problem (Walsh, 1975, p. 7). One approach is to use the method of Lagrange multipliers to remove the constraints. For this, we first write the Lagrangian $\mathcal{L}$:

$$\mathcal{L}(\mathbf{n}, \boldsymbol{\lambda}) = \sum_{i=1}^{N} n_i \mu_i + \sum_{k=1}^{M} \lambda_k \left(b_k - \sum_{i=1}^{N} a_{ki} n_i \right), \qquad (3.5\text{-}2)$$

where $\boldsymbol{\lambda}$ is a vector of M unknown Lagrange multipliers, $\boldsymbol{\lambda} = (\lambda_1, \lambda_2, \ldots, \lambda_M)^T$. Then the necessary conditions provide the following set of $(N + M)$ equations in the $(N + M)$ unknowns $(n_1, n_2, \ldots, n_N, \lambda_1, \lambda_2, \ldots, \lambda_M)$:

$$\left(\frac{\partial \mathcal{L}}{\partial n_i} \right)_{n_{j \neq i}, \boldsymbol{\lambda}} = \mu_i - \sum_{k=1}^{M} a_{ki} \lambda_k = 0, \qquad (n_i > 0) \qquad (3.5\text{-}3)$$

and

$$\left(\frac{\partial \mathcal{L}}{\partial \lambda_k} \right)_{\mathbf{n}, \lambda_{j \neq k}} = b_k - \sum_{i=1}^{N} a_{ki} n_i = 0. \qquad (3.5\text{-}4)$$

As in the stoichiometric formulation, the solution of these equations involves the introduction of an appropriate expression for μ_i.

Example 3.2 Write the set of equations 3.5-3 and 3.5-4 for the system described in Example 2.2.

Solution The system, as represented in Example 2.2, is $\{(NH_3, O_2, NO, NO_2, H_2O), (N, H, O)\}$. Here $N = 5$ and $M = 3$. There are five equations 3.5-3:

$$\mu_{NH_3} - \lambda_N - 3\lambda_H = 0,$$

$$\mu_{O_2} - 2\lambda_O = 0,$$

$$\mu_{NO} - \lambda_N - \lambda_O = 0,$$

$$\mu_{NO_2} - \lambda_N - 2\lambda_O = 0,$$

$$\mu_{H_2O} - 2\lambda_H - \lambda_O = 0.$$

The three equations 3.5-4 are

$$b_N - n_{NH_3} - n_{NO} - n_{NO_2} = 0,$$

$$b_H - 3n_{NH_3} - 2n_{H_2O} = 0,$$

and

$$b_O - 2n_{O_2} - n_{NO} - 2n_{NO_2} - n_{H_2O} = 0.$$

3.6 EQUIVALENCE OF THE TWO FORMULATIONS

The equivalence of the stoichiometric and nonstoichiometric formulations can be shown as follows. From equation 3.5-3, for the nonstoichiometric formulation, we have

$$\mu_i = \sum_{k=1}^{M} a_{ki}\lambda_k; \qquad i = 1, 2, \ldots, N. \tag{3.6-1}$$

Hence, for the quantity on the left side of equation 3.4-5, the stoichiometric formulation, it follows that

$$\sum_{i=1}^{N} \nu_{ij}\mu_i = \sum_{i=1}^{N} \nu_{ij} \left(\sum_{k=1}^{M} a_{ki}\lambda_k \right)$$

$$= \sum_{i=1}^{N} \sum_{k=1}^{M} \lambda_k a_{ki} \nu_{ij}$$

$$= \sum_{k=1}^{M} \lambda_k \sum_{i=1}^{N} a_{ki} \nu_{ij}$$

$$= 0$$

(which is the stoichiometric formulation) since

$$\sum_{i=1}^{N} a_{ki} \nu_{ij} = 0. \tag{2.3-3}$$

3.7 THE CHEMICAL POTENTIAL

3.7.1 Expressions for the Chemical Potential

The structure of chemical thermodynamics, as exemplified by the equations in this chapter to this point, is general and independent of the functional form of the chemical potential μ_i. Although the structure contains derivatives that show how μ_i depends on temperature and pressure (equations 3.2-12 and 3.2-13), thermodynamics itself provides no comparable expressions for the dependence of μ_i on composition. We must then superimpose on the thermodynamic structure, particularly in equations 3.4-5 and 3.5-3, the equilibrium conditions, specific expressions for μ_i to introduce composition explicitly into these equilibrium conditions. A guideline for this is that the expression for μ_i must satisfy the Gibbs-Duhem equation (equation 3.2-17).

The Chemical Potential

We consider expressions for the chemical potential of a pure species first before turning attention to species in solution, in which latter case, composition must be taken into account in addition to T and P.

3.7.1.1 Pure Species

From equation 3.2-13 written for a pure species, we obtain

$$\left(\frac{\partial \mu}{\partial P}\right)_T = v, \tag{3.7-1}$$

where v is molar volume. Integration of this at fixed T from a reference pressure P° to P results in

$$\mu(T, P) - \mu(T, P^\circ) = \int_{P^\circ}^{P} v\, dP. \tag{3.7-2}$$

We apply this to three particular cases: ideal gas; nonideal gas; and liquid or solid.

3.7.1.1.1 Ideal Gas

Introduction into equation 3.7-2 of the equation of state

$$Pv = RT \tag{3.7-3}$$

and a reference or standard-state pressure (P°) of unity results in

$$\mu(T, P) = \mu^\circ(T) + RT \ln P, \tag{3.7-4}$$

where P must be in the same unit of pressure as P°. Thus if P° is chosen to be 1 atm, P must be expressed in atmospheres. We retain this choice in accordance with usual practice, particularly in relation to free-energy data (Denbigh, 1981, p. xxi). In equation 3.7-4 $\mu^\circ(T)$ is called the standard chemical potential that is a function of T only.

3.7.1.1.2 Nonideal Gas

Equation 3.7-2 may be written, on addition and subtraction of $RT\ln(P/P^\circ)$, as

$$\mu(T, P) = \mu(T, P^\circ) - RT \ln P^\circ + RT \ln P + \int_{P^\circ}^{P} \left(v - \frac{RT}{P}\right) dP. \tag{3.7-5}$$

On letting $P^\circ \to 0$ and using equation 3.7-4, since in this limit $\mu(T, P^\circ)$

approaches its ideal value, we have

$$\mu(T, P) = \mu^\circ(T) + RT\ln P + \int_0^P \left(v - \frac{RT}{P}\right) dP. \qquad (3.7\text{-}6)$$

For convenience, it is customary to use the last two terms on the right of equation 3.7-6 to define the fugacity f by means of

$$RT\ln f = RT\ln P + \int_0^P \left(v - \frac{RT}{P}\right) dP. \qquad (3.7\text{-}7)$$

It follows from this definition that

$$\lim_{P \to 0} \frac{f}{P} = 1. \qquad (3.7\text{-}8)$$

3.7.1.1.3 Liquid or Solid

For a pure liquid or solid, it is convenient to take P° to be the vapor pressure p^*, to take advantage of the equilibrium condition for liquid-vapor or solid-vapor equilibrium—equation 3.4-5, in conjunction with equation 3.7-6. Thus from the latter, the chemical potential of the liquid or solid at (T, p^*), which is equal to that of the vapor at (T, p^*), is

$$\mu(T, p^*) = \mu^\circ(T) + RT\ln p^* + \int_0^{p^*} \left(v_g - \frac{RT}{P}\right) dP, \qquad (3.7\text{-}9)$$

where v_g is the molar volume of the vapor. On combining this with equation 3.7-2, we have

$$\mu(T, P) = \mu^\circ(T) + RT\ln p^* + \int_0^{p^*} \left(v_g - \frac{RT}{P}\right) dP + \int_{p^*}^P v\, dP, \qquad (3.7\text{-}10)$$

where v is the molar volume of the liquid or solid. The two integrals on the right of equation 3.7-10 are usually relatively small in value, and hence, for a pure liquid or solid, we obtain

$$\mu(T, P) \simeq \mu^\circ(T) + RT\ln p^*. \qquad (3.7\text{-}10a)$$

Analogous to equation 3.7-7 for the fugacity of a gas, the fugacity of a liquid or solid is given by

$$RT\ln \frac{f}{P} = \int_0^{p^*} \left(v_g - \frac{RT}{P}\right) dP + \int_{p^*}^P \left(v - \frac{RT}{P}\right) dP. \qquad (3.7\text{-}11)$$

3.7.1.2 Species in Solution

3.7.1.2.1 Ideal-Gas Solution

The form of equation 3.7-4 for the chemical potential of a pure, ideal gas suggests the form for a species in an ideal-gas solution (i.e., a solution of ideal gases):

$$\mu_i(T, P, x_i) = \mu_i^\circ(T) + RT \ln p_i, \qquad (3.7\text{-}12)$$

in which pressure P is replaced by the partial pressure p_i, where, by definition,

$$p_i = \left(\frac{n_i}{n_t}\right) P \equiv x_i P, \qquad (3.7\text{-}13)$$

x_i is the mole fraction of species i, and n_t is the total number of moles in the solution. A justification for this form is that application of equation 3.2-13 to equation 3.7-12 leads to the equation of state for an ideal-gas solution:

$$\left(\frac{\partial \mu_i}{\partial P}\right)_{T,\mathbf{n}} = \frac{RT}{P} = \bar{v}_i. \qquad (3.7\text{-}14a)$$

Hence

$$V = \sum n_i \bar{v}_i = \frac{RT}{P} \sum n_i = n_t \frac{RT}{P} \qquad (3.7\text{-}14b)$$

This can be most easily seen if equation 3.7-12 is written as

$$\mu_i(T, P, x_i) = \mu_i^\circ(T) + RT \ln P + RT \ln x_i. \qquad (3.7\text{-}12a)$$

3.7.1.2.2 Ideal Solution

Equation 3.7-12a may be used as the basis for a less restricted type of system —an ideal solution, which may be gaseous (but not necessarily an ideal-gas solution), liquid, or solid. This is accomplished in part by replacing the first two terms on the right by an arbitrary function of T, P, and a standard compositional state x_i^*, $\mu_i(T, P, x_i^*)$, so that

$$\mu_i(T, P, x_i) = \mu_i(T, P, x_i^*) + RT \ln x_i. \qquad (3.7\text{-}15)$$

The definition of an ideal solution in terms of μ_i is completed by specification of the standard state, which then serves to define $\mu_i(T, P, x_i^*)$. With respect to x_i^*, there are two common choices or conventions, each convention leading to a particular type of ideality:

1 The Raoult Convention In this case $x_i^* \to 1$; that is, the standard state is pure species i at (T, P) of the system and in the same physical state. Hence

$$\mu_i(T, P, x_i^*) = \lim_{x_i \to 1} (\mu_i - RT \ln x_i), \qquad (3.7\text{-}16)$$

and equation 3.7-15 is normally written without reference to x_i^* as

$$\mu_i(T, P, x_i) = \mu_i^*(T, P) + RT \ln x_i. \qquad (3.7\text{-}15a)$$

A justification for equation 3.7-15a as a model for a species in an ideal solution is that it can be used to derive the characteristics, including additivity of pure-species enthalpies and Raoult's law, of this type of ideal solution. The quantity $\mu_i^*(T, P)$ is the standard chemical potential of species i that is a function of both T and P. From equation 3.7-16, μ_i^* is the chemical potential of pure species i at (T, P) of the system in the same physical state. We note that an ideal solution based on the Raoult convention is equivalent to the type of ideality to which the Lewis-Randall fugacity rule applies (Prausnitz, 1969, pp. 90–92).

2 The Henry Convention In this case $x_i^* \to 0$; that is, the standard state is the infinitely dilute solution of species i at (T, P) of the system. Hence

$$\mu_i(T, P, x_i^*) \equiv \mu_i^* = \lim_{x_i \to 0} (\mu_i - RT \ln x_i). \qquad (3.7\text{-}17)$$

Since μ_i^* in equation 3.7-17 is different from μ_i^* in equation 3.7-15a, we denote it henceforth by μ_{Hi}^* and write equation 3.7-15 as

$$\mu_i(T, P, x_i) = \mu_{Hi}^*(T, P) + RT \ln x_i. \qquad (3.7\text{-}15b)$$

The Raoult convention is commonly used for all species in a solution in situations in which *no* distinction is made between solute(s) and solvent(s). When this distinction *is* appropriate, the Henry convention is commonly used for the solute species and the Raoult convention for the solvent species.

The composition variable used in equation 3.7-15 need not be the mole fraction. For the Henry convention applied to a solid solute species i dissolved in a liquid solvent, the molality m_i is commonly used, where

$$m_i = 1000 \frac{n_i}{M_s n_s}, \qquad (3.7\text{-}18)$$

n_i is the number of moles of solute i dissolved in n_s moles of solvent, and M_s is the molecular weight of the solvent. In this case we write equation 3.7-15b as

$$\mu_i(T, P, m_i) = \mu_{mi}^*(T, P) + RT \ln m_i \qquad (3.7\text{-}19)$$

The Chemical Potential

and interpret μ^*_{mi} (numerically different from μ^*_{Hi}) by the analog of equation 3.7-17:

$$\mu^*_{mi} = \lim_{m_i \to 0} (\mu_i - RT \ln m_i). \tag{3.7-20}$$

From equations 3.7-15b, 3.7-18, and 3.7-19, μ^*_{Hi} and μ^*_{mi} are related by

$$\mu^*_{Hi} = \mu^*_{mi} + RT \ln m_s, \tag{3.7-21}$$

where m_s is the molality of the solvent, $1000/M_s$.

Another composition variable that is sometimes used in connection with the Henry convention is the molarity C_i, defined by

$$C_i = \frac{n_i}{V}, \tag{3.7-22}$$

the number of moles of species i per unit volume (conventionally in liters) of the system at (T, P). The disadvantage of this variable, in comparison with x_i and m_i, is that it is inherently a function, albeit usually a *weak* function, of T and P. Equations 3.7-19 and 3.7-20 may be rewritten in terms of C_i, with μ^*_{Ci} replacing μ^*_{mi}, and equation 3.7-21 becomes

$$\mu^*_{Hi} = \mu^*_{Ci} + RT \ln C_s, \tag{3.7-23}$$

where C_s is the molarity of the solvent, n_s/V.

When the solute is an electrolyte (e.g., a salt dissolved in water), equation 3.7-19 cannot be used for an individual ionic species since the limiting process of equation 3.7-20 is not operationally possible, because it would violate electrical-charge neutrality. The cation and anion of an electrolyte must be combined to represent the electrolyte as a whole. For this purpose, the mean-ion molality of species i is defined by

$$m^\nu_{\pm i} = m^{\nu_+}_+ m^{\nu_-}_-, \tag{3.7-24}$$

where

$$\nu = \nu_+ + \nu_-. \tag{3.7-25}$$

Here ν_+ and ν_- are the subscripts to the cation and anion, respectively, in the molecular formula of the electrolyte. Then equation 3.7-19 becomes

$$\mu_i(T, P, m_i) = \mu^*_{mi} + RT \ln m^\nu_{\pm i}. \tag{3.7-26}$$

Example 3.3 Calculate the mean-ion molality of $Al_2(SO_4)_3$ in a solution made up by dissolving 0.1 g mole of the salt in 200 g of water.

Solution

$$m = \frac{1000(0.1)}{200} = 0.5$$

$$\nu = \nu_+ + \nu_- = 2 + 3 = 5$$

If we assume that $Al_2(SO_4)_3$ is a "strong" electrolyte (i.e., completely ionized), then

$$m_+ = \nu_+ m = 1.0 \quad \text{and} \quad m_- = \nu_- m = 1.5,$$

and, from equation 3.7-24,

$$m_\pm^5 = m_+^2 \, m_-^3 = (1.0)^2(1.5)^3 = 3.37.$$

$$m_\pm = 3.37^{1/5} = 1.275.$$

3.7.1.2.3 Nonideal Solution

The most general form of expression required for the chemical potential is that for a species in a nonideal solution, whether gaseous, liquid, or solid. To remove the restriction of an ideal solution, we replace the composition variable in equation 3.7-15a by the activity a_i of species i, and to complete the definition of activity, we specify the standard state. As for an ideal solution, there are two common ways of doing the latter; thus

$$\mu_i(T, P, \mathbf{x}) = \mu_i^*(T, P) + RT \ln a_i(T, P, \mathbf{x}), \tag{3.7-27}$$

together with, for the Raoult convention, in terms of mole fraction,

$$\lim_{x_i \to 1} \frac{a_i}{x_i} = 1; \tag{3.7-27a}$$

and for the Henry convention,

$$\lim_{x_i \to 0} \frac{a_i}{x_i} = 1. \tag{3.7-27b}$$

An alternative to the use of activity is the use of the activity coefficient γ_i of species i, where

$$a_i = \gamma_i x_i, \tag{3.7-28}$$

The Chemical Potential

and x_i may be replaced by molality or molarity (with consequent changes for the numerical values of both a_i and γ_i). In this case equations 3.7-27, 3.7-27a, and 3.7-27b become, respectively,

$$\mu_i(T, P, \mathbf{x}) = \mu_i^*(T, P) + RT \ln \gamma_i(T, P, \mathbf{x}) x_i, \qquad (3.7\text{-}29)$$

and

$$\lim_{x_i \to 1} \gamma_i = 1 \quad \text{(Raoult convention)} \qquad (3.7\text{-}29a)$$

or

$$\lim_{x_i \to 0} \gamma_i = 1 \quad \text{(Henry convention)}. \qquad (3.7\text{-}29b)$$

For an electrolyte species in a nonideal solution, the modification of the general forms, following equation 3.7-26 for an ideal solution, involves the introduction of the mean-ion activity or mean-ion activity coefficient, each defined in a manner analogous to the mean-ion molality in equation 3.7-24. In terms of the mean-ion activity coefficient $\gamma_\pm$, equation 3.7-26 becomes

$$\mu_i(T, P, \mathbf{m}) = \mu_{mi}^*(T, P) + RT \ln(\gamma_\pm m_\pm)_i^\nu, \qquad (3.7\text{-}30)$$

together with

$$\lim_{m_i \to 0} \gamma_{\pm i} = 1. \qquad (3.7\text{-}30a)$$

As an alternative to the approach just described, for species in a nonideal solution, the chemical potential may be expressed in terms of the fugacity by using a relation equivalent to equations 3.7-6 and 3.7-7 for a pure species:

$$\mu_i(T, P, \mathbf{x}) = \mu_i^\circ(T) + RT \ln f_i, \qquad (3.7\text{-}31)$$

where the fugacity f_i is defined by (Prausnitz, 1969, p. 30):

$$RT \ln \phi_i = RT \ln \frac{f_i}{x_i P} = \int_0^P \left(\bar{v}_i - \frac{RT}{P} \right) dP, \qquad (3.7\text{-}32)$$

where ϕ_i is the fugacity coefficient of species i, defined by

$$\phi_i = \frac{f_i}{x_i P}. \qquad (3.7\text{-}33)$$

3.7.2 Assigning Numerical Values to the Chemical Potential

Expressions for the chemical potential, and in particular for the chemical potential of a species in a nonideal solution, involve two types of quantity to which numerical values must be assigned: (1) the standard chemical potential ($\mu°$ or μ^*) and (2) the composition and composition-related quantities, such as activity or activity coefficient and fugacity or fugacity coefficient. For the first, we discuss ways in which numerical information is available in Section 3.12, in connection also with the standard free energy of reaction introduced in Section 3.10. For the second, there is a vast literature, and we only point out here some general features, including those related to the temperature and pressure dependence of μ_i given in equations 3.2-12 and 3.2-13, which involve partial molar quantities. We discuss this in more detail in Chapter 7. The types of information required can be listed as follows:

1 *Volumetric (PvTx) Information* This includes information contained in an equation of state, compressibility-factor charts, and tables of densities. This is needed for the determination of fugacity or fugacity coefficient, partial molar volume, and the pressure dependence of the chemical potential, activity coefficient, fugacity, and so on. The partial molar volume is involved in most of these determinations. If the molar volume v of a solution is known as a function of composition at fixed (T, P), the partial molar volume of species i in the solution $\bar{v}_i$ can be determined from v by the relation (cf. Smith and Van Ness, 1975, p. 604)

$$\bar{v}_i = v - \sum_{j \neq i} x_j \left(\frac{\partial v}{\partial x_j} \right)_{T, P, x_{k \neq j}}. \qquad (3.7\text{-}34)$$

2 *Enthalpy Information* This includes data regarding heat capacities, enthalpies of solution and mixing, and enthalpies of formation. This is needed for determination of the temperature dependence of the chemical potential, activity coefficient, and so on. The partial molar enthalpy of species i in a solution $\bar{h}_i$ can be determined from molar enthalpy by means of a relation analogous to equation 3.7-34.

3 *Activity Coefficient Information* This includes information given by correlations of experimental data obtained, such as from phase equilibria for nonelectrolytes and from emf determinations for electrolytes. For the former in particular, many empirical and semiempirical relations, such as the Margules, van Laar, and Wilson equations, have been proposed (Prausnitz, 1969, Chapter 6).

4 *Excess Thermodynamic Function Information* This includes data concerning excess enthalpies and volumes (Missen, 1969). Much of the information required is given in terms of excess functions. An excess function is the difference between the function for a nonideal solution and the same function

Implications of the Nonnegativity Constraint

for an ideal solution based on a specified convention, whether the Raoult or the Henry convention. Thus the excess molar volume of a solution v^E is defined by

$$v^E = v - v^{\text{id}}, \qquad (3.7\text{-}35)$$

where v^{id} is the molar volume of an ideal solution at the same T, P, and $\mathbf{x}$. Expressions involving excess thermodynamic functions are completely analogous to those involving the corresponding thermodynamic functions, except for a few cases of intensive quantities (Missen, 1969). For example, the excess partial molar volume of species i in a solution $\bar{v}_i^E$ may be related to v^E by an equation analogous to equation 3.7-34:

$$\bar{v}_i^E = v^E - \sum_{j \neq i} x_j \left(\frac{\partial v^E}{\partial x_j} \right)_{T, P, x_{k \neq j}}. \qquad (3.7\text{-}36)$$

Finally, we point out that, for numerical work, we frequently use the nondimensional form of the chemical potential μ_i/RT. The equations in Section 3.7.1 could all be rewritten correspondingly in nondimensional form.

3.8 IMPLICATIONS OF THE NONNEGATIVITY CONSTRAINT

To this point we have assumed that the equilibrium conditions discussed in Sections 3.4 and 3.5 have a solution that satisfies $n_i > 0$ for all species. This need not be the case, however, and in this section we show how the equilibrium conditions must be modified to account for the possibility of $n_i = 0$. We then show that this leads to the necessity to develop criteria to test for the presence or absence of an entire phase at equilibrium.

If $n_i = 0$, either $n_t = 0$ or $n_t \neq 0$. In general, for a nonideal solution, from equation 3.7-29, at fixed (T, P) we have

$$\mu_i = \mu_i^* + RT \ln \gamma_i(\mathbf{x}) + RT \ln \frac{n_i}{n_t}, \qquad (3.7\text{-}29)$$

and, from the definition of μ_i,

$$\left(\frac{\partial G}{\partial n_i} \right)_{T, P, n_{j \neq i}} = \mu_i. \qquad (3.2\text{-}9)$$

We assume that γ_i is finite for all possible mole fractions (for an *ideal* solution, this is true, since $\gamma_i = 1$). Then, if $n_i = 0$ and $n_t \neq 0$, $\mu_i \to -\infty$. From equation 3.2-9, it follows that G may be *lowered* by adding an infinitesimal amount of species i, and hence at equilibrium the case $n_i = 0$ and $n_t \neq 0$ is not possible. This, in turn, implies that the only possibility is $n_i = 0$ and $n_t = 0$ at equilibrium. In other words, n_i is zero if, and only if, *all* species in that phase also have zero mole numbers (i.e., the entire phase is absent) (cf. Denbigh, 1981, pp. 160–161).

We have thus shown that, to consider the possibility $n_i = 0$ at equilibrium, we simply focus on establishing whether $n_t = 0$. We examine the three possibilities for a phase: (1) single species; (2) ideal solution; and (3) nonideal solution. We then show how the equilibrium conditions, with equations 3.5-3 for the nonstoichiometric formulation and 3.4-5 for the stoichiometric formulation, must be modified.

3.8.1 Single-Species Phase

For a single-species phase, it is relatively easy to modify the conditions since $\mu_i = \mu_i^*(T, P)$ and is independent of composition. The Kuhn-Tucker conditions are used (Walsh, 1975, pp. 35–39), which are analogous to the Lagrange multiplier conditions when inequality constraints are present. For the species in the single-species phase under consideration, equation 3.5-3 in the nonstoichiometric formulation is replaced by the pair of conditions

$$\left(\frac{\partial \mathcal{L}}{\partial n_i}\right)_{n_{j \neq i}, \lambda} = \mu_i^* - \sum_{k=1}^{M} a_{ki} \lambda_k = 0, \quad (n_i > 0) \qquad (3.8\text{-}1a)$$

and

$$\left(\frac{\partial \mathcal{L}}{\partial n_i}\right)_{n_{j \neq i}, \lambda} = \mu_i^* - \sum_{k=1}^{M} a_{ki} \lambda_k > 0, \quad (n_i = 0). \qquad (3.8\text{-}1b)$$

In the stoichiometric formulation, instead of equation 3.4-5, we have the pair of conditions, stemming from the stoichiometric matrix in canonical form (Section 2.3.3) for the noncomponent species

$$\frac{\partial G}{\partial n_i} \equiv \frac{\partial G}{\partial \xi_j} = \mu_i^* + \sum_{k=1}^{M} \nu_{kj} \mu_k = 0, \quad (n_i > 0) \qquad (3.8\text{-}2a)$$

and

$$\frac{\partial G}{\partial n_i} \equiv \frac{\partial G}{\partial \xi_j} = \mu_i^* + \sum_{k=1}^{M} \nu_{kj} \mu_k > 0, \quad (n_i = 0) \qquad (3.8\text{-}2b)$$

where $i = j + m$.

Relations 3.8-1b and 3.8-2b both essentially state that, if the free energy of the system were to be increased by the formation of species i, the formation would not take place.

3.8.2. Ideal Solution

For a solution, we cannot proceed completely as for a single-species phase because $\partial G/\partial n_i$ is not strictly defined when $n_i = 0$. However, we again suppose that the (rest of the) system is at equilibrium and that the phase under consideration is absent ($n_i = 0$ for all species in that phase) and consider whether a small amount of it could be formed. In the nonstoichiometric formulation for each species in a small amount of the phase, from equation 3.6-1,

$$\mu_i = \mu_i^* + RT \ln x_i = \sum_{k=1}^{M} a_{ki} \lambda_k, \qquad (3.6\text{-}1)$$

or, equivalently,

$$x_i = \exp\left[\left(\frac{1}{RT}\right)\left(-\mu_i^* + \sum_{k=1}^{M} a_{ki} \lambda_k\right)\right]. \qquad (3.8\text{-}3)$$

If $\sum x_i < 1$, the phase is absent; if (by coincidence) $x_i = 1$, the phase is at incipient formation; and if $\sum x_i > 1$, the phase is present in finite amount, and the equilibrium calculation must allow for this.

Thus the test or criterion for the phase to be absent at equilibrium is

$$\sum_i \exp\left[\left(\frac{1}{RT}\right)\left(-\mu_i^* + \sum_{k=1}^{M} a_{ki} \lambda_k\right)\right] < 1, \qquad (3.8\text{-}4)$$

or, in the stoichiometric formulation,

$$\sum_i \exp\left[\left(\frac{1}{RT}\right)\left(-\mu_i^* - \sum_{k=1}^{M} \nu_{kj} \mu_k\right)\right] < 1, \qquad (3.8\text{-}5)$$

where the summations are over all species in the phase. It is readily shown that relations 3.8-4 and 3.8-5 reduce to 3.8-1b and 3.8-2b, respectively, in the case of a single-species phase. Criteria 3.8-4 and 3.8-5 can be shown rigorously to be correct by considering the mathematical dual of the chemical equilibrium problem (Dembo, 1976). However, we have used an heuristic discussion here.

3.8.3 Nonideal Solution

Proceeding as for the case of an ideal solution, we obtain the analog of equation 3.8-3:

$$\gamma_i(\mathbf{x}) x_i = \exp\left[\left(\frac{1}{RT}\right)\left(-\mu_i^* + \sum_{k=1}^{M} a_{ki} \lambda_k\right)\right]. \qquad (3.8\text{-}6)$$

If a solution x to these nonlinear equations satisfies $\sum x_i < 1$, the phase is absent; if it satisfies $\sum x_i > 1$, the phase is present and must be considered in the equilibrium calculation.

3.9 EXISTENCE AND UNIQUENESS OF SOLUTIONS

The conditions for the existence of a solution to a problem in chemical equilibrium have been reviewed by Smith (1980a). We assume that the nonnegativity and element-abundance constraints are satisfied by at least one composition vector n and that all b_k are finite and $b_k \neq 0$ for at least one element. It is also necessary that the function G be continuous in n. This is a potential problem only at $n_i = 0$; by ensuring that $x_i \ln[\gamma_i(\mathbf{x})x_i] = 0$ for all i at $x_i = 0$, we ensure that G is continuous at $x_i = 0$. Then a solution to the equilibrium problem exists. This follows from a theorem in analysis known as the *Weierstrass theorem* (Hadley, 1964, p. 53).

In addition to the existence of a solution, we are interested in the number of solutions, that is, how many possible vectors n satisfy both the element-abundance constraints and the equilibrium conditions. This interest arises because nonuniqueness may occur in several important situations. It is typically connected with incipient formation of a phase. A very simple illustration is provided by the system $\{(H_2O(\ell), H_2O(g)), (H, O)\}$, with $b_1 = 2$ and $b_2 = 1$, at given T and P. There are three possibilities. At the given T, if $P < p^*$, the unique solution is $(n_1, n_2)^T = (0, 1)^T$; if $P > p^*$, the unique solution is $(n_1, n_2)^T = (1, 0)^T$; at $P = p^*$, the solution is not unique, and any $(n_1, n_2)^T$ satisfying $n_1 + n_2 = 1$ ($n_i \geq 0$) is valid. The same type of situation can occur in more complicated multiphase situations involving at least one multispecies phase. The basic reason for the possibility of nonunique solutions lies in the manner in which we have posed the equilibrium problem—in terms of (extensive) mole numbers, in addition to the two intensive parameters T and P.

For the case of a system consisting of a single ideal-solution phase, the chemical equilibrium problem has a unique solution, a proof of which statement follows. A sufficient condition for uniqueness is that G be a strictly convex function of n, subject to the constraints. Then the Kuhn-Tucker conditions are sufficient as well as necessary. For a single phase, convexity thus depends on the quadratic form

$$Q(\delta \mathbf{n}) = \sum_{i=1}^{N} \sum_{j=1}^{N} \left(\frac{\partial^2 G}{\partial n_i \partial n_j} \right) \delta n_i \, \delta n_j, \tag{3.9-1}$$

where $\partial^2 G / \partial n_i \partial n_j$ are the entries of a matrix called the *Hessian matrix of G*. Uniqueness is established if $Q(\delta \mathbf{n}) > 0$ for all allowable compositions n and nonvanishing variations $\delta \mathbf{n}$. From equation 3.7-15a, the entries of the Hessian

Existence and Uniqueness of Solutions

are given by

$$\frac{\partial^2 G}{\partial n_i \partial n_j} = RT\left(\frac{\delta_{ij}}{n_i} - \frac{1}{n_t}\right), \qquad (3.9\text{-}2)$$

where δ_{ij} is the Kronecker delta function. Inserting equation 3.9-2 into 3.9-1, we have

$$\frac{Q(\delta \mathbf{n})}{RT} = \sum_{i=1}^{N} \frac{\delta n_i^2}{n_i} - \frac{1}{n_t}\left(\sum_{j=1}^{N} \delta n_j\right)^2$$

$$= \sum_{i=1}^{N} n_i \left(\frac{\delta n_i}{n_i} - \frac{\sum_{j=1}^{N} \delta n_j}{n_t}\right)^2. \qquad (3.9\text{-}3)$$

Since $n_i > 0$ (which must be true, from our previous discussion), Q is positive unless the quantity in parentheses is zero for each i. In this latter case

$$\frac{\delta n_i}{n_i} = \frac{\sum_{j=1}^{N} \delta n_j}{n_t}. \qquad (3.9\text{-}4)$$

Since $n_i > 0$ and $\delta n_i \neq 0$ (for at least one i), the right side of equation 3.9-4 is nonzero. Multiplying equation 3.9-4 by $a_{ki} n_i$ and summing over i, we have

$$\sum_{i=1}^{N} a_{ki} \delta n_i = \left(\sum_{i=1}^{N} a_{ki} n_i\right)\left(\sum_{j=1}^{N} \frac{\delta n_j}{n_t}\right); \qquad k = 1, 2, \ldots, C. \qquad (3.9\text{-}5)$$

Since the left side of equation 3.9-5 is zero (from equation 2.2-2) and the second factor on the right side is nonzero, the first factor must be zero (for all k). However, this factor is b_k (from equation 2.2-1) and *cannot* be zero for all k. As a result, Q can only be positive.

For a single phase that is an ideal solution, the chemical equilibrium problem then has a unique solution (provided that existence is established). For a single phase that is a *nonideal* solution, we believe that the same result applies, but this has not yet been proved, as far as we are aware.

For a multiphase ideal system, Hancock and Motzkin (1960) have found that uniqueness need not hold. This nonuniqueness is of a degenerate type since it is readily shown that G is convex for such a system. We call this nonuniqueness *degenerate* in the sense that only the relative amount of each phase is not unique, although the mole fractions of the species in each phase are unique (Shapiro and Shapley, 1965). When more than one phase is possible for a *nonideal* system, it has been found that the Gibbs function may possess several local minima; that is, G is not convex (Othmer, 1976; Ceram and Scriven, 1976; Heidemann, 1978; Gautam and Seider, 1979).

3.10 THE STOICHIOMETRIC FORMULATION IN TERMS OF EQUILIBRIUM CONSTANTS

As shown in Section 3.4, the equilibrium conditions may be written as

$$\sum_{i=1}^{N} \nu_{ij}\mu_i = 0; \quad j = 1, 2, \ldots, R. \tag{3.4-5}$$

The traditional way of determining the composition of a system at equilibrium has been to solve equation 3.4-5 by the introduction of equilibrium-constant expressions. There is one such expression for each of the R equations. The form of the equilibrium constant depends on the forms of the expressions introduced for the μ_i in equation 3.4-5. For example, if we introduce the general expression for μ_i in terms of activity,

$$\mu_i = \mu_i^* + RT \ln a_i, \tag{3.7-27}$$

into the jth equation of the set 3.4-5, it becomes

$$\sum_{i=1}^{N} \nu_{ij}\mu_i^* + RT \sum_{i=1}^{N} \nu_{ij} \ln a_i = 0,$$

which can be written as

$$\Delta G_j^* = -RT \sum_{i=1}^{N} \nu_{ij} \ln a_i$$

$$= -RT \ln \prod_{i=1}^{N} a_i^{\nu_{ij}}$$

$$= -RT \ln K_{aj}, \tag{3.10-1}$$

where $\prod$ denotes a product of quantities,

$$\Delta G_j^* = \sum_{i=1}^{N} \nu_{ij}\mu_i^*, \tag{3.10-2}$$

and

$$K_{aj} = \prod_{i=1}^{N} a_i^{\nu_{ij}}. \tag{3.10-3}$$

In equation 3.10-3 K_{aj} is the equilibrium constant for stoichiometric equation j, written in terms of activities, and ΔG_j^* is the standard free-energy change for

The Stoichiometric Formulation in Terms of Equilibrium Constants

the jth equation. Equation 3.10-1 is sometimes referred to as the *reaction isotherm*.

This traditional approach is not used in the algorithms of this book, even though it essentially amounts only to rearrangement of equation 3.4-5. We introduce it here, however, not only because of its widespread description in the literature and its link to the stoichiometric formulation of equation 3.4-5, but also because it provides a source of information about ΔG^* by means of the reaction isotherm.

The dependence of ΔG_j^* and K_{aj} on P and/or T follows from the dependence of the standard chemical potential (μ^* or μ°) on P and/or T (Section 3.7.1). Thus, if they are based on $\mu^*(T, P)$, they depend on both T and P; however, if they are based on $\mu^\circ(T)$, they depend only on T.

The dependence of K_{aj} on T is given by

$$\left(\frac{\partial \ln K_{aj}}{\partial T}\right)_P = \frac{\Delta H_j^*}{RT^2}, \tag{3.10-4}$$

where ΔH_j^* is the standard enthalpy change for the jth stoichiometric equation, corresponding to ΔG_j^*; equation 3.10-4 is known as the *van't Hoff equation* (Denbigh, 1981, p. 144). The dependence of K_{aj} on P is given by

$$\left(\frac{\partial \ln K_{aj}}{\partial P}\right)_T = -\frac{\Delta V_j^*}{RT}, \tag{3.10-5}$$

where ΔV_j^* is the standard volume change for the jth stoichiometric equation, analogous to ΔH_j^* and ΔG_j^*.

As noted previously, the form of the equilibrium-constant expression follows from the form of the chemical potential expression introduced into the equilibrium condition (equation 3.4-5). Thus, if the reacting system can be considered to be an ideal-gas solution, introduction of equation 3.7-12 for each species leads to the equilibrium constant K_P, which depends only on T; similarly, K_x, K_m, or K_C results for an ideal solution, depending on whether equation 3.7-15a or 3.7-19 or the equivalent in terms of molarity is used. For a given equilibrium condition (i.e., a given stoichiometric equation), it is not necessary to use the same expression for μ_i of each species, as shown in the following example.

Example 3.4 Write the equilibrium-constant expression for the stoichiometric equation

$$Pb(s) + PbO_2(s) + 2H_2SO_4(m) = 2PbSO_4(s) + 2H_2O(\ell),$$

which represents the behavior of a lead-acid cell. Here the three species Pb, PbO_2, and $PbSO_4$ are solids, as indicated by the symbol s, and the electrolyte is H_2SO_4, which is at the molality m in an aqueous solution.

Solution For each of the three solid species, the activity is unity, as implied by equation 3.7-27a. For the electrolyte H_2SO_4, we use equation 3.7-30 and the Henry convention, and for water, we use equation 3.7-29 and the Raoult convention. The equilibrium constant may then be written as

$$K_a = \frac{a_{H_2O}^2}{a_{H_2SO_4}^2}$$

$$= \frac{a_{H_2O}^2}{(a_\pm^3)_{H_2SO_4}^2}$$

$$= \frac{\gamma_{H_2O}^2 x_{H_2O}^2}{(\gamma_\pm^3 m_\pm^3)_{H_2SO_4}^2}$$

$$= \frac{\gamma_{H_2O}^2 x_{H_2O}^2}{(\gamma_\pm^3 4m^3)_{H_2SO_4}^2}.$$

3.11 ELECTROCHEMICAL CELLS

For an electrochemical cell, whether galvanic or electrolytic, there is work involved other than that associated with volume change, and relation 3.1-3 is replaced by relation 3.1-4, in which the maximum possible work, under reversible (rev) conditions, is

$$\delta w'_{rev} = zFE\, d\xi. \qquad (3.11\text{-}1)$$

Here z is the number of moles of electrons associated with the cell process (as written; see Example 3.6), F is the Faraday constant [96,487 coulombs (mole electrons)$^{-1}$], E is the emf of the cell, and ξ is the extent-of-reaction variable for the cell process.

If we assume that there is only one cell process, equation 3.4-5 for the stoichiometric formulation becomes

$$\sum_{i=1}^{N} \nu_i \mu_i = -zFE. \qquad (3.11\text{-}2)$$

Introduction of equation 3.7-27 into the left side of equation 3.11-2 leads to the Nernst equation relating the emf of the cell to the activities of the species involved in the cell process:

$$E = E^\circ - \frac{RT}{zF} \ln \prod_{i=1}^{N} a_i^{\nu_i}, \qquad (3.11\text{-}3)$$

where $E°$ is the standard emf of the cell; $E°$ is determined from standard electrode potentials (Section 3.12) and is related to the standard free-energy change $\Delta G°$ for the cell reaction by

$$\Delta G° = -zFE°. \tag{3.11-4}$$

(Here the quantities are represented by $\Delta G°$ and $E°$, rather than by, say, ΔG^* and E^*, to conform to normal practice in electrochemistry.)

3.12 WAYS IN WHICH STANDARD FREE-ENERGY INFORMATION IS AVAILABLE

Regardless of which method is used to determine the composition of a system at equilibrium, that is, by examining the consequences of equation 3.4-5, 3.5-3, 3.10-1, or 3.11-3, it is necessary to assign a numerical value to the standard chemical potential μ_i^* or its equivalent. There are four main ways in which free-energy information is available for this purpose:

1 As standard free energies of formation from the constituent elements ($\Delta G_f°$), based either on the Raoult convention or the Henry convention, or as equilibrium constants.
2 As values of the so-called free-energy function, $(G° - H_0°)/T$ or $(G° - H_{298}°)/T$.
3 As conventional absolute entropies ($S°$) together with enthalpies of formation ($\Delta H_f°$).
4 As standard electrode potentials ($E°$).

We review the use of each of these methods in turn and, following this, list sources of these various forms of free-energy data.

3.12.1 Use of Free Energy of Formation or Equilibrium Constants

Since the standard free-energy change of a reaction (represented, say, by $\Delta G°$) is related to the $\mu_i°$'s, on the one hand, and to the standard free energies of formation $\Delta G_{fi}°$'s, on the other hand, by

$$\Delta G° = \Sigma \nu_i \mu_i° = \Sigma \nu_i \Delta G_{fi}°, \tag{3.12-1}$$

the use of $\Delta G_{fi}°$ is tantamount to identifying it with $\mu_i°$; that is,

$$\mu_i° \equiv \Delta G_{fi}°. \tag{3.12-2}$$

A variation of this occurs when the equilibrium constant of a reaction is given and $\Delta G_{fi}°$'s are not all available. Then $\Delta G°$ is calculated from the reaction isotherm in the generic form

$$\Delta G° = -RT \ln K_a. \qquad (3.10\text{-}1a)$$

A difficulty arises in this case in assigning values to each individual $\mu_i°$. This difficulty is addressed in Section 9.4.

3.12.2 Use of the Free-Energy Function

The free-energy function for a species is defined as $(G° - H_0°)/T$, where $G°$ is the molar standard free energy of the species at T (equivalent to $\mu°$, since the species is normally a pure species), and $H_0°$ is the standard enthalpy of the species at 0 K (298 K may also be used). The standard free-energy change of a reaction is calculated from this function by

$$\Delta G° = \Delta H_0° + T \sum_i \nu_i \left(\frac{G° - H_0°}{T} \right)_i, \qquad (3.12\text{-}3)$$

or by

$$\Delta G° = \Delta H_{298}° + T \sum_i \nu_i \left(\frac{G° - H_{298}°}{T} \right)_i, \qquad (3.12\text{-}3a)$$

where $\Delta H_0°$ and $\Delta H_{298}°$ are the standard enthalpy changes at 0 K and 298 K, respectively, and are evaluated according to the way in which enthalpy data are given (e.g., from standard enthalpies of formation analogous to the use of equation 3.12-1).

Thus, if standard enthalpies of formation $\Delta H_{fi}°$ are provided, together with the enthalpy function $H_T° - H_0°$ or $H_T° - H_{298}°$, then

$$\Delta H_0° = \sum \nu_i \Delta H_{fi,298}° - \sum \nu_i (H_{298}° - H_0°), \qquad (3.12\text{-}4)$$

and

$$\Delta H_{298}° = \sum \nu_i \Delta H_{fi,298}°. \qquad (3.12\text{-}4a)$$

From equations 3.12-1, 3.12-3, and 3.12-4, it follows that

$$\mu_i° \equiv \Delta H_{fi,298}° - (H_{298}° - H_0°)_i + T \left(\frac{G° - H_0°}{T} \right)_i,$$

$$\equiv \Delta H_{fi,0}° + T \left(\frac{G° - H_0°}{T} \right)_i, \qquad (3.12\text{-}5)$$

Ways in Which Standard Free-Energy Information is Available

and from equations 3.12-1, 3.12-3a, and 3.12-4a, that

$$\mu_i^\circ \equiv \Delta H_{fi,298}^\circ + T\left(\frac{G^\circ - H_{298}^\circ}{T}\right)_i. \tag{3.12-5a}$$

3.12.3 Use of Conventional Absolute Entropies

The value of ΔG° may be determined from entropy and enthalpy data by

$$\Delta G^\circ = \Delta H^\circ - T\Delta S^\circ, \tag{3.12-6}$$

where

$$\Delta S^\circ = \Sigma \nu_i S_i^\circ, \tag{3.12-7}$$

and S_i° is the conventional absolute entropy of species i determined by means of either the third law of thermodynamics or statistical mechanics.

From equations 3.12-1, 3.12-6, and 3.12-7 and $\Delta H^\circ = \Sigma \nu_i \Delta H_{fi}^\circ$, it follows that

$$\mu_i^\circ \equiv \Delta H_{fi}^\circ - TS_i^\circ. \tag{3.12-8}$$

Example 3.5 Calculate μ_i° for each species in, and ΔG° for,

$$2CH_4(g) + N_2(g) = 2HCN(g) + 3H_2(g)$$

at 1500 K by means of each of the three methods described, from the data of Table 3.1, which have been recalculated from the JANAF tables (JANAF, 1971).

Solution From equation 3.12-1, it follows that

$$\Delta G^\circ = \Sigma \nu_i \Delta G_{fi}^\circ$$

$$= -2(74.72) - 1(0) + 2(85.55) + 3(0)$$

$$= 21.66 \text{ kJ}$$

The value of μ° for each species, which is equivalent to ΔG_{fi}° in each case (from equation 3.12-2), is given in column 7 in Table 3.1 under $\mu^\circ(1)$.

From equation 3.12-3a and the enthalpy analog of equation 3.12-1, we obtain

$$\Delta G^\circ = -2(-74.87) - 1(0) + 2(135.14) + 3(0) + \frac{1500}{1000}[-2(-227.56)$$

$$-1(-216.17) + 2(-236.61) + 3(-154.54)]$$

$$= 21.70 \text{ kJ}$$

Table 3.1 Data and Calculated μ°'s for Example 3.5

Species	$\Delta G_f^\circ(1500)$	$\Delta H_f^\circ(298)$	$\Delta H_f^\circ(1500)$	S_{1500}°	$[(G^\circ - H_{298}^\circ)/T]_{1500}$	$\mu^\circ(1)$	$\mu^\circ(2)$	$\mu^\circ(3)$
	kJ mole^{-1}			J mole^{-1} K^{-1}		kJ mole^{-1}		
CH$_4$(g)	74.72	−74.87	−92.48	279.66	−227.56	74.72	−416.21	−511.97
N$_2$(g)	0	0	0	241.77	−216.17	0	−324.26	−362.66
HCN(g)	85.55	135.14	132.13	274.99	−236.61	85.55	−219.78	−280.36
H$_2$(g)	0	0	0	178.72	−154.54	0	−231.81	−268.08

Ways in Which Standard Free-Energy Information is Available

The value of $\mu°$ for each species, calculated from equation 3.12-5a, is given in column 8 in Table 3.1 under $\mu°(2)$.

From equations 3.12-6 and 3.12-7, and the enthalpy analog of equation 3.12-1, it follows that

$$\Delta G° = -2(-92.48) - 1(0) + 2(132.13) + 3(0) - \frac{1500}{1000}$$

$$\cdot [-2(279.66) - 1(241.77) + 2(274.99) + 3(178.72)]$$

$$= 21.65 \text{ kJ}$$

The value of $\mu°$ for each species, calculated from equation 3.12-8, is given in column 9 in Table 3.1 under $\mu°(3)$.

Although the values of $\Delta G°$ calculated by the three methods are (virtually) the same, the values of $\mu_i°$ differ, as they are based on different *zero points*. All the data would give the same equilibrium composition, however, regardless of how they are calculated. This is because $N^T \mu°$ is the same for the three sets of $\mu°$ data, where N is any complete stoichiometric matrix for the system. The proof of this requirement is left as a problem.

3.12.4 Use of Standard Electrode Potentials

$\Delta G°$ for a reaction may be obtained from the standard emf $E°$ of a chemical cell in which the given reaction takes place. In turn, $E°$ is obtained from the standard electrode potentials for the two electrode processes (oxidation at the anode and reduction at the cathode) that constitute the overall cell reaction. The electrode potentials are conventionally given for the electrode processes written as reduction processes. The procedure is illustrated in Example 3.6.

Example 3.6 Calculate $\Delta G°$ for the reaction in the lead-acid cell written in Example 3.4 (which is for the cell as a galvanic cell on discharge). For the anode, the electrode process and standard reduction electrode potential are

$$PbSO_4 + 2e = Pb + SO_4^{2-}; \quad E° = -0.356 \text{ V}$$

and for the cathode,

$$PbO_2 + SO_4^{2-} + 4H^+ + 2e = PbSO_4 + 2H_2O; \quad E° = +1.685 \text{ V}.$$

Solution The overall cell process, as given in Example 3.4, is obtained by reversing the anode equation (since the anode process is actually an oxidation) and adding it to the cathode equation. For such an addition, the coefficient of e, which represents one mole of electrons, must be the same for each electrode, for the charge balance to be maintained; in this example this coefficient, which is z in the equations in Section 3.11, is 2. Since reversal of the anode equation

Table 3.2 Sources of Data[a]

Species Group	$PvTx$ (Including p^*)	ΔH_f° or ΔH_c°	$H - H^\circ$, $H - H_{298}^\circ$, etc.	h^E, ΔH_s°, etc.	C_P
Elements	3, 15, 17, 35	1, 2, 11, 12, 15, 16, 23, 24, 26, 32, 34	11, 13, 14, 17, 29, 32, 34	15, 26	1, 11–15, 26, 29, 32
Common gases, substances	5, 17, 27, 35, 36	1, 2, 11, 12, 16, 26, 27, 32, 36	5, 11, 17, 27, 32, 36	26	1, 5, 11, 26, 27, 32, 36
Organic compounds (general)	3, 17, 30, 31, 35	1, 12, 17, 19, 24, 26, 29, 32	17, 29, 32, 35	26, 31	1, 12, 15, 26, 29, 30, 32, 35
Organic compounds (hydrocarbons)	3, 17, 27, 30, 31, 36	12, 24, 27, 29, 36	27, 29, 36	31	12, 27, 29, 30, 31, 36
Inorganic compounds (pure)	15, 17, 35	1, 2, 11, 12, 15–17, 23, 24, 26, 32, 34	11, 13, 17, 32, 34	—	1, 11–15, 26, 32
Inorganic compounds (solutions)	31	2, 12, 16, 19, 23, 24, 26, 32	32	9, 17, 26, 31	9, 12, 17, 26, 31, 32

	$S°, \Delta S,$ etc.	$\Delta G_f°$	$(G - H_0°)/T$ $(G - H_{298}°)/T$	$E°$	$\gamma, \gamma_\pm,$ etc.	g^E, Solubility, Phase Equilibrium
Elements	1,2,11–16,26, 29,32,34	1,2,11,12,15, 16,26,32,34	10,17,29	—	—	6,7,15,18,28,31,33
Common gases, substances	1,2,5,11,12,16, 26,27,32,36	1,2,11,12, 16,26,27,32,36	11,17,27,36	—	—	—
Organic compounds (general)	1,12,19,26,29, 32,35	1,12,19,26, 29,32	17,29,35	21,22	20	6,7,18,28,31,33
Organic compounds (hydrocarbons)	12,27,29,36	12,27,29,36	17,27,29,36	21	—	6,7,28,31,33
Inorganic compounds (pure)	1,2,11–16,26, 32,34	1,2,11,12,15, 16,26,32,34	11,17	—	—	—
Inorganic compounds (solutions)	2,12,16,19,26, 32	2,12,16,19, 26,32	—	4,9,16,22, 25	8,9,10,16, 17,25	6,7,18,28,31,33

[a] The numbers in the columns refer to the following references: 1—Barin and Knacke (1973); 2—Barner and Scheuerman (1978); 3—Boublik et al. (1973); 4—Charlot (1958); 5—Din (1956, 1961); 6—Hala et al. (1967); 7—Hala et al. (1968); 8—Hamer (1968); 9—Harned and Owen (1958); 10—Helgeson et al. (1969); 11—JANAF (1971); 12—Karapet'yants and Karapet'yarts (1970); 13—Kelley (1960); 14—Kelley and King (1961); 15—Kubaschewski and Alcock (1979); 16—Latimer (1952); 17—Lewis and Randall (1961); 18—Linke (1958, 1965); 19—Martell and Smith (1974–1977); 20—Mash and Pemberton (1980); 21—Meites and Zuman (1974); 22—Milazzo and Caroli (1978); 23—Pedley (1972); 24—Pedley and Rylance (1977); 25—Robinson and Stokes (1965); 26—Rossini et al. (1952); 27—Rossini et al. (1953); 28—Stephen and Stephen (1979); 29—Stull et al. (1969); 30—Timmermans (1950, 1965); 31—Timmermans (1959, 1960); 32—Wagman et al. (1965–1973); 33—Wichterle et al. (1973, 1976, 1979); 34—Wicks and Block (1963); 35—Zwolinski et al. (1963); 36—Zwolinski et al. (1974).

causes reversal of the sign of the electrode potential, then

$$E° = 1.685 + 0.356 = 2.041 \text{ V},$$

and from equation 3.11-4,

$$\Delta G° = -2(96,487)2.041$$

$$= -393.86 \text{ kJ}.$$

The method described in this section differs from the methods described in Sections 3.12.1 to 3.12.3 in one important respect, and this involves consideration of free energy of formation of a species in solution, as opposed to in the pure state. The standard chemical potentials in the first three methods (Sections 3.12.1 to 3.12.3) are appropriate for free-energy considerations based on the Raoult convention, such as in equations 3.7-12 and 3.7-15a. For free-energy considerations and standard chemical potentials based on the Henry convention, such as in equations 3.7-15b and 3.7-19, the dissolved state is involved; this is inherent for at least one of the species in the fourth method (this section). A species involved in the use of the Henry convention need not be an electrolyte, and the reaction need not be one in a chemical cell.

To illustrate the difference and the methods of determination of free energies of formation of dissolved species, consider chlorine (Cl_2) both as a gas and as dissolved in water, and chloride ion (Cl^-) as dissolved in water. For chlorine gas, the standard free energy of formation is zero. For chlorine dissolved in water, the standard free energy of formation at 25°C on the molality scale, as determined from solubility data (Denbigh, 1981, pp. 296, 327, 481), is 6900 J mole^{-1}. For chloride ion dissolved in water, the standard free energy of formation at 25°C, also on the molality scale, as determined from the emf of the chlorine electrode ($E° = 1.3595$ V), is -131.17 kJ mole^{-1}.

3.13 SOURCES OF DATA

Table 3.2 provides a list of sources of data (volumetric, enthalpy, free energy, etc.) of use in determining free-energy quantities and their temperature and pressure derivatives. The species are grouped into elements, common gases, organic compounds, and inorganic compounds. Each entry in Table 3.2 refers to a numbered citation following the table, and each citation is contained in the list of references at the end of the book.

In Table 3.2, in addition to symbols already defined, $\Delta H_c°$ is the standard enthalpy of combustion; h^E and g^E are the molar excess enthalpy and free energy, respectively, of a solution; $\Delta H_s°$ is the standard enthalpy of solution; and C_P is the molar heat capacity at constant pressure.

Table 3.2 is intended to be reasonably comprehensive but, of course, is not exhaustive. Much of the information on properties of solutions is scattered throughout the literature. Comprehensive bibliographies on phase equilibria

Problems

(e.g., Wichterle, et al. 1973, 1976, 1979) contain data that relate directly or indirectly to quantities such as g^E or γ. Collections of phase diagrams for systems of elements (mostly metals) (Hansen and Anderko, 1958; Elliott, 1965; Shunk, 1969; Moffatt, 1977, et seq.), and for ceramic (inorganic) systems (Levin, et al., 1964, 1969, 1975), not included in Table 3.2, may also contain information in the original sources enabling thermodynamic properties to be derived.

In the use of free-energy data from any source, care must be taken to ensure consistency with respect to the basis for the standard state of a given element, for example, whether it relates to solid, liquid, or ideal-gas state. This is especially the case for elements such as sulfur and phosphorus.

PROBLEMS

3.1 If ΔG_f° for $C_6H_6(l)$ is 124.50 kJ mole^{-1} at 25°C and 1 atm (a stable state) and the vapor pressure of $C_6H_6(l)$ is 0.1253 atm at 25°C, what is the value of ΔG_f° for $C_6H_6(g)$ at 25°C and 1 atm (a metastable state)?

3.2 If the free-energy function $(G - H_0^\circ)/T$ for ClO_2 is -234.72 J mole^{-1} K^{-1} at 500 K, $\Delta H_{f(298)}^\circ$ is 104.60 kJ mole^{-1}, and $H_{298}^\circ - H_0^\circ$ is 10.78 kJ mole^{-1}, what is the value of the standard chemical potential μ° at 500 K for use, for example, in association with equation 3.7-12? (Data taken from Lewis and Randall, 1961, p. 683.)

3.3 If the conventional absolute entropy (S°) for SO_2 is 299.96 J mole^{-1} K^{-1} at 900 K and the enthalpy of formation is -362.24 kJ mole^{-1} at 900 K, what is the value of the standard chemical potential μ° at 900 K for use, for example, in association with equation 3.7-12? (Data taken from JANAF, 1971.)

3.4 Write equations 3.4-5, the equilibrium conditions for the stoichiometric formulation, for the system in Problem 2.3a.

3.5 Repeat Problem 3.4 for equations 3.5-3 and 3.5-4, the equilibrium conditions for the nonstoichiometric formulation.

3.6 For any two sets of standard chemical potentials $\mu^\circ(1)$ and $\mu^\circ(2)$ (see Example 3.5) for a given system (with any number of species N and elements M), show that the two sets must satisfy

$$\mathbf{N}^T(\boldsymbol{\mu}^\circ(2) - \boldsymbol{\mu}^\circ(1)) = \mathbf{0},$$

where $\mathbf{N}$ is any complete stoichiometric matrix for the system. Illustrate this using the three sets of data in Example 3.5.

3.7 Given the standard electrode potential (E°) for $Cd^{2+} + 2e = Cd$ is -0.403 V at 25°C relative to a hydrogen electrode, on a molality basis (and the Henry convention), calculate

(a) the standard chemical potential of the cadmium ion Cd^{2+};

(b) the standard chemical potential and the standard electrode potential on a molarity basis (the density of water is 0.9971 kg liter^{-1} at 25°C).

3.8 Calculate the standard free energy of formation of N_2 in water at 75°C, based on the Henry convention and the molality scale. Assume that the solubility of N_2 at a partial pressure of 1 atm corresponds to a mole fraction of 8.3×10^{-6} (Prausnitz, 1969, p. 358).

3.9 The mean-ion activity coefficient $\gamma_\pm$ for H_2SO_4 in water is 0.257 on the molality scale (Henry convention) at 25°C and $m = 6.0$ (Robinson and Stokes, 1965, p. 477). Calculate the value on (a) the molarity scale (Henry convention) and (b) the mole fraction scale (Henry convention). The density of the 6-m solution is 1.273 kg liter^{-1}, and that of water is 0.9971 at 25°C.

3.10 Suppose that it is desired to work in terms of T and V as independent variables, rather than in terms of T and P, as in most of Chapter 3. What are the equations corresponding to equations 3.2-10 to 3.2-17, 3.3-1, 3.4-1 to 3.4-5, 3.5-1 to 3.5-4, 3.7-12, 3.7-15a, ad 3.7-29?

3.11 Show that $\partial^2 G/\partial \xi^2$ is positive definite for a single ideal-solution phase; that is, show that $Q(\delta\xi)$ corresponding to equation 3.9-1 is positive for all $\delta\xi \neq 0$.

CHAPTER FOUR

Computation of Chemical Equilibrium for Relatively Simple Systems

We are now in a position to consider actual examples of equilibrium analysis, having developed the equilibrium conditions in Chapter 3 in terms of two formulations, examined the nature of the constraints, and introduced expressions for the chemical potential. We develop algorithms for the two formulations for relatively simple systems prior to the development of general-purpose algorithms in later chapters.

Initially we define a relatively simple system and then comment on factors that affect the choice of formulation to use. We subsequently develop first the stoichiometric formulation and then the nonstoichiometric formulation, in special forms applicable to such systems. Each approach is illustrated by examples. For these examples, T and P are fixed, and we defer consideration of the effect of changes in T and/or P to Chapter 8.

4.1 RELATIVELY SIMPLE SYSTEMS AND THEIR TREATMENT

For the purpose of this chapter, a relatively simple system consists of a single phase that is an ideal solution of two or more species (including the case of an ideal-gas solution) and involves a relatively small number M of elements or a relatively small difference $(N - M)$ between the number of species and the number of elements. [We continue to assume in this chapter, for convenience, that $M = \text{rank}(\mathbf{A}) \equiv C$.] These restrictions are related to the means by which the calculations are actually performed—by "hand" (i.e., by means of a nonprogrammable calculator or graphically), by means of a programmable calculator, or by means of a small computer. The devices used are then characterized by having either no storage memory or a memory of a size of up to perhaps 64K bytes. Recent developments in both programmable calculators and in computers have meant that the difference between a calculator and a computer has narrowed, resulting in an almost continuous spectrum of capability, from the smallest programmable calculator to the largest mainframe computer.

More precisely, in terms of M and N, a relatively simple system is characterized by relatively small values of NM and $M(N - M)$; the latter is a measure of the size of the matrix that must be manipulated in the stoichiometric formulation (i.e., a measure of the size of the computer memory required), and the former is a similar measure for the nonstoichiometric formulation. Consideration of relatively complex systems involving nonideality, more than one phase, and relatively large values of NM or $M(N - M)$ requires more storage than is available on many small machines, and we defer discussion of such systems to later chapters, which describe general-purpose algorithms for use with large computers.

In the examples given in this chapter we illustrate three levels of increasing problem complexity, along with corresponding levels of computational capability. The most primitive of the latter, by hand, involves values of M or $(N - M)$ of 1 or 2; that is, we consider systems for hand calculation to consist of two nonlinear equations at most, for the solution of which the Newton-Raphson or another procedure can be used (Ralston and Rabinowitz, 1978, Chapter 8). Recent developments in programmable calculators allow a significant increase in the size of system that can be considered relative to that for calculation by hand. We use an HP-41C calculator for this purpose and in Appendix B present algorithms for both stoichiometric and nonstoichiometric formulations of the equilibrium problem. Finally, recent developments in small computers allow a further increase in the size of system that can be considered simple. In Appendix B, we also present algorithms written in BASIC for each of the two problem formulations.

4.2 REMARKS ON CHOICE OF FORMULATION

The simplest case for the stoichiometric formulation is when there is only one stoichiometric equation ($R = 1$), which is the case when $(N - M) = 1$. The simplest case for the nonstoichiometric formulation is when there is only one element ($M = 1$). These simplest cases illustrate the determining characteristics for relatively small systems for the two formulations. Comparison of $(N - M)$ with M is a useful guide as to which formulation to use for a relatively simple system. For the stoichiometric case to be preferred, $(N - M)$ is smaller, and for the nonstoichiometric case to be preferred, M is smaller. More precisely, if $(N - M) < M$ ($N < 2M$), the stoichiometric formulation is preferable; if $(N - M) > M$ ($N > 2M$), the nonstoichiometric formulation is preferable.

4.3 STOICHIOMETRIC FORMULATION FOR RELATIVELY SIMPLE SYSTEMS

4.3.1 System Involving One Stoichiometric Equation ($R = 1$)

We consider first the simplest case of a system that can be represented by one stoichiometric equation to illustrate the stoichiometric approach, both numerically and graphically. The dissociation of hydrogen is used in the following paragraphs as an example of this situation.

In general, for a system represented by the stoichiometric equation

$$\sum_i \nu_i A_i = 0, \qquad (2.3\text{-}8)$$

equation 2.3-1a relates n_i to ζ, the extent-of-reaction variable. Numerically, the solution is obtained from equation 3.4-5, the equilibrium condition, and equation 2.3-1a, together with appropriate chemical potential expressions. The solution of equation 3.4-5 in terms of ξ provides the equilibrium value of ξ, from which the composition can be calculated. Graphically, the solution occurs at the minimum of the function $G(\xi)$, which is constructed from equations 3.4-1 and 2.3-1a, together with the chemical potential expressions.

Example 4.1 For the system $\{(H, H_2), (H)\}$, calculate the equilibrium composition at 4000 K and 1 atm (1) numerically, and (2) graphically, if the system is composed initially of an equimolar mixture of H and H_2. At 4000 K, the standard free energy of formation of H is $-15{,}480$ J mole^{-1} (Zwolinski et al., 1974).

Solution Numerically, the system may be represented by the stoichiometric equation

$$H_2 = 2H \quad \text{or} \quad 2H - H_2 = 0. \qquad (A)$$

Since H is species 1 and H_2 is species 2, $\nu_1 = 2$ and $\nu_2 = -1$. The equilibrium criterion, from equation 3.4-5, is

$$\mu_2 = 2\mu_1, \qquad (B)$$

and equation 2.3-1a applied to each species is

$$n_1 = n_1^\circ + 2\xi, \qquad (C)$$

$$n_2 = n_2^\circ - \xi. \qquad (D)$$

If we assume that the system is an ideal-gas solution, so that the chemical potential expression is given by equation 3.7-12a, then

$$\mu_1 = \mu_1^\circ + RT \ln \frac{n_1}{n_t} + RT \ln P \tag{E}$$

and

$$\mu_2 = \mu_2^\circ + RT \ln \frac{n_2}{n_t} + RT \ln P. \tag{F}$$

We also set $n_1^\circ = n_2^\circ = 1$. On substitution of equations C to F and the data ($\mu_1^\circ = -15{,}480$; $\mu_2^\circ = 0$; $R = 8.314$; $T = 4000$ K; $P = 1$ atm) in equation B, we have the following equation for ξ at equilibrium:

$$\frac{(1 + 2\xi)^2}{(1 - \xi)(2 + \xi)} = 2.537,$$

from which the relevant solution is $\xi = 0.4345$. This results in $n_1 = 1.869$ and $n_2 = 0.565$ moles; from these, the composition, expressed in mole fractions, is $x_1 = 0.768$ and $x_2 = 0.232$.

Graphically, the solution may be obtained by either minimizing $G(\xi)$ or solving the nonlinear equation $\Delta G(\xi) \equiv \Sigma v_i \mu_i = 0$. Here we illustrate the former, which is shown in Figure 4.1, a plot of $G(\xi)$ against ξ. Beginning with equation 3.2-16, $G(\xi)$ is constructed as follows:

$$G = n_1 \mu_1 + n_2 \mu_2$$

$$= -15480 - 30960\xi + 33257$$

$$\times [(1 + 2\xi)\ln(1 + 2\xi) + (1 - \xi)\ln(1 - \xi) - (2 + \xi)\ln(2 + \xi)]. \tag{H}$$

Figure 4.1 is a plot of equation H and shows that G is a minimum at $\xi = 0.434$, which leads to essentially the same results as in the numerical solution (preceding paragraph). The minimum value of G is $-72{,}800$ J relative to the datum implied by the μ_i° values.

4.3.2 System Involving Two Stoichiometric Equations ($R = 2$)

We consider here only the graphical method of solution for a system represented by two stoichiometric equations. The numerical method should be implemented by the algorithm developed in the following section. As for $R = 1$, we may consider either the minimization or the nonlinear equation point of view. For $R = 2$, the former involves finding the minimum point on a three-dimensional surface, and the latter involves finding the intersection of

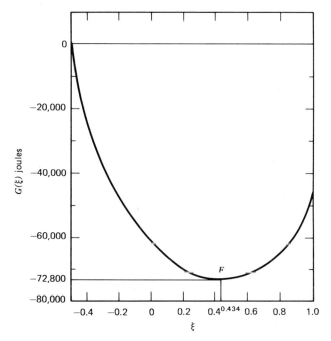

Figure 4.1 Graphical solution for Example 4.1 showing minimum in $G(\xi)$ at equilibrium (point E).

two curves in the (ξ_1, ξ_2) plane. In Example 4.1 we used the minimization point of view, and here we illustrate the use of the alternative. This graphical solution involves first establishing two nonlinear equations in ξ_1 and ξ_2, the extents of reaction for the two stoichiometric equations, from the equilibrium criteria, the chemical potential expressions, and equation 2.3-1a. We use the system involving gaseous polymeric forms of carbon at high temperature to illustrate the procedure.

Example 4.2 For the system $\{(C_1, C_2, C_3), (C)\}$, calculate the equilibrium distribution of the three species at 4200 K and 1 atm, given that $\mu°/RT$ is 1.695 for C_1 (species 1), 1.119 for C_2 (species 2), and 0.171 for C_3 (species 3) (JANAF, 1971). Also assume that the system behaves as an ideal-gas solution.

Solution The system may be represented by the following two stoichiometric equations with corresponding extent-of-reaction variables as indicated:

$$2C_1 = C_2; \quad \xi_1, \tag{A}$$

$$3C_1 = C_3; \quad \xi_2. \tag{B}$$

Applying equation 2.3-1a and taking, for convenience, $n_1^\circ = 3$, $n_2^\circ = n_3^\circ = 0$, we have

$$n_1 = 3 - 2\xi_1 - 3\xi_2, \qquad \text{(C)}$$

$$n_2 = \xi_1, \qquad \text{(D)}$$

and

$$n_3 = \xi_2. \qquad \text{(E)}$$

The equilibrium conditions, from equations A, B, and 3.4-5, are

$$2\mu_1 = \mu_2, \qquad \text{(F)}$$

$$3\mu_1 = \mu_3. \qquad \text{(H)}$$

Substituting chemical potential expressions for μ_1, μ_2, and μ_3 from equation 3.7-12a into equations F and H, together with the use of equations C to E to eliminate n_1, n_2, n_3, and the use of the numerical data given and rearranging, we have (from equation F)

$$f_1(\xi_1, \xi_2) = \frac{\xi_1(3 - \xi_1 - 2\xi_2)}{(3 - 2\xi_1 - 3\xi_2)^2} - 9.689 \qquad \text{(J)}$$

$$= 0,$$

and from equation H

$$f_2(\xi_1, \xi_2) = \frac{\xi_2(3 - \xi_1 - 2\xi_2)^2}{(3 - 2\xi_1 - 3\xi_2)^3} - 136.18 \qquad \text{(K)}$$

$$= 0.$$

It is mathematically convenient to replace equation K by J/K (which is equivalent to replacing equation B by A − B or $C_3 = C_1 + C_2$). This results in

$$f_2'(\xi_1, \xi_2) = \frac{\xi_1(3 - 2\xi_1 - 3\xi_2)}{\xi_2(3 - \xi_1 - 2\xi_2)} - 0.07115 \qquad \text{(L)}$$

$$= 0.$$

Values of ξ_1 may be calculated from specified values of ξ_2 for each of equations J and L. Figure 4.2 is a plot of the two sets of values of ξ_1 against ξ_2. The solution lies at the intersection of the two curves, which then gives the equilibrium values of ξ_1 and ξ_2, 0.315 and 0.723, respectively. From these,

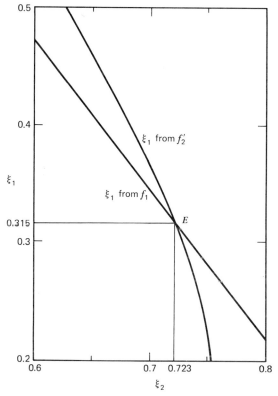

Figure 4.2 Graphical solution for Example 4.2 showing equilibrium values of ξ_1 and ξ_2 at point E.

the equilibrium mole fractions as a measure of the distribution are $x_1 = 0.162$, $x_2 = 0.254$, and $x_3 = 0.584$.

4.3.3 Stoichiometric Algorithm

To consider the general case of any number of stoichiometric equations for relatively simple systems, we begin with the equilibrium conditions

$$\sum_{i=1}^{N} \nu_{ij}\mu_i(\boldsymbol{\xi}) = 0; \quad j = 1, 2, \ldots, R. \tag{3.4-5}$$

From an estimate $\mathbf{n}^{(m)}$ of the solution of equation 3.4-5, mole numbers at the next iteration are obtained by means of (see equation 2.3-1a)

$$n_i^{(m+1)} = n_i^{(m)} + \omega^{(m)} \sum_{j=1}^{R} \nu_{ij}\delta\xi_j^{(m)}, \tag{4.3-1}$$

where $\omega^{(m)}$ is a positive step-size parameter, which is usually set to unity or less (see Section 5.4.1 for general discussion).

Expanding equation 3.4-5 about $\mathbf{n}^{(m)}$ in a Taylor series, neglecting the second- and higher-order terms, and setting the result to zero, we obtain the Newton-Raphson method (see Section 5.3.1 for general discussion). This gives

$$\sum_{l=1}^{R}\sum_{k=1}^{N}\sum_{i=1}^{N} \nu_{ij}\left(\frac{\partial \mu_i}{\partial n_k}\right)^{(m)}\left(\frac{\partial n_k}{\partial \xi_l}\right)^{(m)} \delta\xi_l^{(m)} = -\sum_{n=1}^{N} \nu_{ij}\mu_i^{(m)};$$

$$j = 1, 2, \ldots, R, \quad (4.3\text{-}2)$$

where superscript (m) denotes evaluation at $\mathbf{n}^{(m)}$. For an ideal solution, we introduce the chemical potential expression from equation 3.7-15a, which is rewritten as

$$\mu_i = \mu_i^* + RT \ln \frac{n_i}{n_t}. \tag{4.3-3}$$

From this, it follows that

$$\frac{\partial \mu_i}{\partial n_k} = RT\left(\frac{\delta_{ik}}{n_i} - \frac{1}{n_t}\right), \tag{4.3-4}$$

where δ_{ik} is the Kronecker delta. Substituting equations 4.3-4 and 2.3-6 in equation 4.3-2, we have

$$\sum_{l=1}^{R} \delta\xi_j^{(m)}\left(\sum_{i=1}^{N} \frac{\nu_{ij}\nu_{il}}{n_i^{(m)}} - \frac{\bar{\nu}_j\bar{\nu}_l}{n_t^{(m)}}\right) = -\sum_{i=1}^{N} \frac{\nu_{ij}\mu_i^{(m)}}{RT};$$

$$j = 1, 2, \ldots, R \quad (4.3\text{-}5)$$

where

$$\bar{\nu}_j = \sum_{i=1}^{N} \nu_{ij}. \tag{4.3-6}$$

Equations 4.3-5 are solved for $\delta\xi^{(m)}$, and the result is used in equation 4.3-1 to determine $\mathbf{n}^{(m+1)}$. The procedure is repeated until convergence is attained. [This approach is essentially that suggested by Hutchison (1962), Stone (1966), and Bos and Meerschoek (1972).] A flow chart for this algorithm is given in Figure 4.3. Computer program listings for the HP-41C and in BASIC are provided in Appendix B.

Stoichiometric Formulation for Relatively Simple Systems

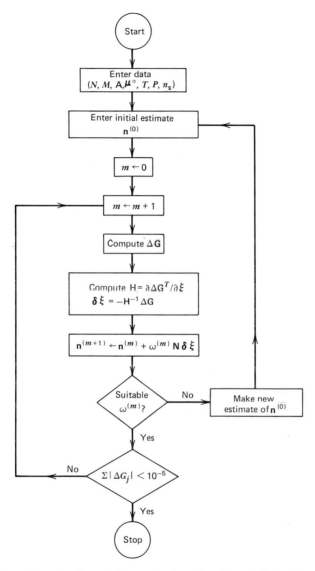

Figure 4.3 Flow chart for the stoichiometric algorithm for relatively simple solutions.

Example 4.3 Calculate the equilibrium mole numbers for the system $\{(CO_2, N_2, H_2O, CO, O_2, NO, H_2), (C, H, O, N)\}$ at 2200 K and 40 atm, resulting from the combustion of one mole of propane in air with the stoichiometric amount of oxygen (for complete combustion); assume that air consists of N_2 and O_2 in a 4:1 ratio. [This is a simplified version of a problem originally considered by Damköhler and Edse (1943), in which the presence of the species H, O and OH is neglected here.]

Table 4.1   Summary of Input Data and Results for Example 4.3

Species	Formula Vector	$\mu°$, kJ mole^{-1}	$\mathbf{n}^{(0)}$	$\mathbf{n}^{(9)}$
CO_2	1 0 2 0	−396.125	2.0	2.923
N_2	0 0 0 2	0	19	1.999 × 10
H_2O	0 2 1 0	−123.93	1.5	3.980
CO	1 0 1 0	−302.65	1.0	7.667 × 10^{-2}
O_2	0 0 2 0	0	0.75	3.471 × 10^{-2}
NO	0 0 1 1	62.51	2.0	2.731 × 10^{-2}
H_2	0 2 0 0	0	2.5	2.005 × 10^{-2}

Solution   The stoichiometric algorithm is appropriate in this case since $N < 2M$. For illustration, we use the HP-41C program given in Appendix B. From the statement of the problem, $\mathbf{b} = (3, 8, 10, 40)^T$. We enter data and execute the program in accordance with the *User's Guide* in Appendix B. A summary of the input data and the results is given in Table 4.1. We have ordered the species in column 1 in accordance with the note at the end of the *User's Guide*. The $\mu°$ in column 3 is taken from JANAF (1971). The initial estimate $\mathbf{n}^{(0)}$ in column 4 has been arbitrarily set to satisfy $\mathbf{b}$. The solution, obtained after nine iterations, is given in column 5. The dominant species are CO_2 and H_2O as reaction products and N_2 as relatively inert. If the combustion were indeed stoichiometrically complete, the amounts of these species would be 3, 4, and 20, respectively.

Since $N = 7$ and rank ($\mathbf{A}$) is 4, $R = 3$. The three chemical equations used by the algorithm are

1. $2CO_2 - 2CO = O_2$,
2. $CO_2 + \tfrac{1}{2}N_2 - CO = NO$, and
3. $-CO_2 + H_2O + CO = H_2$.

$\Delta G/RT$ for these equations at $\mathbf{n}^{(9)}$ is $(-1.12 \times 10^{-7}, -5.70 \times 10^{-8}, -1.20 \times 10^{-8})^T$.

4.4   NONSTOICHIOMETRIC FORMULATION FOR RELATIVELY SIMPLE SYSTEMS

4.4.1   System Consisting of One Element ($M = 1$)

We consider the simplest case of a system consisting of a single element to illustrate the minimization problem given in equation 3.5-1, subject to the constraints of equation 2.2-1. First, to provide geometric insight into the nature of the nonstoichiometric formulation in terms of the Lagrange multipliers, we use the case of $N = 2$. Then we consider a procedure for arbitrary N that can be generalized to the numerical algorithm given in the following section.

We note, however, that the computer programs of Appendix B.2 do not allow the case $M = 1$, although they could be suitably modified to do so.

4.4.1.1 Geometric Illustration for $N = 2$

Consider a system of species 1 and 2 involving one element. The problem is to minimize

$$G(n_1, n_2) = n_1\mu_1 + n_2\mu_2 \qquad (4.4\text{-}1)$$

at given T and P such that

$$a_1 n_1 + a_2 n_2 = b, \qquad (4.4\text{-}2)$$

where b is the number of moles of the element in the (closed) system. The solution is obtained from equations 3.2-8 and 3.3-1, with

$$dG = \mu_1 \, dn_1 + \mu_2 \, dn_2 = 0, \qquad (4.4\text{-}3)$$

from which

$$\frac{dn_2}{dn_1} = -\frac{\mu_1}{\mu_2}. \qquad (4.4\text{-}4)$$

Since, from equation 4.4-2,

$$\frac{dn_2}{dn_1} = -\frac{a_1}{a_2}, \qquad (4.4\text{-}5)$$

it follows that, at equilibrium,

$$\frac{\mu_1}{a_1} = \frac{\mu_2}{a_2} \, (= \lambda), \qquad (4.4\text{-}6)$$

where the parameter λ has been introduced to represent the common ratio. These two equations can be rearranged as

$$\mu_1 = a_1 \lambda, \qquad (4.4\text{-}7)$$

$$\mu_2 = a_2 \lambda, \qquad (4.4\text{-}8)$$

which we recognize as the equilibrium conditions of equation 3.5-3, with λ as the (single) Lagrange multiplier.

The quantity dn_2/dn_1 in equation 4.4-4 is the slope of a tangent to the curve $G = $ constant. Similarly, dn_2/dn_1 in equation 4.4-5 is the slope of a tangent to the constraint (which is coincident with the constraint itself in this case). Equations 4.4-7 and 4.4-8 express the equality of these slopes. This condition, coupled with the requirement that the solution lie on the constraint, means that graphically the constraint itself must be tangent to a contour of constant G. For a linear constraint, the solution occurs graphically where the element-abundance constraint line (equation 4.4-2) is tangent to the $G(n_1, n_2)$ surface (equation 4.4-1). This can be illustrated by constructing contours of fixed G values and showing tangency of one of the contours to the constraint line.

Example 4.4 Use the system described in Example 4.1 to illustrate the Lagrange multiplier method graphically.

Solution Equations 4.4-1 and 4.4-2 are, respectively,

$$G = 33257[n_1 \ln n_1 + n_2 \ln n_2 - 0.4655 n_1 - (n_1 + n_2)\ln(n_1 + n_2)], \quad (A)$$

where G is in joules and

$$n_1 + 2n_2 = 3, \quad (B)$$

based on a system containing one mole of each species initially.

The graphical construction is shown in Figure 4.4, which is a plot of $n_1(n_H)$ against $n_2(n_{H_2})$, showing the constraint line of equation B together with contours of constant G calculated from equation A. Figure 4.4 shows the constraint line tangent to the contour $G = -72{,}800$ J at the equilibrium point E. The coordinates of this point are $n_1 = 1.87$ and $n_2 = 0.56$, in essential agreement with the result given in Example 4.1. The value of G at point E is consistent with the minimum value of G in Example 4.1.

4.4.1.2 General Case ($N \geq 2$)

Consider the general system for $M = 1\{(A_1, A_2,\ldots,A_N), (A)\}$. The $N + 1$ conditions at equilibrium from equations 3.5-3 and 3.5-4 are

$$\mu_i = i\lambda; \quad i = 1, 2,\ldots,N \quad (4.4\text{-}9)$$

and

$$\sum_{i=1}^{N} i n_i = b. \quad (4.4\text{-}10)$$

Using equation 3.7-15a for μ_i in equation 4.4-9, we obtain

$$n_i = n_t \exp\left(\frac{i\lambda - \mu_i^*}{RT}\right); \quad i = 1, 2,\ldots,N. \quad (4.4\text{-}11)$$

Nonstoichiometric Formulation for Relatively Simple Systems

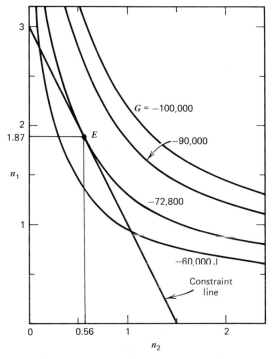

Figure 4.4 Graphical solution for Example 4.4; equilibrium is at point E.

Substituting equation 4.4-11 in equation 4.4-10 and summing equation 4.4-11, we have, respectively,

$$n_t \sum_{i=1}^{N} i \exp\left(\frac{i\lambda - \mu_i^*}{RT}\right) = b \quad (4.4\text{-}12)$$

and

$$\sum_{i=1}^{N} \exp\left(\frac{i\lambda - \mu_i^*}{RT}\right) = 1. \quad (4.4\text{-}13)$$

Equations 4.4-12 and 4.4-13 are two equations in the two unknowns λ and n_t. Since equation 4.4-13 contains only the unknown λ, n_t and the mole fractions of the species can be obtained by solving this equation and substituting the result in equations 4.4-12 and 4.4-11, respectively.

The general problem for $M = 1$ is thus equivalent to solving the single Nth-degree polynomial equation

$$\sum_{i=1}^{N} \sigma_i z^i - 1 = 0, \quad (4.4\text{-}14)$$

where

$$\sigma_i = \exp\left(\frac{-\mu_i^*}{RT}\right) \qquad (4.4\text{-}15)$$

and

$$z = \exp\left(\frac{\lambda}{RT}\right). \qquad (4.4\text{-}16)$$

From Descarte's rule of signs (Wilf, 1962, p. 94), equation 4.4-14 has a unique, positive, real root.

Example 4.5 Repeat Example 4.2, using the nonstoichiometric formulation and the Lagrange multiplier method.

Solution The solution involves the polynomial equation of degree 3 given by equation 4.4-14, which, on substitution of the data given in Example 4.2, becomes

$$0.1836 \exp\left(\frac{\lambda}{RT}\right) + 0.3266\left[\exp\left(\frac{\lambda}{RT}\right)\right]^2 + 0.8428\left[\exp\left(\frac{\lambda}{RT}\right)\right]^3 - 1 = 0.$$

This equation may be solved analytically or graphically. The result is $\lambda/RT = -0.123$. The mole fractions calculated from this result, with the use of equation 4.4-11, are $x_1 = 0.162$, $x_2 = 0.255$, and $x_3 = 0.583$, essentially the same as in Example 4.2.

In cases where not all species A_i are present in the system, equation 4.4-14 becomes

$$\sum_{i=1}^{N} \sigma_i z^{a_{1i}} - 1 = 0, \qquad (4.4\text{-}17)$$

where a_{1i} is the subscript to species A_i (i.e., its formula vector) and N is the number of species present. In equation 4.4-11, $i\lambda$ is replaced by $a_{1i}\lambda$.

4.4.2 Nonstoichiometric Algorithm

In this section we describe an algorithm for the computation of equilibrium in a system consisting of a single phase that is an ideal solution, based on the minimization problem stated in Section 3.5, for which the solution is given in general by equations 3.5-3 and 3.5-4.

For an ideal solution, we introduce the appropriate chemical potential expression (equation 3.7-15a), written as

$$\mu_i = \mu_i^* + RT \ln \frac{n_i}{n_t}, \qquad (4.3\text{-}3)$$

into the first equilibrium condition

$$\mu_i - \sum_{k=1}^{M} a_{ki}\lambda_k = 0; \quad i = 1, 2, \ldots, N. \tag{3.5-3}$$

This results in

$$n_i = n_t \sigma_i \prod_{l=1}^{M} z_l^{a_{li}}; \quad i = 1, 2, \ldots, N, \tag{4.4-18}$$

where

$$z_l = \exp\left(\frac{\lambda_l}{RT}\right), \tag{4.4-19}$$

and we have replaced the dummy index k by l to avoid two dummy indices in the following equation being denoted by the same symbol. We substitute equation 4.4-18 into the second equilibrium condition, equation 3.5-4, to give

$$n_t \sum_{i=1}^{N'} a_{ki}\sigma_i \prod_{l=1}^{M} z_l^{a_{li}} = b_k; \quad \begin{array}{l} k = 1, 2, \ldots, M, \\ l = 1, 2, \ldots, M, \end{array} \tag{4.4-20}$$

where the sum to N' excludes inert species.* The total number of moles is

$$n_t = \sum_{i=1}^{N'} n_i + n_z, \tag{4.4-21}$$

where n_z is the total number of moles of inert species. Substituting equation 4.4-18 into equation 4.4-21, we obtain

$$n_t\left(1 - \sum_{i=1}^{N'} \sigma_i \prod_{l=1}^{M} z_l^{a_{li}}\right) = n_z. \tag{4.4-22}$$

From equation 4.4-20, with $k = 1$ and $b_1 \neq 0$,

$$\sum_{i=1}^{N'} a_{1i}\sigma_i \prod_{l=1}^{M} z_l^{a_{li}} = \frac{b_1}{n_t}. \tag{4.4-23}$$

---

*Henceforth we frequently distinguish reacting species from inert species in order to reduce the number of nonlinear equations that must be solved. The number of reacting species is N' (cf. N, the total number of species, including inert species).

We combine equations 4.4-20 and 4.4-23 to eliminate n_t:

$$\sum_{i=1}^{N'} a_{ki}\sigma_i \prod_{l=1}^{M} z_l^{a_{li}} = r_k \sum_{i=1}^{N'} a_{1i}\sigma_i \prod_{l=1}^{M} z_l^{a_{li}};$$

$$k = 2, 3, \ldots, M, \quad (4.4\text{-}24)$$

where

$$r_k = \frac{b_k}{b_1}; \quad k = 2, 3, \ldots, M. \quad (4.4\text{-}25)$$

Similarly, equations 4.4-22 and 4.4-23 yield

$$\sum_{i=1}^{N'} \sigma_i(1 + r_1 a_{1i}) \prod_{l=1}^{M} z_l^{a_{li}} = 1, \quad (4.4\text{-}26)$$

where

$$r_1 = \frac{n_z}{b_1}. \quad (4.4\text{-}27)$$

Finally, equations 4.4-24 and 4.4-26 may be written as

$$\sum_{i=1}^{N'} \beta_{ki}\sigma_i \prod_{l=1}^{M} z_l^{a_{li}} = \delta_{k1}; \quad k = 1, 2, \ldots, M, \quad (4.4\text{-}28)$$

where

$$\beta_{1i} = 1 + r_1 a_{1i}, \quad (4.4\text{-}29)$$

$$\beta_{ki} = a_{ki} - r_k a_{1i}; \quad k = 2, 3, \ldots, M. \quad (4.4\text{-}30)$$

Equations (4.4-28) are M equations in the M unknown z's (or λ's). These equations have been considered previously by Brinkley (1966), White (1967), and Vonka and Holub (1971), except that they did not incorporate inert species.

Equations 4.4-28 require storage of both a_{li} and β_{li} or recalculation of one from the other at each iteration. For better efficiency, we use only β_{li} by transforming a_{li} into β_{li} from equation 4.4-30:

$$a_{li} = \beta_{li} + r_l a_{1i}; \quad l = 2, 3, \ldots, M. \quad (4.4\text{-}31)$$

Substituting this result into equation 4.4-28 and rearranging, we have

$$\sum_{i=1}^{N'} \beta_{ki}\sigma_i \prod_{l=2}^{M} z_l^{\beta_{li}} \left(z_1 \prod_{l=2}^{M} z_l^{r_l} \right)^{a_{1i}} = \delta_{k1}, \quad (4.4\text{-}32)$$

or

$$\sum_{i=1}^{N'} \beta_{ki}\sigma_i \prod_{l=1}^{M} \theta_l^{\alpha_{li}} = \delta_{k1}, \qquad k = 1, 2, \ldots, M \qquad (4.4\text{-}33)$$

where

$$\theta_l = z_l, \qquad l = 2, 3, \ldots, M \qquad (4.4\text{-}34)$$

and

$$\theta_1 = z_1 \prod_{l=2}^{M} z_l^{r_l}, \qquad (4.4\text{-}35)$$

where

$$\alpha_{li} = \beta_{li}; \qquad l = 2, 3, \ldots, M \qquad (4.4\text{-}36)$$

and

$$\alpha_{1i} = a_{1i}. \qquad (4.4\text{-}37)$$

Hence we need store only the β matrix and the α_{1i}'s. If desired, the total number of moles can be determined from equation 4.4-23, which can be written, together with equation 4.4-18, as

$$n_t = \frac{b_1}{\sum_{i=1}^{N'} a_{1i} x_i}. \qquad (4.4\text{-}38)$$

The mole fractions are determined from the solution of equation 4.4-33 and

$$x_i = \sigma_i \prod_{l=1}^{M} z_l^{\alpha_{li}}. \qquad (4.4\text{-}39)$$

The Newton-Raphson method (see Section 5.3.1 for general discussion) for solving equations 4.4-33 for $\ln \theta_l$ (we use $\ln \theta_l$ to ensure that θ_l remains positive) is given by

$$\sum_{l=1}^{M} \left(\frac{\partial f_k}{\partial \ln \theta_l} \right)_{\boldsymbol{\theta}^{(m)}} \delta(\ln \theta_l)^{(m)} = -f_k; \qquad k = 1, 2, \ldots, M, \qquad (4.4\text{-}40)$$

$$\ln \theta_l^{(m+1)} = \ln \theta_l^{(m)} + \omega^{(m)} \delta(\ln \theta_l)^{(m)}; \qquad m = 0, 1, 2, \ldots, \qquad (4.4\text{-}41)$$

where equation 4.4-33 has been written as $\mathbf{f} = \mathbf{0}$, and ω is a step-size parameter. From equation 4.4-33, we obtain

$$\frac{\partial f_k}{\partial \ln \theta_l} = \sum_{i=1}^{N'} \beta_{ki}\alpha_{li}\sigma_i \prod_{l=1}^{M} \theta_l^{\alpha_{li}}$$

$$= \sum_{i=1}^{N'} \beta_{ki}\alpha_{li} x_i. \qquad (4.4\text{-}42)$$

A flow chart for this algorithm is given in Figure 4.5. Computer program listings for the HP-41C and in BASIC are provided in Appendix B.

We illustrate this procedure first by a very simple system involving only two elements, to show explicitly the structure of equations 4.4-33. We then present a more complex example to illustrate the use of the BASIC computer program in Appendix B.

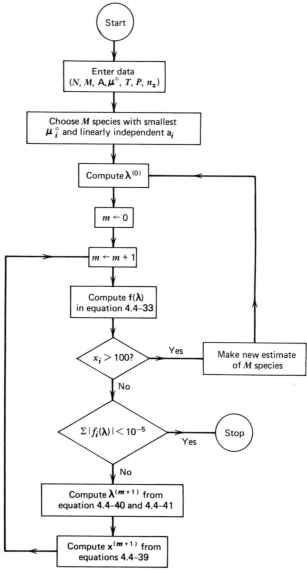

Figure 4.5 Flow chart for the nonstoichiometric algorithm for relatively simple systems.

Nonstoichiometric Formulation for Relatively Simple Systems

Example 4.6 Calculate the fraction of SO_2 converted at equilibrium to SO_3 on oxidation with O_2 at 900 K and 1 atm. Assume SO_2 and O_2 are present initially in an equimolar ratio and that the reacting system is an ideal-gas solution. At 900 K, $\mu°/RT$ is 0 for O_2 (species 1), -39.603 for SO_2 (species 2), and -41.509 for SO_3 (species 3) (JANAF, 1971).

Solution The system is represented by

$$\{(O_2, SO_2, SO_3), (S, O)\}; \quad \mathbf{b} = (1,4)^T.$$

$$\mathbf{A} = \begin{pmatrix} a_{11} & a_{12} & a_{13} \\ a_{21} & a_{22} & a_{23} \end{pmatrix} = \begin{pmatrix} 0 & 1 & 1 \\ 2 & 2 & 3 \end{pmatrix}$$

$$r_2 = \frac{b_2}{b_1} = 4$$

$$\beta_{11} = \beta_{12} = \beta_{13} = 1$$

$$\beta_{21} = a_{21} - r_2 a_{11} = 2 - 4(0) = 2$$

$$\beta_{22} = a_{22} - r_2 a_{12} = 2 - 4(1) = -2$$

$$\beta_{23} = a_{23} - r_2 a_{13} = 3 - 4(1) = -1$$

$$\begin{pmatrix} \alpha_{11} & \alpha_{12} & \alpha_{13} \\ \alpha_{21} & \alpha_{22} & \alpha_{23} \end{pmatrix} = \begin{pmatrix} 0 & 1 & 1 \\ 2 & -2 & -1 \end{pmatrix}$$

$$\sigma_1 = \exp\left(\frac{-\mu_1^*}{RT}\right) = 1$$

$$\sigma_2 = \exp\left(\frac{-\mu_2^*}{RT}\right) = 1.5826 \times 10^{17}$$

$$\sigma_3 = \exp\left(\frac{-\mu_3^*}{RT}\right) = 1.0645 \times 10^{18}$$

Equations 4.4-33 become, for $k = 1$,

$$\beta_{11}\sigma_1\theta_1^{\alpha_{11}}\theta_2^{\alpha_{21}} + \beta_{12}\sigma_2\theta_1^{\alpha_{12}}\theta_2^{\alpha_{22}} + \beta_{13}\sigma_3\theta_1^{\alpha_{13}}\theta_2^{\alpha_{23}} = 1,$$

and for $k = 2$,

$$\beta_{21}\sigma_1\theta_1^{\alpha_{11}}\theta_2^{\alpha_{21}} + \beta_{22}\sigma_2\theta_1^{\alpha_{12}}\theta_2^{\alpha_{22}} + \beta_{23}\sigma_3\theta_1^{\alpha_{13}}\theta_2^{\alpha_{23}} = 0.$$

Inserting values in these two equations for β_{ki}, σ_i, and α_{li}, we have

$$f_1(\theta_1, \theta_2) = \theta_2^2 + 1.5826 \times 10^{17}\theta_1\theta_2^{-2} + 1.0645 \times 10^{18}\theta_1\theta_2^{-1} - 1$$

$$= 0, \qquad (A)$$

and

$$f_2(\theta_1, \theta_2) = 2\theta_2^2 - 2(1.5826 \times 10^{17})\theta_1\theta_2^{-2} - 1.0645 \times 10^{18}\theta_1\theta_2^{-1}$$
$$= 0 \tag{B}$$

for $k = 1$ and $k = 2$, respectively. These two equations may be solved by means of one of the two nonstoichiometric computer programs in Appendix B, or graphically. For illustration, a graphical solution is shown in Figure 4.6, which is a plot of θ_1, calculated from each of equations A and B for assigned values of θ_2, against θ_2. The intersection provides the equilibrium values of $\theta_1 = 2.90 \times 10^{-19}$ and $\theta_2 = 0.612$. From these results and equation 4.4-39, the mole fractions are

$$x_1 = \theta_2^2 = 0.374,$$

$$x_2 = \frac{\sigma_2\theta_1}{\theta_2^2} = 0.122,$$

$$x_3 = \frac{\sigma_3\theta_1}{\theta_2} = 0.504.$$

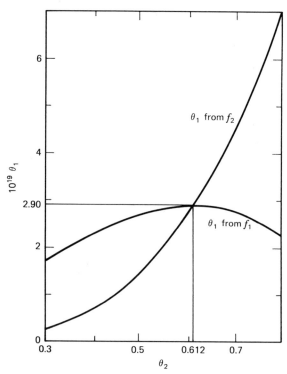

Figure 4.6 Graphical solution for Example 4.6.

Nonstoichiometric Formulation for Relatively Simple Systems

From equation 4.4-38, $n_t = 1/(x_2 + x_3) = 1.594$. Thus the fraction of SO_2 converted at equilibrium is $[1 - 0.122(1.594)]/1 = 0.806$.

Example 4.7 We reconsider the system described in Example 4.3, but now assume that N_2 is inert and that the species OH, H, and O, previously neglected, are present. The first assumption means that the species NO is omitted.

```
ENTER DEVICE NUMBER FOR PRINTING
61

ENTER NUMBER OF SPECIES AND ELEMENTS:
8,3

TYPE OF CHEM. POT. :KJ, KCAL OR MU/RT
ENTER 1, 2 OR 3 RESPECTIVELY
1

HOW MANY SIGNIFICANT FIGURES?
8

ENTER SPECIES NAME
CO2

ENTER FORMULA VECTOR,CHEM POTENTIAL 4 NUMBERS
1,0,2,-396.41

ENTER SPECIES NAME
CO

ENTER FORMULA VECTOR,CHEM POTENTIAL 4 NUMBERS
1,0,1,-302.65

ENTER SPECIES NAME
H2O

ENTER FORMULA VECTOR,CHEM POTENTIAL 4 NUMBERS
0,2,1,-123.93

ENTER SPECIES NAME
H2

ENTER FORMULA VECTOR,CHEM POTENTIAL 4 NUMBERS
0,2,0,0

ENTER SPECIES NAME
O2

ENTER FORMULA VECTOR,CHEM POTENTIAL 4 NUMBERS
0,0,2,0

ENTER SPECIES NAME
OH

ENTER FORMULA VECTOR,CHEM POTENTIAL 4 NUMBERS
0,1,1,6.954

ENTER SPECIES NAME
H

ENTER FORMULA VECTOR,CHEM POTENTIAL 4 NUMBERS
0,1,0,94.81
```

Figure 4.7 Computer input display for Example 4.7 with use of BASIC program in Appendix B.

```
ENTER SPECIES NAME
O

ENTER FORMULA VECTOR,CHEM POTENTIAL  4 NUMBERS
0,0,1,108.30

ENTER ELEMENTAL ABUNDANCES:
3,8,10

ARE THERE ANY INERT SPECIES PRESENT    (Y/N)
Y

ENTER NO OF MOLES OF INERT SPECIES
20

ENTER ELEMENT 1
C

ENTER ELEMENT 2
H

ENTER ELEMENT 3
O

ENTER TEMPERATURE IN K, PRESSURE IN ATM
2200,40

ENTER TITLE
PROPANE COMBUSTION

DO YOU WANT TO MAKE YOUR OWN INITIAL ESTIMATE FOR THE LAGR.MULT. (Y/N)
N

DO YOU WANT TO PRINT INTERMEDIATE RESULTS (Y/N)
N
```

Figure 4.7 (*Continued*)

The data for the problem are entered as indicated, in accordance with the BASIC computer program in Appendix B.2. Figure 4.7 shows the input procedure as displayed on the computer (a Tektronix 4052). Each pair of lines consists of a prompting statement by the program, followed by a user response. The penultimate prompting statement is a question concerning the procedure to be used to determine an initial estimate of the Lagrange multipliers λ/RT. In this case the program generates these internally using a procedure discussed in Section 9.1.2.

The computer output is shown in Figure 4.8. The effects of the changes made with respect to Example 4.3 are very small. The mole fractions of the species are the same to within 0.001. Thus the assumption that N_2 is inert is a good approximation (environmental considerations notwithstanding). Also, the neglect of H and O in Example 4.3 is an excellent approximation. The mole fraction of OH or of NO (0.001), although not negligible, is very small.

PROBLEMS

4.1 Would nitrogen (N_2) be a suitable inert gas in the study of the dissociation of hydrogen (H_2) at 4000 K and moderate pressure?

Problems

```
PROPANE COMBUSTION

8 SPECIES           3 ELEMENTS

PRESSURE    40 ATM

TEMPERATURE  2200 K

ELEMENTAL ABUNDANCES    C          3
                        H          8
                        O         10

SPECIES   FORMULA VECTOR              STAN. CHEM. POT.
          C H O
CO2       1 0 2                            -396.410
CO        1 0 1                            -320.650
H2O       0 2 1                            -123.930
H2        0 2 0                               0.000
O2        0 0 2                               0.000
OH        0 1 1                               6.954
H         0 1 0                              94.810
O         0 0 0                             108.300
STAN. CHEM. POT. IS IN KJ/MOLE

MACHINE ESTIMATE

INITIAL ESTIMATES FOR THE LAGRANGE MULTIPLIERS
-1.018672146E+001
 2.730364543E-002
-4.901918402E+000

NUMBER OF ITERATIONS            8

SPECIES          MOLE FRACTION        MOLES

CO2              1.082671128E-001     2.928661397E+000
CO               2.637254212E-003     7.133860326E+002
H2O              1.468279573E-001     3.971745062E+000
H2               6.872857093E-004     1.859130694E-002
O2               1.487093281E-003     4.022636767E-002
OH               6.912399157E-004     1.869826954E-002
H                2.325214550E-005     6.289782666E-004
O                1.635950485E-005     4.425300453E-004

INERTS           7.393624466E-001     2.000000000E+001

TOTAL NUMBER OF MOLES:       2.705033247E+001

FINAL LAGRANGE MULTIPLIERS (LAMBDA/RT)

-1.738408412E+001
-1.796940507E+000
-1.411026218E+000

G/RT =  -8.667737905E+001
```

Figure 4.8 Computer output for Example 4.7 from BASIC program in Appendix B.

Consider an equimolar mixture of H_2 and N_2 initially and a pressure of 0.5 atm, together with a reasonable criterion for the maximum allowable amount of hydrogen reacted with nitrogen. The following free-energy data are taken from JANAF (1971):

Species:	H	N	NH	NH_2	NH_3
ΔG_f°, kJ mole^{-1}	−15.36	210.77	258.66	331.16	411.06

4.2 Consider the steam reforming of natural gas (say, CH_4) for the production of hydrogen (CO and CO_2 are also formed) (Shreve and Brink, 1977, pp. 99–102). Calculate the composition (mole fractions) of the system at equilibrium at 1100 K and 2 atm, resulting from the reaction of H_2O and CH_4 in a molar ratio of 3.5 : 1. Free energies of formation for the species in the system $\{(CH_4, H_2O, H_2, CO, CO_2), (C, H, O)\}$ at 1100 K are $(30.32, -187.08, 0, -209.04, -396.05)^T$ in kJ mole^{-1} (JANAF, 1971).

4.3 Consider the system $\{(CH_4, S_2, CS_2, H_2S, H_2), (C, H, S)\}$, involved in the production of CS_2 from CH_4 and S_2 (Shreve and Brink, 1977, p. 711). Calculate the composition at equilibrium at 1000 K and 1.5 atm, if $\mathbf{b} = (1, 4, 4)^T$ and free energies of formation (JANAF, 1971), in kJ mole^{-1}, are $(19.35, 0, -17.42, -41.17, 0)^T$.

4.4 Acrylonitrile is produced by reaction of propylene, ammonia, and oxygen (from air—assume this to be 79 mole % N_2 and 21% O_2) in a fluidized-bed catalytic reactor at about 700 K and 2 atm (Shreve and Brink, 1977, pp. 614–615). Calculate the composition at equilibrium for the system $\{(C_3H_6, NH_3, O_2, C_3H_3N, H_2O, CH_3CN, HCN, N_2), (C, H, N, O, N_2)\}$ if the feed consists of C_3H_6 and NH_3 in the stoichiometric ratio for formation of C_3H_3N and, say, 50% excess air. Comment on the need for catalyst selectivity, if formation of CH_3CN and/or HCN is also to be favored or not favored. Standard free energies of formation at 700 K, in kJ mole^{-1} (Stull et al., 1969) are $(128.03, 27.41, 0, 212.63, -208.87, 133.55, 106.40)^T$. Note the assumption that N_2 is inert.

4.5 Consider the partial oxidation of natural gas (say, CH_4) for the production of synthesis gas (CO + H_2), with H_2O and CO_2 as other products (Shreve and Brink, 1977, p. 103). Calculate the composition of the product gas at equilibrium at 1500 K and 30 atm, if the feed contains CH_4 and O_2 in the stoichiometric ratio for the formation of CO and H_2. Standard free energies of formation for the system $\{(CH_4, O_2, H_2, CO, H_2O, CO_2), (C, H, O)\}$ are $(74.72, 0, 0, -243.68, -164.42, -396.34)^T$ (JANAF, 1971).

4.6 The chlorination of methane (CH_4) in stages is used to produce CH_3Cl, CH_2Cl_2, $CHCl_3$, and CCl_4; HCl is also a product (Shreve and Brink, 1977, pp. 699–701). Compare the calculated equilibrium composition for the first (vapor-phase) stage at 600 K and 2 atm, resulting from a feed of Cl_2 and CH_4 in the molar ratio 0.6 : 1, with the indicated actual [molar (?)] distribution of $CH_3Cl : CH_2Cl_2 : CHCl_3 : CCl_4 = 6 : 3 : 1 : 0.25$; Cl_2 conversion is essentially 100% and CH_4 conversion about 65%. Standard free energies of formation for the system $\{(CH_4, Cl_2, CH_3Cl, CH_2Cl_2, CHCl_3, CCl_4, HCl), (C, H, Cl)\}$ are $(-22.98, 0, -36.17, -40.05, -36.55, -11.85, -97.99)^T$ (JANAF, 1971), in kJ mole^{-1}.

Problems

4.7 The refrigerants CCl_3F and CCl_2F_2 are made by the reaction between CCl_4 and HF; HCl is a by-product (Shreve and Brink, 1977, pp. 322–323). The reaction uses a catalyst of molten $SbCl_5$ at, say, 350 K and 1 atm. Investigate the minimum (molar) feed ratio of CCl_4 to HF so that there is virtually no HF at equilibrium. Free-energy data in the ideal-gas state are as follows (JANAF, 1971):

Species	ΔG_f°, kJ mole^{-1}			
	200 K	300 K	400 K	500 K
CCl_4	−67.70	−53.41	−39.34	−25.49
HF	−273.94	−274.65	−275.34	−275.97
CCl_3F	−262.35	−249.14	−236.02	−223.03
CCl_2F_2	−465.46	−452.48	−439.43	−426.42
HCl	−94.29	−95.32	−96.28	−97.17

4.8 As a variation of Problem 4.2, consider the production of hydrogen by the steam cracking of ethane at 1000 K and 1 atm (Balzhiser et al., 1972, pp. 513–527). If the feed contains 4 moles of H_2O per mole of C_2H_6 and no C(gr) is formed, calculate the composition of the following system at equilibrium: $\{(C_2H_6, H_2O, H_2, CO, CO_2, O_2, CH_4, C_2H_2, C_2H_4), (C, H, O)\}$, with standard free energies of formation, in kJ mole^{-1} (Balzhiser et al., 1972, p. 514): $(109.33, -192.59, 0, -200.59, -395.85, 0, 19.29, 169.89, 118.19)^T$.

4.9 Show that equation 4.4-17 becomes
$$\sum_{i=1}^{N} \sigma_i(1 + a_{1i}r)z^{a_{1i}} - 1 = 0,$$
where $r = n_z/b$, for the system $\{(A_{m_1}, A_{m_2}, \ldots, A_{m_N}, X), (A, X)\}$.

4.10 Calculate the equilibrium composition in the system $\{(C_2H_5OH(\ell), CH_3COOH(\ell), CH_3COOC_2H_5(\ell), H_2O(\ell)), (C, H, O)\}$, relating to the production of ethyl acetate by the esterification reaction between ethyl alcohol and acetic acid at 358 K and a pressure sufficiently high to keep the system liquid. Assume that the feed is equimolar in C_2H_5OH and CH_3COOH and that the reacting system is an ideal solution. [The effect of nonideality in the liquid phase for this system is explored in a problem in Chapter 7, the ideal two-phase (liquid-vapor) system is considered in Chapter 6, and the nonideal, two-phase system is considered in Chapter 9.] For the purpose of the calculation, show that the use of Newton's interpolation formula on the data given by Stull et al. (1969), together with equation 3.7-10a in conjunction with Antoine constants (for vapor pressure) provided by Suzuki et al. (1970), leads to the following standard free energies of formation at 358 K:

Species:	$C_2H_5OH(\ell)$	$CH_3COOH(\ell)$	$CH_3COOC_2H_5(\ell)$	$H_2O(\ell)$
ΔG_f°, kJ mole^{-1}:	−153.90	−365.10	−303.06	−227.59

(Sanderson and Chien, 1973; George et al., 1976).

CHAPTER FIVE

Survey of Numerical Methods

In this chapter we provide some elementary background on numerical methods prior to considering the general-purpose algorithms discussed in Chapter 6. In Chapter 3 we provided an introduction to the two main approaches used throughout this book, the stoichiometric and the nonstoichiometric approaches, primarily from the thermodynamic point of view. In Chapter 4 we applied these to relatively simple (ideal, single-phase) systems and for this purpose developed special-purpose algorithms to be used in conjunction with small computers.

Numerous general-purpose algorithms have appeared in the literature relating to the solution of the chemical equilibrium problem, motivated by one of the two equivalent mathematical formulations described in Chapter 3 (cf. Smith, 1980a). Most of these algorithms can be classified as to whether they are methods for function minimization or for solving sets of nonlinear equations. In this chapter we outline some of the relations between these two types of method and some numerical ways of using them, as a prelude to detailed discussion of existing algorithms. We do not attempt to give an exhaustive discussion of such numerical methods. For example, we focus primarily on necessary conditions and consider only problems and approaches that are particularly appropriate to the formulations of the equilibrium problem discussed in Chapter 3. We also discuss only methods that use analytical expressions for first and second derivatives since these are usually readily available for chemical equilibrium problems. Recent treatments of general numerical aspects of optimization and nonlinear equations are given by, for example, Walsh (1975), Ralston and Rabinowitz (1978), Bazaraa and Shetty (1979), and Fletcher (1980).

We believe that the general discussion given here is useful for two reasons: (1) it is important for an understanding of existing chemical equilibrium algorithms in terms of these general techniques; and (2) it should enable recognition of the basis of other equilibrium algorithms that may be encountered.

5.1 TWO CLASSES OF NUMERICAL PROBLEM

The approaches we discuss for solving the chemical equilibrium problem involve consideration of two classes of numerical problem. These are (1) the minimization of a function $f(\mathbf{x})$, perhaps subject to certain constraints, and (2) the solution of a set of nonlinear algebraic or transcendental equations. Thus we wish to solve either

$$\min_{\mathbf{x} \in \Omega} f(\mathbf{x}), \tag{5.1-1}$$

or

$$\mathbf{g}(\mathbf{x}) = \mathbf{0}, \tag{5.1-2}$$

where Ω is a constraint set and $\mathbf{g}$ and $\mathbf{x}$ are N-vectors. In the following, we assume that f and $\mathbf{g}$ are twice continuously differentiable.

Figure 5.1 shows the methods discussed in this chapter for solving these two types of problem. The numbers indicate the sections in which the various methods are treated, and the dashed lines indicate interrelationships or links, with precedence indicated by arrows. There are five such links shown, indicated by the letters A, B, C, D, and E.

Link A in Figure 5.1 reflects the fact that, when $\Omega = R^N$, the first-order necessary conditions for $f(\mathbf{x})$ to take on a local minimum at $\mathbf{x}^*$ are that $\mathbf{x}^*$ satisfy the nonlinear equations

$$\left(\frac{\partial f}{\partial \mathbf{x}}\right)_{\mathbf{x}^*} = \mathbf{0} \tag{5.1-3}*$$

Conversely, if $\mathbf{x}^*$ is a solution of equation 5.1-2, it is also a solution of the minimization problem

$$\min \sum_{i=1}^{N} g_i^2(\mathbf{x}^*). \tag{5.1-4}$$

A sufficient condition for $\mathbf{x}^*$ to be a local minimum of f is that equation 5.1-3 holds and that the Hessian matrix $(\partial^2 f / \partial x_i \, \partial x_j)$ is positive definite at $\mathbf{x}^*$.

At the outset we note that we cannot in general solve such nonlinear problems in closed form except in the simplest of special cases. However, the solution of sets of linear algebraic equations on a digital computer is usually a relatively straightforward task. Hence most numerical algorithms for solving nonlinear problems proceed by solving a sequence of problems whose degree of difficulty is no greater than that of solving sets of linear equations. Each step in this sequence is called an *iteration*. The sequence is usually terminated

---
*$\partial / \partial \mathbf{x}$ is a column vector with entries $\partial / \partial x_i$.

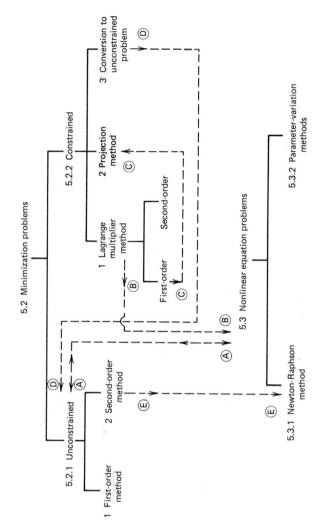

Figure 5.1 Some methods for solving two classes of numerical problem.

Minimization Problems

when the left and right sides of equation 5.1-2 agree to within some specified tolerance or the components of x change from iteration to iteration by less than some specified small amount. The success of any such numerical algorithm depends on two main factors. The first is the complexity of each iteration (the amount of calculation that must be performed), and the second is the convergence properties of the algorithm (the number of iterations required).

Algorithms thus proceed from an initial estimate $\mathbf{x}^{(0)}$ of a solution and calculate a sequence by means of

$$\mathbf{x}^{(m+1)} = \mathbf{x}^{(m)} + \omega^{(m)} \delta \mathbf{x}^{(m)}; \quad m = 0, 1, 2, \ldots, \quad (5.1\text{-}5)$$

where m is an iteration index. The scalar quantity ω is called a *step-size parameter*, which determines the distance between successive iterations in the direction defined by $\delta \mathbf{x}^{(m)}$. Equation 5.1-5 is applied in general, and algorithms differ by the way in which $\delta \mathbf{x}^{(m)}$ is determined. We discuss ways of choosing $\omega^{(m)}$ [the so-called line-search algorithm (Fletcher, 1980, p. 17)] in Section 5.4.1.

5.2 MINIMIZATION PROBLEMS

We consider here the minimization of a function $f(\mathbf{x})$, where $\mathbf{x} \in R^N$, perhaps subject to linear equality and nonnegativity constraints. In the absence of constraints, we have the unconstrained problem

$$\min_{\mathbf{x} \in R^N} f(\mathbf{x}). \quad (5.2\text{-}1)$$

The constrained minimization problem that we consider is

$$\min_{\mathbf{x} \in \Omega} f(\mathbf{x}), \quad (5.2\text{-}2)$$

where the constraint set Ω is defined by

$$\Omega = \{\mathbf{x} : \mathbf{A}\mathbf{x} = \mathbf{b}, x_i \geqslant 0\}. \quad (5.2\text{-}3)$$

Here $\mathbf{A}$ is an $M \times N$ matrix and $\mathbf{b}$ is an M-vector of real constants.

In equation 5.1-5, $\delta \mathbf{x}^{(m)}$ is usually chosen at each iteration so that

$$\left(\frac{df}{d\omega^{(m)}}\right)_{\omega^{(m)}=0} = \sum_{i=1}^{N} \left(\frac{\partial f}{\partial x_i}\right)_{\mathbf{x}^{(m)}} \delta x_i^{(m)} < 0, \quad (5.2\text{-}4)$$

unless $(\partial f/\partial x_i)_{\mathbf{x}^{(m)}} = 0$. An algorithm satisfying equation 5.2-4 is called a *descent method*. Different algorithms use different ways of choosing $\delta \mathbf{x}^{(m)}$, but all have the common property that, once $\delta \mathbf{x}^{(m)}$ is chosen, the (positive) step-size parameter $\omega^{(m)}$ is chosen so that $f(\mathbf{x}^{(m)} + \omega^{(m)}\delta\mathbf{x}^{(m)})$ is smaller than $f(\mathbf{x}^{(m)})$. Equation 5.2-4 ensures that this is possible.

5.2.1 Unconstrained Minimization Methods

5.2.1.1 First-Order Method

An intuitively appealing choice of $\delta\mathbf{x}^{(m)}$ is the vector along which $f(\mathbf{x})$ decreases most rapidly at $\mathbf{x}^{(m)}$. This is the gradient vector $\nabla\mathbf{f}$, with entries $\partial f/\partial x_i$. This choice of $\delta\mathbf{x}^{(m)}$ yields a first-order method usually called the *gradient method* (also referred to as the *method of steepest descent* or the *first-variation method*), which is defined by

$$\delta\mathbf{x}^{(m)} = -\left(\frac{\partial f}{\partial \mathbf{x}}\right)_{\mathbf{x}^{(m)}} \equiv -(\nabla\mathbf{f})_{\mathbf{x}^{(m)}}. \tag{5.2-5}$$

The rate of change of f at $\mathbf{x}^{(m)}$ in the direction defined by equations 5.1-5 and 5.2-5 is

$$\left(\frac{df}{d\omega^{(m)}}\right)_{\omega^{(m)}=0} = -\sum_{i=1}^{N}\left(\frac{\partial f}{\partial x_i}\right)_{\mathbf{x}^{(m)}}^{2}. \tag{5.2-6}$$

This satisfies relation 5.2-4, unless we are at a value of $\mathbf{x}^{(m)}$ that makes the gradient vector vanish [in which case $\mathbf{x}^{(m)}$ satisfies the first-order necessary conditions for a minimum]. Unfortunately, the gradient method can be quite slow to converge, especially near the minimum $\mathbf{x}^*$. The reason for this is that the behavior of f near $\mathbf{x}^*$ is determined largely by its second derivatives since the first derivatives become vanishingly small.

5.2.1.2 Second-Order Method

We may use information concerning second derivatives by approximating f near each $\mathbf{x}^{(m)}$ by a quadratic function and then finding the minimum of that approximation. This is sometimes called the *second-variation method*. The algorithm is based on minimizing the local quadratic approximation to $f(\mathbf{x})$ at $\mathbf{x}^{(m)}$, given by

$$Q(\mathbf{x}) = f(\mathbf{x}^{(m)}) + \sum_{i=1}^{N}\left(\frac{\partial f}{\partial x_i}\right)_{\mathbf{x}^{(m)}}(x_i - x_i^{(m)})$$

$$+ \frac{1}{2}\sum_{i=1}^{N}\sum_{j=1}^{N}\left(\frac{\partial^2 f}{\partial x_i \partial x_j}\right)_{\mathbf{x}^{(m)}}(x_i - x_i^{(m)})(x_j - x_j^{(m)}). \tag{5.2-7}$$

Here $Q(\mathbf{x})$ is the quadratic function that agrees with the first two terms of the Taylor series expansion of $f(\mathbf{x})$ about $\mathbf{x}^{(m)}$.

Minimization Problems

The necessary conditions that Q be a minimum with respect to $\mathbf{x}$ are

$$\frac{\partial Q}{\partial x_i} = 0; \quad i = 1, 2, \ldots, N. \tag{5.2-8}$$

This yields

$$\left(\frac{\partial f}{\partial x_i}\right)_{\mathbf{x}^{(m)}} + \sum_{j=1}^{N}\left(\frac{\partial^2 f}{\partial x_i \partial x_j}\right)_{\mathbf{x}^{(m)}}\left(x_j - x_j^{(m)}\right) = 0; \quad i = 1, 2, \ldots, N. \tag{5.2-9}$$

Equation 5.2-9 is a set of N linear equations in the N unknown elements of the vector $\delta\mathbf{x}^{(m)} = \mathbf{x} - \mathbf{x}^{(m)}$. Thus the second-variation method is formally given by equation 5.1-5 with

$$\delta\mathbf{x}^{(m)} = -\left(\frac{\partial^2 f}{\partial \mathbf{x}^2}\right)^{-1}_{\mathbf{x}^{(m)}}\left(\frac{\partial f}{\partial \mathbf{x}}\right)_{\mathbf{x}^{(m)}}, \tag{5.2-10}*$$

where superscript (-1) denotes a matrix inverse.

The rate of change of f at $\mathbf{x}^{(m)}$ in the direction defined by equations 5.1-5 and 5.2-10 is

$$\left(\frac{df}{d\omega^{(m)}}\right)_{\omega^{(m)}=0} = -\sum_{i=1}^{N}\sum_{j=1}^{N}\frac{\left(\dfrac{\partial f}{\partial x_i}\right)\left(\dfrac{\partial f}{\partial x_j}\right)}{\left(\dfrac{\partial^2 f}{\partial x_i \partial x_j}\right)}, \tag{5.2-11}$$

where all quantities on the right side are evaluated at $\mathbf{x}^{(m)}$. If the Hessian matrix is positive definite at $\mathbf{x}^{(m)}$, the criterion of equation 5.2-4 is satisfied.

5.2.2 Constrained Minimization Methods

5.2.2.1 Lagrange Multiplier Method

When the constraints on $\mathbf{x}$ are

$$\mathbf{A}\mathbf{x} = \mathbf{b} \tag{5.2-12}$$

(cf. equation 2.2-3), the classical method of Lagrange multipliers may be used (Walsh, 1975, p. 7), as was done in Chapter 3. (For simplicity, we assume that

---

*$\partial^2 f/\partial \mathbf{x}^2 = (\partial/\partial \mathbf{x}(\partial f/\partial \mathbf{x})^T)^T$

A is of full rank M.) We form the Lagrangian function

$$\mathcal{L}(\mathbf{x}, \boldsymbol{\lambda}) = f(\mathbf{x}) + \sum_{k=1}^{M} \lambda_k \left(b_k - \sum_{i=1}^{N} A_{ki} x_i \right) \quad (5.2\text{-}13)$$

and then minimize $\mathcal{L}$ with respect to $\mathbf{x}$, while ensuring that the constraints are satisfied. This results in the set of nonlinear equations

$$\left. \begin{array}{l} \dfrac{\partial \mathcal{L}}{\partial x_i} = \dfrac{\partial f}{\partial x_i} - \sum_{k=1}^{M} A_{ki} \lambda_k = 0; \quad i = 1, 2, \ldots, N \\[2mm] \dfrac{\partial \mathcal{L}}{\partial \lambda_j} = b_j - \sum_{i=1}^{N} A_{ji} x_i = 0; \quad j = 1, 2, \ldots, M. \end{array} \right\} \quad (5.2\text{-}14)$$

Equations 5.2-14 correspond to link B in Figure 5.1. As in the case of unconstrained minimization methods, there are first- and second-order implementations of the Lagrange multiplier method.

5.2.2.1.1 First-Order Method

If we use a first-order method analogous to the gradient method employed in equation 5.2-5, we obtain

$$\delta x_i^{(m)} = -\left(\dfrac{\partial \mathcal{L}}{\partial x_i} \right)_{\mathbf{x}^{(m)}, \boldsymbol{\lambda}^{(m)}}$$

$$= -\left(\dfrac{\partial f}{\partial x_i} \right)_{\mathbf{x}^{(m)}} + \sum_{k=1}^{M} \lambda_k^{(m)} A_{ki}; \quad i = 1, 2, \ldots, N, \quad (5.2\text{-}15)$$

where the Lagrange multipliers $\boldsymbol{\lambda}^{(m)}$ are determined by using equation 5.2-15 in conjunction with equation 5.2-12 to yield the set of M linear equations:

$$\sum_{k=1}^{M} \lambda_k^{(m)} \sum_{i=1}^{N} A_{ji} A_{ki} = \sum_{i=1}^{N} A_{ji} \left(\dfrac{\partial f}{\partial x_i} \right)_{\mathbf{x}^{(m)}}; \quad j = 1, 2, \ldots, M. \quad (5.2\text{-}16)$$

We have assumed here that $\mathbf{x}^{(m)}$ satisfies the constraints of equation 5.2-12. Equations 5.2-16 ensure that $\mathbf{x}^{(m+1)}$, as ultimately determined by equation 5.1-5, also satisfies these constraints. If the constraints are not satisfied at $\mathbf{x}^{(m)}$, that is, if

$$\mathbf{A}\mathbf{x}^{(m)} \equiv \mathbf{b}^{(m)} \neq \mathbf{b}, \quad (5.2\text{-}17)$$

then equation 5.2-16 is modified to become

$$\sum_{k=1}^{M} \lambda_k^{(m)} \sum_{i=1}^{N} A_{ji} A_{ki} = \sum_{i=1}^{N} A_{ji} \left(\dfrac{\partial f}{\partial x_i} \right)_{\mathbf{x}^{(m)}} + b_j - b_j^{(m)}; \quad j = 1, 2, \ldots, M.$$

$$(5.2\text{-}18)$$

5.2.2.1.2 Second-Order Method

A second-order method analogous to the second-variation method of equation 5.2-9 results when $\delta x^{(m)}$ is determined from the $(N + M)$ linear equations

$$\sum_{k=1}^{M} \lambda_k^{(m)} A_{ki} - \sum_{j=1}^{N} \left(\frac{\partial^2 f}{\partial x_i \partial x_j} \right)_{x^{(m)}} \delta x_j^{(m)} = \left(\frac{\partial f}{\partial x_i} \right)_{x^{(m)}} ; i = 1, 2, \ldots, N \quad (5.2\text{-}19)$$

and

$$\sum_{j=1}^{N} A_{lj} \delta x_j^{(m)} = 0; \quad l = 1, 2, \ldots, M. \quad (5.2\text{-}20)$$

Again, if equation 5.2-17 applies, equation 5.2-20 becomes

$$\sum_{j=1}^{N} A_{lj} \delta x_j^{(m)} = b_l - b_l^{(m)}; \quad l = 1, 2, \ldots, M. \quad (5.2\text{-}21)$$

5.2.2.2 Projection Methods

Projection methods (Walsh, 1975, pp. 146–148) are based on the use of an $N \times N$ matrix $\mathbf{P}$, such that

$$\mathbf{AP} = \mathbf{0}, \quad (5.2\text{-}22)$$

where $\mathbf{A}$ is the matrix of the linear constraints of equation 5.2-12. For any vector $\mathbf{y} \in R^N$, the direction defined by

$$\delta \mathbf{x}^{(m)} = \mathbf{P}\mathbf{y} \quad (5.2\text{-}23)$$

satisfies

$$\mathbf{A} \delta \mathbf{x}^{(m)} = \mathbf{0}. \quad (5.2\text{-}24)$$

Thus, from equation 5.1-5, if $\mathbf{x}^{(m)}$ satisfies the constraints, so also does $\mathbf{x}^{(m+1)}$. The matrix $\mathbf{P}$ can be regarded as "projecting" the direction $\mathbf{y}$ onto the linear constraint set. Several choices are possible for $\mathbf{y}$, including the right sides of equations 5.2-5 and 5.2-10.

One way of obtaining the matrix $\mathbf{P}$ is by considering the first-order Lagrange multiplier method used in equations 5.2-15 and 5.2-16. These equations may be written in vector-matrix form as

$$\delta \mathbf{x} = -\nabla \mathbf{f} + \mathbf{A}^T \boldsymbol{\lambda} \quad (5.2\text{-}25)$$

and

$$(AA^T)\lambda = A \nabla f, \qquad (5.2\text{-}26)$$

where, for ease of notation, we have dropped the iteration index m. Solving equation 5.2-26 for λ and substituting the result in equation 5.2-25, we have

$$\delta x = -\left(I - A^T(AA^T)^{-1}A\right)\nabla f. \qquad (5.2\text{-}27)$$

This way of determining δx is a projection method since it satisfies equation 5.2-23, with

$$P = I - A^T(AA^T)^{-1}A \qquad (5.2\text{-}28)$$

and

$$y = -\nabla f. \qquad (5.2\text{-}29)$$

The quantity P given by equation 5.2-28 may be used in conjunction with any direction y; with y given by equation 5.2-29, this method is called the *gradient-projection method*. Link C in Figure 5.1 reflects the way in which this projection method has been derived.

5.2.2.3 Method of Conversion to Unconstrained Problem

Constrained minimization problems may be converted to unconstrained problems in several ways (Walsh, 1975, Chapter 5), as indicated by link D in Figure 5.1. In the special case when the constraints are linear, as in equation 5.2-12, a linear transformation of variables may be used to obtain an unconstrained problem. We have already explored the chemical implications of such a transformation in Chapter 2 as it relates to chemical stoichiometry and have developed some preliminary ideas for a "stoichiometric" algorithm in Chapter 3 and a special-purpose algorithm of this type in Chapter 4. Here we briefly review those results in relation to a general numerical algorithm for minimizing a nonlinear function subject to a set of linear equality constraints, which we refer to as *the method of stoichiometric elimination*.

The stoichiometric elimination technique focuses on a set of independent variables ξ related to x through the linear transformation

$$\delta x = N \delta \xi \qquad (5.2\text{-}30)$$

and seeks to minimize $f(\xi)$. The matrix N satisfies

$$AN = 0, \qquad (5.2\text{-}31)$$

and has $R = (N - M)$ linearly independent columns. Equation 5.2-31 ensures

Nonlinear Equation Problems

that

$$\mathbf{A}\,\delta\mathbf{x} = \mathbf{0}. \qquad (5.2\text{-}32)$$

Thus, from equation 5.1-5, if $\mathbf{x}^{(m)}$ satisfies equation 5.2-12, so also does $\mathbf{x}^{(m+1)}$. The matrix $\mathbf{N}$ is arbitrary apart from equation 5.2-31 and may be redefined on each iteration, if required. However, a convenient way of forming $\mathbf{N}$ is first to choose a set of M linearly independent columns of $\mathbf{A}$ and then express the remaining R columns as a linear combination of these. Formation of this particular $\mathbf{N}$ matrix thus entails the solution of $(N - M)$ sets of M linear algebraic equations.

As before, we may employ either a first- or a second-order method for choosing the $\delta\boldsymbol{\xi}$ in equation 5.2-30. The first-order method sets

$$\delta\xi_j^{(m)} = -\left(\frac{\partial f}{\partial \xi_j}\right)_{\mathbf{x}^{(m)}}$$

$$-- \sum_{i=1}^{N} v_{ij}\left(\frac{\partial f}{\partial x_i}\right)_{\mathbf{x}^{(m)}}; \quad j = 1, 2, \ldots, R. \qquad (5.2\text{-}33)$$

As in Chapter 2, v_{ij} denotes entry (i, j) of $\mathbf{N}$. The second-order method sets

$$\delta\boldsymbol{\xi}^{(m)} = -\left(\frac{\partial^2 f}{\partial \boldsymbol{\xi}^2}\right)_{\mathbf{x}^{(m)}}^{-1}\left(\frac{\partial f}{\partial \boldsymbol{\xi}}\right)_{\mathbf{x}^{(m)}}. \qquad (5.2\text{-}34)$$

The gradient vector $\partial f/\partial \boldsymbol{\xi}$ in equation 5.2-34 is expressed in terms of $\partial f/\partial \mathbf{x}$ in equation 5.2-33. The Hessian matrix $\partial^2 f/\partial \boldsymbol{\xi}^2$ is related to $\partial^2 f/\partial \mathbf{x}^2$ by

$$\frac{\partial^2 f}{\partial \xi_i \, \partial \xi_j} = \sum_{k=1}^{N}\sum_{l=1}^{N}\left(\frac{\partial^2 f}{\partial x_k \partial x_l}\right) v_{ki} v_{lj}. \qquad (5.2\text{-}35)$$

Equation 5.2-34 essentially entails the solution of a set of $R = (N - M)$ linear equations on each iteration. This should be contrasted with the second-order Lagrange multiplier method described in Section 5.2.2, which requires the solution of a set of $(N + M)$ linear equations in general.

5.3 NONLINEAR EQUATION PROBLEMS

As noted previously, the minimization problem is associated with the problem of solving sets of nonlinear equations. Here we first recapitulate various forms of these equations and then describe two methods for solving sets of nonlinear equations.

The nonlinear equations associated with the unconstrained minimization of $f(\mathbf{x})$ are

$$\frac{\partial f}{\partial \mathbf{x}} = \mathbf{0}. \tag{5.3-1}$$

For the constrained minimization problem given by equations 5.2-2 and 5.2-12, if Lagrange multipliers are used to incorporate the constraints, we have the set of $(N + M)$ nonlinear equations (cf. equation 5.2-14)

$$\left.\begin{aligned}\mathbf{A}^T \boldsymbol{\lambda} &= \frac{\partial f}{\partial \mathbf{x}}, \\ \mathbf{A}\mathbf{x} &= \mathbf{b}.\end{aligned}\right\} \tag{5.3-2}$$

When the stoichiometric elimination technique is used to eliminate the constraints specified by equation 5.2-12, we have the set of nonlinear equations

$$\frac{\partial f}{\partial \boldsymbol{\xi}} = \mathbf{0}. \tag{5.3-3}$$

This may be written in terms of the original $\mathbf{x}$ variables through the chain rule for differentiation to yield (cf. equation 5.2-33)

$$\mathbf{N}^T \frac{\partial f}{\partial \mathbf{x}} = \mathbf{0}. \tag{5.3-4}$$

As we have seen in Chapter 3, for chemical equilibrium problems, this yields the so-called classical form of the chemical equilibrium conditions.

5.3.1 Newton-Raphson Method

The Newton-Raphson method (Ralston and Rabinowitz, 1978, p. 360) is one of the oldest and still most widely used numerical techniques for solving the N nonlinear equations

$$\mathbf{g}(\mathbf{x}) = \mathbf{0}. \tag{5.1-2}$$

The technique sets to zero the local linear approximation to $\mathbf{g}(\mathbf{x})$ at $\mathbf{x}^{(m)}$, $\mathbf{l}^{(m)}(\mathbf{x})$, given by

$$l_i^{(m)}(\mathbf{x}) = g_i(\mathbf{x}^{(m)}) + \sum_{j=1}^{N} \left(\frac{\partial g_i}{\partial x_j}\right)_{\mathbf{x}^{(m)}} \left(x_j - x_j^{(m)}\right); \quad i = 1, 2, \ldots, N. \tag{5.3-5}$$

The resulting equations for $\delta \mathbf{x}^{(m)} = \mathbf{x} - \mathbf{x}^{(m)}$ are

$$\sum_{j=1}^{N} \left(\frac{\partial g_i}{\partial x_j}\right)_{\mathbf{x}^{(m)}} \delta x_j^{(m)} = -g_i(\mathbf{x}^{(m)}); \quad i = 1, 2, \ldots, N. \tag{5.3-6}$$

Equation 5.3-6 is identical to the second-variation method for minimizing $f(\mathbf{x})$ when $\mathbf{g}(\mathbf{x}) = \nabla f(\mathbf{x})$, as given by equation 5.2-9 (link E in Figure 5.1). The equivalence results from the fact that the Hessian matrix of f appearing in equation 5.2-9 is identical to the Jacobian matrix of $\mathbf{g}$ given by $\partial \mathbf{g}^T/\partial \mathbf{x}$, appearing in equation 5.3-6.

The Newton-Raphson method is a descent method for the objective function

$$S = \frac{1}{2} \sum_{i=1}^{N} g_i^2(\mathbf{x}). \tag{5.3-7}$$

This is demonstrated by differentiation of equation 5.3-7 to yield

$$\left(\frac{dS}{d\omega^{(m)}}\right)_{\omega^{(m)}=0} = \sum_{i=1}^{N} \sum_{j=1}^{N} g_i(\mathbf{x}^{(m)}) \left(\frac{\partial g_i}{\partial x_j}\right)_{\mathbf{x}^{(m)}} \delta x_j^{(m)}$$

$$= - \sum_{i=1}^{N} g_i^2(\mathbf{x}^{(m)}) \leq 0. \tag{5.3-8}$$

Equation 5.3-8 allows us to choose the step-size parameter $\omega^{(m)}$ in equation 5.1-5 so as to minimize approximately $S(\mathbf{x}^{(m+1)})$ at each iteration.

We may also apply the Newton-Raphson method to the nonlinear equations 5.3-2. This yields (cf. equations 5.2-19 and 5.2-21)

$$\left.\begin{aligned}
\sum_{k=1}^{M} A_{ki} \delta \lambda_k^{(m)} - \sum_{j=1}^{N} \left(\frac{\partial^2 f}{\partial x_i \partial x_j}\right)_{\mathbf{x}^{(m)}} \delta x_j^{(m)} \\
= \left(\frac{\partial f}{\partial x_i}\right)_{\mathbf{x}^{(m)}} - \sum_{k=1}^{M} A_{ki} \lambda_k^{(m)}; \qquad i = 1,2,\ldots,N \\
\sum_{j=1}^{N} A_{lj} \delta x_j^{(m)} = b_l - \sum_{j=1}^{N} A_{lj} x_j^{(m)}; \qquad l = 1,2,\ldots,M
\end{aligned}\right\} \tag{5.3-9}$$

which are used in conjunction with

$$\boldsymbol{\lambda}^{(m+1)} = \boldsymbol{\lambda}^{(m)} + \omega^{(m)} \delta \boldsymbol{\lambda}^{(m)}$$

and

$$\mathbf{x}^{(m+1)} = \mathbf{x}^{(m)} + \omega^{(m)} \delta \mathbf{x}^{(m)}.$$

5.3.2 Parameter-Variation Methods

The general approach of parameter-variation methods has been of interest recently in numerical analysis (Ralston and Rabinowitz, 1978, p. 363). It attempts to solve a set of N nonlinear equations $\mathbf{g}(\mathbf{x}) = \mathbf{0}$ by introducing one or more auxiliary parameters $\boldsymbol{\alpha}$ and then solving the equations

$$\mathbf{h}(\mathbf{x}, \boldsymbol{\alpha}) = \mathbf{0} \tag{5.3-10}$$

at a sequence of values of $\boldsymbol{\alpha}$ that approach zero. The parameters $\boldsymbol{\alpha}$ must be incorporated in $\mathbf{h}$ in such a way that

$$\mathbf{h}(\mathbf{x}, \mathbf{0}) \equiv \mathbf{g}(\mathbf{x}). \tag{5.3-11}$$

In the use of this method there are the two important questions of the choices of $\mathbf{h}$ and the sequence of $\boldsymbol{\alpha}$ values. One possibility is to choose

$$\mathbf{h}(\mathbf{x}, \boldsymbol{\alpha}) = \mathbf{g}(\mathbf{x}) - \boldsymbol{\alpha}, \tag{5.3-12}$$

and an initial value

$$\boldsymbol{\alpha}^{(0)} = \mathbf{g}(\mathbf{x}^{(0)}), \tag{5.3-13}$$

where $\mathbf{x}^{(0)}$ is arbitrarily chosen. At $\mathbf{x} = \mathbf{x}^{(0)}$ we have

$$\mathbf{h}(\mathbf{x}^{(0)}, \boldsymbol{\alpha}^{(0)}) = \mathbf{0}. \tag{5.3-14}$$

We then gradually change $\boldsymbol{\alpha}$ to $\mathbf{0}$ through some sequence of values and at the same time solve the sequence of problems

$$\mathbf{h}(\mathbf{x}^{(m)}, \boldsymbol{\alpha}^{(m)}) = \mathbf{0}; \quad m = 1, 2, \ldots. \tag{5.3-15}$$

The philosophy of the method is that if $\boldsymbol{\alpha}^{(m+1)}$ differs only slightly from $\boldsymbol{\alpha}^{(m)}$, the solutions $\mathbf{x}^{(m)}$ and $\mathbf{x}^{(m+1)}$ should also differ only slightly. Thus we might expect that $\mathbf{x}^{(m)}$ should be a good initial estimate of the solution of

$$\mathbf{h}(\mathbf{x}^{(m+1)}, \boldsymbol{\alpha}^{(m+1)}) = \mathbf{0}. \tag{5.3-16}$$

One possibility for choosing the sequence $\{\boldsymbol{\alpha}^{(m)}\}$ is to regard equation 5.3-15 as defining $\mathbf{x}(\boldsymbol{\alpha})$ and then to differentiate this equation to obtain the differential equation

$$\sum_{j=1}^{N} \left(\frac{\partial g_i}{\partial x_j}\right) \left(\frac{\partial x_j}{\partial \alpha_k}\right) = \delta_{ik}; \quad i, k = 1, 2, \ldots, N, \tag{5.3-17}$$

where δ_{ik} is the Kronecker delta. Equation 5.3-17 may be rewritten as

$$\frac{\partial \mathbf{x}^T}{\partial \boldsymbol{\alpha}} = \mathbf{J}^{-1}(\mathbf{x}), \tag{5.3-18}$$

where **J** is the Jacobian matrix of **g**. Equation 5.3-18 is a matrix set of ordinary differential equations, which we can integrate from $\mathbf{x}(\boldsymbol{\alpha}^{(0)}) = \mathbf{x}^{(0)}$ to $\boldsymbol{\alpha} = \mathbf{0}$ along an appropriate path. This can be performed by using a computer algorithm for solving initial-value problems for ordinary differential equations. The value of $\mathbf{x}(\mathbf{0})$ is then the desired solution to $\mathbf{g}(\mathbf{x}) = \mathbf{0}$.

Another way of choosing **h** in equation 5.3-10 is to incorporate a single auxiliary parameter t and to write

$$\mathbf{g}(\mathbf{x}, t) = \mathbf{g}(\mathbf{x}) - t\mathbf{g}(\mathbf{x}^{(0)}) = \mathbf{0}, \quad (5.3\text{-}19)$$

where $\mathbf{x}^{(0)}$ is arbitrarily chosen. Differentiating equation 5.3-19 with respect to t, we have

$$\sum_{j=1}^{N} \left(\frac{\partial g_i}{\partial x_j} \right)\left(\frac{dx_j}{dt} \right) = tg_i(\mathbf{x}^{(0)}); \quad i = 1, 2, \ldots, N. \quad (5.3\text{-}20)$$

This may be rewritten as

$$\frac{d\mathbf{x}}{dt} = t\mathbf{J}^{-1}\mathbf{g}(\mathbf{x}^{(0)}). \quad (5.3\text{-}21)$$

We then integrate equation 5.3-21 from $t = 1$, where $\mathbf{x}(1) = \mathbf{x}^{(0)}$, to $t = 0$. The value of $\mathbf{x}(0)$ is the desired solution.

In practice, these differential equation methods fail if the Jacobian matrix $\mathbf{J}(\mathbf{x})$ becomes singular at any stage of the integration. In that case, one must resort to additional techniques for computing a "path" of **x** values that terminates at the solution to the problem.

5.4 STEP-SIZE PARAMETER AND CONVERGENCE CRITERIA

5.4.1 Computation of the Step-Size Parameter

All the previous methods for minimization and for solving sets of nonlinear equations, except for the parameter-variation technique, involve computation of new values of **x** from current ones by means of

$$\mathbf{x}^{(m+1)} = \mathbf{x}^{(m)} + \omega^{(m)} \delta\mathbf{x}^{(m)}, \quad (5.1\text{-}5)$$

where $\delta\mathbf{x}^{(m)}$ is determined by the particular algorithm used and $\omega^{(m)}$ is a positive step-size parameter. In this section we discuss the computation of $\omega^{(m)}$.

If the problem can be posed in the form of a minimization problem for **x** of the form

$$\min G(\mathbf{x}), \quad (5.4\text{-}1)$$

a convenient way of choosing $\omega^{(m)}$ is by finding the value of ω that approximately minimizes $G(\mathbf{x}^{(m)} + \omega\, \delta\mathbf{x}^{(m)})$ on each iteration. All minimization problems are naturally of the form of problem 5.4-1, and we have seen that

$$G(\mathbf{x}) = \sum_{i=1}^{N} g_i^2(\mathbf{x}) \qquad (5.4\text{-}2)$$

is a function whose minimum yields the solution of the nonlinear equations

$$\mathbf{g}(\mathbf{x}) = \mathbf{0}. \qquad (5.4\text{-}3)$$

Thus the determination of a step-size parameter is of general importance in the practical application of most of the numerical methods previously discussed.

We have seen that the concept of a descent method is especially important since for such a method, G is a decreasing function of $\omega^{(m)}$ at $\mathbf{x}^{(m)}$; that is, the method yields $\delta\mathbf{x}^{(m)}$ satisfying

$$\left(\frac{dG}{d\omega^{(m)}}\right)_{\omega^{(m)}=0} \equiv \sum_{i=1}^{N} \left(\frac{\partial G}{\partial x_i}\right)_{\mathbf{x}^{(m)}} \delta x_i^{(m)} < 0, \qquad (5.4\text{-}4)$$

provided that $\partial G/\partial \mathbf{x} \neq \mathbf{0}$. Equation 5.4-4 ensures that a positive value of $\omega^{(m)}$ in equation 5.1-5 can be found so that $G(\mathbf{x}^{(m+1)}) < G(\mathbf{x}^{(m)})$. Determination of the optimal step-size parameter on each iteration is thus equivalent to the one-dimensional optimization problem

$$\min_{\omega>0} G(\mathbf{x}^{(m)} + \omega\, \delta\mathbf{x}^{(m)}). \qquad (5.4\text{-}5)$$

In the solution of this problem, care must be taken that too much computation time is not spent searching for the *exact* minimizing value of ω. Usually it is preferable to determine this value only approximately and then proceed to the next iteration.

Methods for solving this one-dimensional optimization problem are of two types. The first type brackets the minimum in smaller and smaller intervals. Techniques such as *interval halving, golden-section search*, and *Fibonacci search* may be used (Fletcher, 1980, pp. 25–29). These methods use values of $G(\omega)$ for comparison purposes only and do not use $G(\omega)$ values explicitly. The second type fits $G(\omega)$ to a suitable low-order polynomial, whose minimum is then found analytically. For example, the parabola fitted to three values may be used. Davidon, as cited by Walsh (1975, pp. 97–101), fits a cubic polynomial to two points and the derivatives at these two points.

We now discuss a very simple procedure, when $\omega = 1$ is known to provide an estimate of the optimum value [e.g., when $\delta\mathbf{x}^{(m)}$ is determined from the

Step-Size Parameter and Convergence Criteria

Newton-Raphson method]. First, the value of

$$\left(\frac{dG}{d\omega}\right)_{\omega=1} = \sum_{i=1}^{N}\left(\frac{\partial G}{\partial x_i}\right)_{\omega=1} \delta x_i^{(m)} \tag{5.4-6}$$

is calculated. If this quantity is negative or zero, we assume that we have not passed the minimizing value of ω, and we proceed to the next iteration, with $\omega^{(m)} = 1$ in equation 5.1-5. If the quantity in equation 5.4-6 is positive, we set

$$\omega^{(m)} = \frac{(dG/d\omega)_{\omega=0}}{(dG/d\omega)_{\omega=0} - (dG/d\omega)_{\omega=1}}. \tag{5.4-7}$$

Equation 5.4-7 ensures that $0 < \omega^{(m)} < 1$ since we assume that we have passed a minimum in $G(\omega)$ at $\omega = 1$, and $\delta \mathbf{x}^{(m)}$ defines a descent method. This technique has been used with some success in a simple optimization algorithm (Smith and Missen, 1967) and is employed in the general-purpose computer programs given in Appendixes C and D.

Finally, if it is known that all x_i of the solution of equation 5.4-1 are positive, we must also choose ω to ensure that all x_i remain positive. A convenient way of doing this is to ensure that ω satisfies

$$\omega \leq \max_{1 \leq i \leq N}\left\{1, -\frac{\delta x_i^{(m)}}{x_i^{(m)}}(1-\varepsilon)\right\}, \tag{5.4-8}$$

where ε is a small number (e.g., 0.01).

5.4.2 Convergence Criteria

The iterative procedure defined by equation 5.1-5 is ideally terminated when

$$|x_i^{(m)} - x_i^*| \leq \varepsilon; \quad i = 1, 2, \ldots, N, \tag{5.4-9}$$

where $\mathbf{x}^*$ is the solution and ε is some small positive number. Since $\mathbf{x}^*$ is not known, practical criteria are often chosen as one or more of the following:

$$\max_{1 \leq i \leq N} |\delta x_i^{(m)}| \leq \varepsilon, \tag{5.4-10}$$

$$\max_{1 \leq i \leq N} \left|\frac{\delta x_i^{(m)}}{x_i^{(m)}}\right| \leq \varepsilon, \tag{5.4-11}$$

$$\max_{1 \leq i \leq N} \left|\left(\frac{\partial f}{\partial x_i}\right)_{\mathbf{x}^{(m)}}\right| \leq \varepsilon \tag{5.4-12}$$

and

$$\max_{1 \leq i \leq N} |g_i(\mathbf{x}^{(m)})| \leq \varepsilon. \tag{5.4-13}$$

Criteria 5.4-10 and/or 5.4-11 may be used for both optimization and nonlinear equation problems. Criterion 5.4-11 is relevant only when it is known that $x_i^{(m)} \neq 0$. Criterion 5.4-12 is relevant to minimizing $f(\mathbf{x})$ and criterion 5.4-13, to solving $\mathbf{g}(\mathbf{x}) = \mathbf{0}$. In the programs presented in Appendixes B, C, and D, criteria 5.4-11 and 5.4-12 are used, with the former for nonstoichiometric algorithms and the latter for stoichiometric algorithms.

CHAPTER SIX

Chemical Equilibrium Algorithms for Ideal Systems

In Chapter 4 we developed special-purpose algorithms for use on small computers to treat single-phase equilibrium problems for ideal systems with a relatively small number of species and elements. For these problems, the chemical potential of each species is given by the ideal-solution form of equation 3.7-15a, which we rewrite as

$$\mu_i = \mu_i^*(T, P) + RT \ln \frac{n_i}{n_t}. \tag{6.1-1}$$

In this chapter we discuss general-purpose algorithms to treat problems with any number of phases, species, and elements. We continue to assume that equation 6.1-1 holds for each species. The quantity n_t is the total number of moles in the phase in which species i is a constituent. Thus when a phase contains only species i, the logarithmic term vanishes. Composition variables other than the mole fraction, which is indicated in equation 6.1-1, can be used for an ideal solution, and we discuss this at the end of the chapter.

Computer programs for two selected general-purpose algorithms developed in this chapter are given in Appendixes C and D. In the literature such algorithms have been applied primarily to equilibrium problems involving a single gas phase, with perhaps pure condensed phases also present. Gas-phase reactors and metallurgical problems involving gases and condensed solids are examples of these situations.

We derive all the algorithms on the assumption that a solution to the equilibrium problem exists and is unique. We recall from Chapter 3 that, for ideal systems, this is guaranteed only in general in the case of problems consisting of one phase. Existence seldom presents practical difficulties, but the mathematical possibility of nonuniqueness can cause difficulties in the implementation of certain equilibrium algorithms, as can the nonnegativity constraints on the equilibrium mole numbers. In the ensuing discussion we occasionally refer to these potential difficulties, but a complete discussion of them is postponed to Chapter 9.

Reviews of equilibrium algorithms have been given by Zeleznik and Gordon (1968), Van Zeggeren and Storey (1970), Klein (1971), Holub and Vonka (1976), Seider et al. (1980), and Smith (1980a, 1980b). We are concerned here primarily with a detailed critical analysis of the most important algorithms themselves and do not attempt an exhaustive review.

6.1 CLASSIFICATIONS OF ALGORITHMS

Many algorithms for calculating chemical equilibrium have appeared in the literature. It is useful to classify them into groups with common characteristics to understand relations between them. Any such classification is, however, not unique, and in what follows we discuss algorithms in the context of four alternative classification schemes.

1. One broad way of classifying equilibrium algorithms from a numerical point of view is according to whether they are based on minimization methods or on methods for solving sets of nonlinear equations. This classification may sometimes be an artificial one, as we have seen in Chapter 5.

2. A second way of classifying algorithms is with respect to their incorporation of the element-abundance constraints and the equilibrium conditions, as described in Chapters 2 and 3. Some algorithms satisfy the element-abundance constraints at every iteration of the calculation and proceed to a solution of the equilibrium conditions. Conversely, some algorithms satisfy the equilibrium conditions at every iteration and proceed to a solution of the element-abundance constraints. Still other algorithms satisfy neither condition at each iteration and proceed to satisfy both simultaneously. This classification scheme has been suggested by Johansen (1967).

3. A third classification scheme that has been used is *equilibrium-constant methods* versus *free-energy-minimization methods*. We believe that this classification is often misleading, and its use in the past has had the historical result of obscuring basic similarities between certain algorithms.

4. Finally, as a fourth way, we may classify algorithms as to the particular way in which the element-abundance constraints are utilized in the calculations. As in Chapter 4, we refer to algorithms that eliminate these constraints by means of the technique discussed in Section 5.2.2.3 as *stoichiometric algorithms*. Such methods essentially treat the number of unknown independent variables as $(N' - M)$. Also, we refer to algorithms that explicitly utilize the element-abundance constraints in the form of equation 2.2-3 as *nonstoichiometric algorithms*. For these algorithms, the number of variables is $(N' + M)$, although for ideal systems,

this number is usually effectively reduced to $(M + \pi)$, where π is the number of phases in the system.

In summary, equilibrium algorithms can be examined from several points of view. This chapter is structured to focus on the fourth classification, but reference is also made to the others where appropriate. We have adopted the philosophy that, by taking various points of view into account and by studying the structures of some representative algorithms, we can better understand the basic features of *any* equilibrium algorithm.

6.2 STRUCTURE OF CHAPTER

The presentation in this chapter approximately parallels the discussion of numerical methods in Chapter 5 and is outlined in Figure 6.1. We consider nonstoichiometric algorithms first (Section 6.3 as indicated) and then stoichiometric algorithms (Section 6.4). Within the former, and following the development of Section 5.2.2 (on constrained minimization methods), we discuss first-order methods (Section 6.3.1) and then the Brinkley, NASA, and RAND algorithms, which are essentially variations of the same second-order method (Section 6.3.2); some other approaches are also mentioned (Section 6.3.3). Within the latter, and following the development of Section 5.2.1 (on uncon-

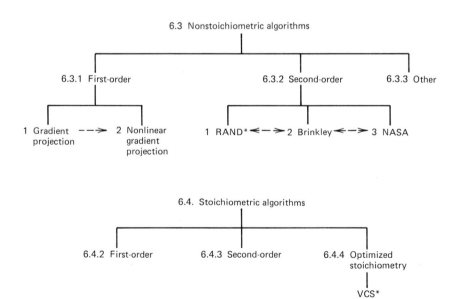

*General—purpose algorithms for which computer programs are given in Appendices C and D.

Figure 6.1 Chemical equilibrium algorithms.

strained minimization methods), we also discuss first- and second-order methods (Sections 6.4.2 and 6.4.3); an important method related to the second-order method is developed separately using the concept of *optimized stoichiometry* (Section 6.4.4). We derive all the algorithms primarily in the case of a single ideal-solution phase and indicate any extensions required to treat other types of problems.

6.3 NONSTOICHIOMETRIC ALGORITHMS

6.3.1 First-Order Algorithms

6.3.1.1 Gradient Projection

The gradient-projection algorithm results from equation 5.2-23. Mole-number changes from a given estimate $\mathbf{n}^{(m)}$ are computed by means of

$$\delta\mathbf{n}^{(m)} = -\mathbf{P}\left(\frac{\partial G}{\partial \mathbf{n}}\right)_{\mathbf{n}^{(m)}} = -\mathbf{P}\boldsymbol{\mu}^{(m)}, \tag{6.3-1}$$

$$\mathbf{n}^{(m+1)} = \mathbf{n}^{(m)} + \omega^{(m)}\delta\mathbf{n}^{(m)}. \tag{6.3-2}$$

The projection matrix $\mathbf{P}$ is given by

$$\mathbf{P} = \mathbf{I} - \mathbf{A}^T(\mathbf{A}\mathbf{A}^T)^{-1}\mathbf{A}. \tag{5.2-28}$$

It is assumed that $\mathbf{n}^{(m)}$ satisfies the element-abundance constraints; equations 6.3-1 and 6.3-2 are used iteratively to minimize the Gibbs function of the system. This method has not appeared in the literature, although it has some useful computational features. For example, only a single matrix inversion is required (in equation 5.2-28), which need be performed only once at the beginning of the algorithm.

6.3.1.2 Nonlinear Gradient Projection

A related first-order algorithm has been proposed by Storey and Van Zeggeren (1964). The nonnegativity constraints on the mole numbers are incorporated by means of the logarithmic transformation

$$y_i = \ln n_i. \tag{6.3-3}$$

This results in the transformed problem

$$\min G(\mathbf{y}),$$

such that

$$\sum_{i=1}^{N} a_{ki} \exp(y_i) = b_k; \quad k = 1, 2, \ldots, M. \tag{6.3-4}$$

Since the constraint equations are now nonlinear, the gradient-projection method is not strictly applicable. However, if we use the local linear Taylor series approximation to the constraints, we obtain, assuming that $y^{(m)}$ satisfies equation 6.3-4,

$$\sum_{i=1}^{N} a_{ki} n_i^{(m)} \delta y_i^{(m)} = 0; \quad k = 1, 2, \ldots, M. \tag{6.3-5}$$

We can utilize the gradient-projection method for minimizing $G(\delta y)$ subject to the linear constraints of equation 6.3-5, which may be expressed in the form

$$\mathbf{AD}^{(m)} \delta \mathbf{y}^{(m)} = \mathbf{0}, \tag{6.3-6}$$

where $\mathbf{D}^{(m)}$ is the diagonal matrix with entries $n_i^{(m)}$. The resulting algorithm computes changes to $\mathbf{y}^{(m)}$ by means of

$$\delta \mathbf{y}^{(m)} = -\mathbf{P}\left(\frac{\partial G}{\partial \mathbf{y}}\right)_{\mathbf{y}^{(m)}} = -\mathbf{PD}\boldsymbol{\mu}^{(m)}, \tag{6.3-7}$$

where we have omitted the superscript (m) on $\mathbf{P}$ and $\mathbf{D}$ for ease of notation. The projection matrix $\mathbf{P}$, which must be recalculated on each iteration, results from replacing $\mathbf{A}$ in equation 5.2-28 by $\mathbf{AD}$, thus yielding

$$\mathbf{P} = \mathbf{I} - \mathbf{D}^T \mathbf{A}^T (\mathbf{ADD}^T \mathbf{A}^T)^{-1} \mathbf{AD}. \tag{6.3-8}$$

Storey and Van Zeggeren (1964) originally derived the preceding algorithm (equations 6.3-7 and 6.3-8) in a quite different manner. We can see the connection with their approach by considering the Lagrange multiplier formulation of the gradient-projection algorithm discussed in Section 5.2.2. We first define Lagrange multipliers $\boldsymbol{\lambda}$ by means of the linear equations (cf. equation 5.2-26)

$$\mathbf{ADD}^T \mathbf{A}^T \boldsymbol{\lambda} = \mathbf{AD}^2 \boldsymbol{\mu}^{(m)}. \tag{6.3-9}$$

Then we set (cf. equation 5.2-25)

$$\delta \mathbf{y}^{(m)} = -\mathbf{D}\boldsymbol{\mu}^{(m)} + \mathbf{D}^T \mathbf{A}^T \boldsymbol{\lambda}. \tag{6.3-10}$$

Equations 6.3-9 and 6.3-10 are the working equations used by Storey and Van Zeggeren (1964). They are simply a minor rearrangement of equations 6.3-7 and 6.3-8.

One practical difficulty with the preceding method is that the satisfaction of the element-abundance constraints of equation 6.3-4 tends to deteriorate as the iterations proceed, unless the step-size parameter ω is very small. It is clear from the derivation that this is a consequence of the linear approximation to the nonlinear constraints. This "drifting" phenomenon may also occur to a minor extent in the use of the gradient-projection algorithm in equation 6.3-1. However, in that case the drifting occurs solely due to the accumulation of computer rounding errors.

The drifting phenomenon may be alleviated by using the modification discussed in Section 5.2.2.1.1. Equation 6.3-7 then becomes

$$\delta \mathbf{y}^{(m)} = -\mathbf{PD}\boldsymbol{\mu}^{(m)} + \beta^{(m)}\mathbf{D}^T\mathbf{A}^T(\mathbf{ADD}^T\mathbf{A}^T)^{-1}\delta\mathbf{b}, \quad (6.3\text{-}11)$$

where

$$\delta \mathbf{b} = \mathbf{b} - \mathbf{A}\mathbf{n}^{(m)}; \quad (6.3\text{-}12)$$

β is an additional step-size parameter, which is usually set to unity. An approach equivalent to that of equation 6.3-11 was proposed by Storey and Van Zeggeren (1970).

This modification may also be applied to equation 6.3-1 to minimize the effects of computer rounding errors. Equation 6.3-1 then becomes

$$\delta \mathbf{n}^{(m)} = -\mathbf{P}\boldsymbol{\mu}^{(m)} + \beta^{(m)}\mathbf{A}^T(\mathbf{A}\mathbf{A}^T)^{-1}\delta\mathbf{b}. \quad (6.3\text{-}13)$$

We note that equations 6.3-11 and 6.3-13 in principle permit the use of initial-solution estimates $\mathbf{n}^{(0)}$ that do not satisfy the element-abundance constraints.

We remark in passing that these projection methods can also be viewed as types of stoichiometric techniques, which we discuss in detail in Section 6.4. This is due to the fact that the projection matrix used in each case can be viewed as a stoichiometric matrix. Thus, for $\mathbf{P}$ defined in equation 5.2-28, $\mathbf{AP}$ vanishes. We recall from Chapter 2 that this means that the columns of $\mathbf{P}$ are stoichiometric vectors. However, $\mathbf{P}$ is not a *complete* stoichiometric matrix since the number of columns N is larger than $R = (N - M)$.

We finally note that the two algorithms discussed in this section make no special assumptions as to the algebraic form of $\boldsymbol{\mu}$. Thus they can also be utilized for nonideal systems (Chapter 7).

6.3.2 Second-Order Algorithms—the Brinkley-NASA-RAND (BNR) Algorithm

We consider here the nonstoichiometric formulation (discussed in Chapter 3) on which the Brinkley algorithm (Brinkley, 1947), the NASA algorithm (Huff et al., 1951), and the RAND algorithm (White et al., 1958) are based. This views the problem as one of solving a set of nonlinear equations.

The equilibrium conditions (equation 3.5-3), with the ideal-solution chemical potential incorporated, are

$$\frac{\mu_i^*}{RT} + \ln n_i - \ln n_t - \sum_{k=1}^{M} \psi_k a_{ki} = 0; \quad i = 1,2,\ldots,N', \quad (6.3\text{-}14)$$

where

$$\psi_k = \frac{\lambda_k}{RT}. \quad (6.3\text{-}15)$$

Equations 6.3-14 are linear in the logarithms of the mole numbers n_i and the logarithm of the total number of moles n_t, where

$$n_t = \sum_{i=1}^{N'} n_i + n_z. \quad (4.4\text{-}21)$$

In contrast to this, equation 4.4-21 and the element-abundance constraints

$$\sum_{i=1}^{N'} a_{ki} n_i - b_k = 0; \quad k = 1,2,\ldots,M, \quad (6.3\text{-}16)$$

are linear in n_i and n_t.

The three variations (Brinkley, NASA, and RAND) of the basic algorithm discussed in this section differ essentially only in the way in which they numerically treat the mole-number variables. The RAND version uses n_i as variables, and employs the Newton-Raphson method on equations 6.3-14, 4.4-21, and 6.3-16, which is equivalent to linearizing the logarithmic terms in equation 6.3-14. The Brinkley and NASA versions use $\ln n_i$ as variables and employ the Newton-Raphson method on the same set of equations, which is equivalent to linearizing the resulting exponential terms in equations 4.4-21 and 6.3-16 (cf. Section 6.3.1.2). We discuss the RAND variation first and then the Brinkley and NASA variations and show how all three algorithms are intimately related. We emphasize that in the following discussion we explicitly include the possibility of inert species through equation 4.4-21. This has not previously been considered in the literature, although Apse (1965) discussed their effect on the RAND variation of the algorithm.

6.3.2.1 The RAND Variation

We consider problems consisting of a single multispecies phase first and then generalize to multiphase problems. At the outset we allow the phase to be nonideal and then show the simplifications that ideality introduces. Linearization of equation 6.3-14 about an arbitrary estimate of the solution $(\mathbf{n}^{(m)}, \boldsymbol{\psi}^{(m)})$

yields, after rearrangement,

$$-\frac{1}{RT}\sum_{j=1}^{N'}\left(\frac{\partial \mu_i}{\partial n_j}\right)_{\mathbf{n}^{(m)}} \delta n_j^{(m)} + \sum_{k=1}^{M} a_{ki}\delta\psi_k^{(m)} = \frac{\mu_i^{(m)}}{RT} - \sum_{k=1}^{M} a_{ki}\psi_k^{(m)};$$

$$i = 1, 2, \ldots, N', \quad (6.3\text{-}17)$$

where

$$\delta\psi_k^{(m)} = \psi_k - \psi_k^{(m)} \quad (6.3\text{-}18)$$

and

$$\delta n_j^{(m)} = n_j - n_j^{(m)}. \quad (6.3\text{-}19)$$

As before, superscript (m) denotes evaluation at $(\mathbf{n}^{(m)}, \boldsymbol{\psi}^{(m)})$. The quantities $\mathbf{n}$ and $\mathbf{n}^{(m)}$ are related through the element-abundance constraints (equation 6.3-16) by

$$\sum_{j=1}^{N'} a_{kj}\delta n_j^{(m)} = b_k - b_k^{(m)}; \quad k = 1, 2, \ldots, M, \quad (6.3\text{-}20)$$

where

$$b_k^{(m)} = \sum_{j=1}^{N'} a_{kj}n_j^{(m)}; \quad k = 1, 2, \ldots, M. \quad (6.3\text{-}21)$$

Equations 6.3-17 and 6.3-20 are a set of $(N' + M)$ linear equations in the unknowns $\boldsymbol{\delta n}^{(m)}$ and $\boldsymbol{\delta\psi}^{(m)}$. These linear equations are solved, and new estimates of $(\mathbf{n}, \boldsymbol{\psi})$ are obtained from

$$\boldsymbol{\psi}^{(m+1)} = \boldsymbol{\psi}^{(m)} + \omega^{(m)}\boldsymbol{\delta\psi}^{(m)} \quad (6.3\text{-}22)$$

and

$$\mathbf{n}^{(m+1)} = \mathbf{n}^{(m)} + \omega^{(m)}\boldsymbol{\delta n}^{(m)}. \quad (6.3\text{-}2)$$

The process is then repeated, using these new solution estimates until convergence is achieved.

The usual working equations of the RAND algorithm in the literature are those for an ideal solution, although the preceding description applies to nonideal systems in general. For ideal systems, the number of linear equations

Nonstoichiometric Algorithms

to be solved on each iteration of the procedure may be reduced from $(N' + M)$ to $(M + 1)$ by eliminating the variables $\delta \mathbf{n}^{(m)}$ in equations 6.3-17 and 6.3-20. This can be done because of the special form of equation 6.1-1. Thus equation 6.1-1 gives

$$\frac{1}{RT}\left(\frac{\partial \mu_i}{\partial n_j}\right) = \frac{\delta_{ij}}{n_j} - \frac{1}{n_t}, \qquad (6.3\text{-}23)$$

where δ_{ij} is the Kronecker delta. Substitution of equation 6.3-23 in 6.3-17 allows $\delta \mathbf{n}^{(m)}$ to be obtained explicitly in terms of ψ in equation 6.3-18:

$$\delta n_j^{(m)} = n_j^{(m)}\left(\sum_{k=1}^{M} a_{kj}\psi_k + u - \frac{\mu_j^{(m)}}{RT}\right); \qquad j = 1, 2, \ldots, N', \quad (6.3\text{-}24)$$

where the additional variable u is defined by

$$u - \frac{\sum_{j=1}^{N'} \delta n_j^{(m)}}{n_t^{(m)}} = \frac{\delta n_t^{(m)}}{n_t^{(m)}}. \qquad (6.3\text{-}25)$$

Substitution of equation 6.3-24 in 6.3-20 yields the M linear equations

$$\sum_{i=1}^{M}\left(\sum_{k=1}^{N'} a_{ik} a_{jk} n_k^{(m)}\right)\psi_i + b_j^{(m)} u$$

$$= \sum_{k=1}^{N'} a_{jk} n_k^{(m)} \frac{\mu_k^{(m)}}{RT} + b_j - b_j^{(m)}; \qquad j = 1, 2, \ldots, M. \quad (6.3\text{-}26)$$

A further equation is obtained by using equation 6.3-25 and summing equation 6.3-24 over j to give

$$\sum_{i=1}^{M} b_i^{(m)}\psi_i - n_z u = \sum_{k=1}^{N'} n_k^{(m)} \frac{\mu_k^{(m)}}{RT}. \qquad (6.3\text{-}27)$$

Each iteration of the RAND algorithm consists of solving the set of $(M + 1)$ linear equations 6.3-26 and 6.3-27 and using equation 6.3-24 to determine $\delta \mathbf{n}^{(m)}$. The values of $\mathbf{n}$ used on the next iteration are obtained from

$$\mathbf{n}^{(m+1)} = \mathbf{n}^{(m)} + \omega^{(m)} \delta \mathbf{n}^{(m)}, \qquad (6.3\text{-}2)$$

where ω is a step-size parameter.

Several minor modifications of the RAND algorithm appear in the literature. Although we have derived it as a method for solving nonlinear equations, it was originally formulated (White et al., 1958) as a second-variation method for minimizing G subject to the element-abundance and nonnegativity constraints (Section 5.2.2.1). The original formulation requires that each $\mathbf{n}^{(m)}$ satisfy the element-abundance constraints. This removes the quantity $(b_j - b_j^{(m)})$ from the right side of equation 6.3-26. Another modification of the algorithm consists of the reduction of the number of working equations in the case of a single ideal-solution phase from $(M + 1)$ to M. This modification has been presented several times in the literature (Brinkley, 1966; White, 1967; Vonka and Holub, 1971) and is essentially the algorithm discussed in Section 4.4.2.

Equations 6.3-26 and 6.3-27 are due to Zeleznik and Gordon (1962) and Levine (1962), apart from our treatment of inerts. As Levine pointed out, even when $\mathbf{n}^{(m)}$ satisfies the element-abundance constraints, it is useful numerically to include the quantities $(b_j - b_j^{(m)})$ on the right side of equation 6.3-26 since this prevents the accumulation of computer rounding errors.

The RAND algorithm is easily extended to any number of single-species phases (Kubert and Stephanou, 1960; Oliver et al., 1962; Core et al., 1963; Eriksson, 1971), and to more than one multispecies phase (Boynton, 1960; Raju and Krishnaswami, 1966; Eriksson and Rosen, 1973; Eriksson, 1975). In this general case, when there are π_m multispecies phases and π_s single-species phases, equations 6.3-24, 6.3-26, and 6.3-27 become, respectively,

$$\delta n_j^{(m)} = \begin{cases} n_j^{(m)} \left(\sum_{i=1}^{M} a_{ij} \psi_i + u_\alpha - \frac{\mu_j^{(m)}}{RT} \right) & \text{(for species in multispecies phases)} \\ u_\alpha n_j^{(m)} & \text{(for species in single-species phases)} \end{cases}$$

(6.3-28)

$$\sum_{i=1}^{M} \sum_{k=1}^{N'} a_{ik} a_{jk} n_k^{(m)} \psi_i + \sum_{\alpha=1}^{\pi} b_{j\alpha}^{(m)} u_\alpha = \sum_{k=1}^{N'} a_{jk} n_k^{(m)} \frac{\mu_k^{(m)}}{RT} + b_j - b_j^{(m)};$$

$$j = 1, 2, \ldots, M, \quad (6.3\text{-}29)$$

and

$$\sum_{i=1}^{M} b_{i\alpha}^{(m)} \psi_i - n_{z\alpha} u_\alpha = \sum_{k=1}^{N'} n_{k\alpha}^{(m)} \frac{\mu_{k\alpha}^{(m)}}{RT}; \quad \alpha = 1, 2, \ldots, \pi_s + \pi_m,$$

(6.3-30)

where subscript α refers to a phase. We thus see that, in general, the RAND algorithm consists of iteratively solving the set of $(M + \pi)$ linear equations 6.3-29 and 6.3-30, where

$$\pi = \pi_m + \pi_s. \quad (6.3\text{-}31)$$

Equations 6.3-26 and 6.3-27 are the special case $\pi = \pi_m = 1$.

In spite of the straightforward way in which we have generalized to the multiphase situation, nontrivial numerical problems may sometimes be encountered in the use of equations 6.3-29 and 6.3-30. These problems arise when the coefficient matrix of the linear equations becomes singular at some point in the calculations. It can be shown that in principle this is not possible in problems consisting of only a single ideal phase but can occur whenever there is more than one phase. Such difficulties have been only briefly alluded to in the literature (Oliver et al., 1962; Barnhard and Hawkins, 1963; Samuels, 1971; Gordon and McBride, 1971, 1976; Madeley and Toguri, 1973a, 1973b; Eriksson, 1975). We discuss these in detail in Section 9.2.

We observe from our discussion of classification schemes at the beginning of this chapter that the RAND algorithm, as originally formulated by White et al. (1958), is a minimization method. At each iteration the element-abundance constraints are satisfied, and the algorithm iteratively minimizes the Gibbs free energy. We have also shown that the same algorithm may be considered to be a method of solving the nonlinear equations 6.3-14 and 6.3-16. We have seen that the mole numbers and chemical potentials on each iteration need not necessarily satisfy either equation 6.3-14 or 6.3-16, and the algorithm may iterate to satisfy both these conditions simultaneously. It is usually called a *free-energy-minimization method*. Finally, the RAND algorithm solves a numerical problem in which there are essentially $(M + \pi)$ variables that must be ultimately determined. These are the M Lagrange multipliers and the π values of the total number of moles in each phase. This is the case, however, only when all phases are ideal and is due to the fact that only then are we able to reduce the $(N' + M)$ equations 6.3-17 and 6.3-20 to the $(M + 1)$ equations 6.3-26 and 6.3-27. We have demonstrated the reduction for the case $\pi = 1$. In general, for nonideal systems (Chapter 7), we cannot reduce the number of equations in the set.

In Figure 6.2 a flow chart is shown for the RAND algorithm as developed here. In view of the discussion in the following two sections, we also refer to this as the BNR algorithm. In Appendix C we present a FORTRAN computer program that implements this algorithm.

6.3.2.2 *The Brinkley Variation*

Although this variation was historically the earliest (Brinkley, 1947, 1951, 1956, 1960, 1966; Kandiner and Brinkley, 1950a, 1950b), it has been displaced by the RAND variation. This has been partly due to the use of the apparently appealing term "free-energy-minimization method" used to describe it, but also because Brinkley chose to discuss his algorithm by using notation that made it *appear* to be quite different from the RAND variation. In this section we show that the Brinkley algorithm differs from the RAND algorithm in only a minor way. This observation was apparently first made by Zeleznik and Gordon (1960).

We again start from equations 6.3-14 and 6.3-16, but we now use $\ln n_i$ as independent variables, rather than n_i. In the RAND variation $\mathbf{n}^{(m)}$ usually

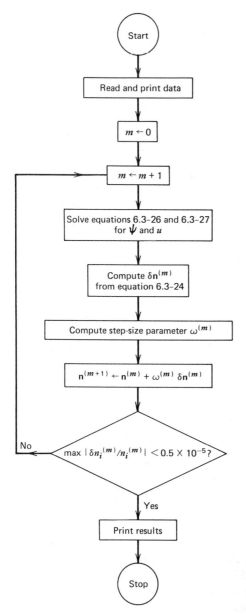

Figure 6.2 Flow chart for the RAND algorithm.

satisfies the element-abundance constraints, and iterations proceed until the equilibrium conditions are satisfied. In contrast, in the Brinkley variation, $\mathbf{n}^{(m)}$ satisfies the equilibrium conditions on each iteration, and iterations proceed until the element-abundance constraints are satisfied. Since $\ln n_i$ are the independent variables, it is convenient to set (cf. Section 6.3.1.2)

$$y_i = \ln n_i. \tag{6.3-3}$$

Then for a single ideal phase equations 6.3-14 and 6.3-16 become, respectively,

$$\exp(y_i) = n_i = \exp\left(\sum_{k=1}^{M} a_{ki}\psi_k - \frac{\mu_i^*}{RT} + \ln n_t\right); \quad i = 1, 2, \ldots, N',$$

(6.3-32)

$$\sum_{i=1}^{N'} a_{ji}\exp(y_i) = b_j; \quad j = 1, 2, \ldots, M. \quad (6.3\text{-}33)$$

Substitution of equation 6.3-32 into 6.3-33 yields

$$\sum_{i=1}^{N'} a_{ji}\exp\left(\sum_{k=1}^{M} a_{ki}\psi_k - \frac{\mu_i^*}{RT} + \ln n_t\right) = b_j; \quad j = 1, 2, \ldots, M. \quad (6.3\text{-}34)$$

Finally, we also have, from equations 4.4-21 and 6.3-32,

$$\sum_{i=1}^{N'} \exp\left(\sum_{k=1}^{M} a_{ki}\psi_k - \frac{\mu_i^*}{RT}\right) = 1 - \frac{n_z}{n_t}. \quad (6.3\text{-}35)$$

Equations 6.3-34 and 6.3-35 are a set of $(M + 1)$ nonlinear equations in the $(M + 1)$ unknowns ψ and n_t. Note that the mole fractions obtained from equation 6.3-32 for an arbitrary set of Lagrange multipliers ψ define an equilibrium composition for some hypothetical set of element abundances $\mathbf{b}^*$. Thus the Brinkley variation of the BNR algorithm iteratively modifies $\mathbf{b}^*$ until it coincides with $\mathbf{b}$ specified by the right side of equation 6.3-33.

If we choose estimates $(\psi^{(m)}, n_t^{(m)})$ and determine $\mathbf{n}^{(m)}$ from equation 6.3-32, the Newton-Raphson iteration equations obtained from linearizing equations 6.3-34 and 6.3-35 are

$$\sum_{i=1}^{M}\left(\sum_{k=1}^{N'} a_{ik}a_{jk}n_k^{(m)}\right)\delta\psi_i^{(m)} + b_j^{(m)}v = b_j - b_j^{(m)}; \quad j = 1, 2, \ldots, M$$

(6.3-36)

and

$$\sum_{i=1}^{M} b_i^{(m)} \delta\psi_i^{(m)} - n_z v = n_t^{(m)} - \sum_{k=1}^{N'} n_k^{(m)} - n_z, \quad (6.3\text{-}37)$$

where

$$v = \ln\frac{n_t}{n_t^{(m)}}, \quad (6.3\text{-}38)$$

and $\delta\boldsymbol{\psi}$ is defined by equation 6.3-18. On each iteration the linear equations 6.3-36 and 6.3-37 are solved for the $(M+1)$ unknowns $\delta\boldsymbol{\psi}^{(m)}$ and v. Then new values of $\boldsymbol{\psi}$ and $\ln n_t$ are obtained from

$$\boldsymbol{\psi}^{(m+1)} = \boldsymbol{\psi}^{(m)} + \omega^{(m)}\delta\boldsymbol{\psi}^{(m)} \qquad (6.3\text{-}22)$$

and

$$\ln n_t^{(m+1)} = \ln n_t^{(m)} + \omega^{(m)}v. \qquad (6.3\text{-}39)$$

The resulting values of $\boldsymbol{\psi}^{(m+1)}$ and $\ln n_t^{(m+1)}$ are used in equations 6.3-32 to determine $\mathbf{n}^{(m+1)}$, and the iteration is repeated.

Note the similarity of the linear equations of the RAND variation (equations 6.3-26 and 6.3-27) to those of the Brinkley variation (equations 6.3-36 and 6.3-37). The coefficient matrices of the two sets of linear equations are identical. Only the right sides differ slightly. This is because in Brinkley's variation $\mathbf{n}^{(m)}$ satisfies the equilibrium conditions, whereas in the RAND variation this is not the case.

We call equations 6.3-32, 6.3-36, and 6.3-37 the *Brinkley algorithm* here, although in Brinkley's earlier papers it appeared in a somewhat different, but completely equivalent, form. The main differences are twofold: (1) Brinkley chose to express the element-abundance constraints of equation 6.3-16 in terms of stoichiometric coefficients (this was primarily because he discussed two other methods for solving the resulting equations—intended for use in hand calculations and not discussed here—for which this form of the constraints was essential); and (2) he chose to express the equilibrium conditions of equation 6.3-32 in terms of equilibrium constants. We now examine the progression from the form in equations 6.3-32, 6.3-36, and 6.3-37 to the form in Brinkley's papers.

When the equilibrium conditions and the element-abundance constraints are expressed in stoichiometric form, equations 6.3-32, 6.3-34, and 6.3-35 become, respectively,

$$n_i = \exp\left(\sum_{k=1}^{M} \nu_{ki}\psi_k - \frac{\mu_i^*}{RT} + \ln n_t \right); \qquad i = 1, 2, \ldots, N', \qquad (6.3\text{-}40)$$

$$\sum_{i=1}^{N'} \nu_{ji} \exp\left(\sum_{k=1}^{M} \nu_{ki}\psi_k - \frac{\mu_i^*}{RT} + \ln n_t \right) = q_j; \qquad j = 1, 2, \ldots, M, \qquad (6.3\text{-}41)$$

and

$$\sum_{i=1}^{N'} \exp\left(\sum_{k=1}^{M} \nu_{ki}\psi_k - \frac{\mu_i^*}{RT} \right) = 1 - \frac{n_z}{n_t}, \qquad (6.3\text{-}42)$$

where

$$q_j = \sum_{i=1}^{N'} \nu_{ji} n_i; \quad j = 1, 2, \ldots, M. \quad (6.3\text{-}43)$$

When i is greater than M, ν_{ki} are the negatives of the stoichiometric coefficients in the stoichiometric equation in which one mole of species i is formed from a set of M component species. However, when i is less than or equal to M, we define, for the component species,

$$\nu_{ki} = \delta_{ki}; \quad i, k = 1, 2, \ldots, M, \quad (6.3\text{-}44)$$

where δ_{ki} is the Kronecker delta. In equations 6.3-40 to 6.3-42, ψ denotes the chemical potentials divided by RT of the component species used in the stoichiometric equations. Note the formal similarity between the two sets of equations 6.3-32, 6.3-34, and 6.3-35 and 6.3-40 to 6.3-42.

The Newton-Raphson iteration equations resulting from equations 6.3-41 and 6.3-42 are

$$\sum_{i=1}^{M} \sum_{k=1}^{N'} \nu_{ik} \nu_{jk} n_k^{(m)} \delta \psi_i^{(m)} + q_j^{(m)} v = q_j - q_j^{(m)}; \quad j = 1, 2, \ldots, M, \quad (6.3\text{-}45)$$

and

$$\sum_{i=1}^{M} q_i^{(m)} \delta \psi_i^{(m)} - n_z v = n_t^{(m)} - \sum_{k=1}^{N'} n_k^{(m)} - n_z, \quad (6.3\text{-}46)$$

where

$$q_j^{(m)} = \sum_{i=1}^{N'} \nu_{ji} n_i^{(m)}; \quad j = 1, 2, \ldots, M. \quad (6.3\text{-}47)$$

Equations 6.3-22 and 6.3-38 to 6.3-40 are then used to determine $\psi^{(m+1)}$, $\ln n_t^{(m+1)}$, and $\mathbf{n}^{(m+1)}$. As in the case of the RAND variation, the Brinkley variation is readily extended to consider more than one phase (see Problem 6.2).

Brinkley's papers differ in some minor ways from this description. In his earlier papers he replaced v by u since

$$v = \ln \frac{n_t}{n_t^{(m)}} = \ln\left(1 + \frac{\delta n_t}{n_t^{(m)}}\right) \approx \frac{\delta n_t}{n_t^{(m)}} = u, \quad (6.3\text{-}48)$$

which resulted in a minor modification of equation 6.3-39. Also, instead of using equation 6.3-40 for the component species, Brinkley used the approximation

$$\frac{n_i^{(m+1)}}{n_t^{(m+1)}} = \frac{n_i^{(m)}}{n_t^{(m)}} \exp(\delta\psi_i^{(m)}) \approx \frac{n_i^{(m)}}{n_t^{(m)}} \left(1 + \delta\psi_i^{(m)}\right) \tag{6.3-49}$$

and set

$$\delta\psi_k^{(m)} \approx \ln\left(1 + \delta\psi_k^{(m)}\right). \tag{6.3-50}$$

The modifications of equations 6.3-45 to 6.3-50 are minor. Equations 6.3-32 and 6.3-34 to 6.3-37 are respectively equivalent to equations 6.3-40 to 6.3-42, 6.3-45, and 6.3-46. However, in his earlier papers Brinkley's equations *appear* very different from the latter set of equations. This is due only to his *notation*. Brinkley's notation also had the effect of causing his algorithm to be regarded as an equilibrium-constant method when, in fact, it is essentially equivalent to the RAND algorithm. This notation, involving equilibrium constants, is the second main way in which his algorithm differs from equations 6.3-32 and 6.3-34 to 6.3-37.

Brinkley expressed equation 6.3-40 in terms of equilibrium constants. Equations 6.3-45 and 6.3-46, the basic working equations of his algorithm, appear in the literature essentially as we have given them here (Brinkley, 1947). Since $RT\psi_k$ in equations 6.3-40 to 6.3-42 is the chemical potential of component species k, we have

$$\psi_k = \frac{\mu_k^*}{RT} + \ln n_k - \ln n_t; \quad k = 1, 2, \ldots, M. \tag{6.3-51}$$

For a mixture of ideal gases, equation 6.3-51 becomes

$$\psi_k = \frac{\mu_k^\circ}{RT} + \ln n_k + \ln \frac{P}{n_t}; \quad k = 1, 2, \ldots, M. \tag{6.3-52}$$

Using equation 6.3-52, we may rewrite equation 6.3-40 for *all* the species as

$$n_i = K_{pi} \left(\frac{P}{n_t}\right)^{\bar{\nu}_i} \prod_{k=1}^{M} n_k^{\nu_{ki}}; \quad i = 1, 2, \ldots, N', \tag{6.3-53}$$

where

$$K_{pi} = \exp\left(\sum_{k=1}^{M} \nu_{ki} \frac{\mu_k^\circ}{RT} - \frac{\mu_i^\circ}{RT}\right) \tag{6.3-54}$$

and

$$\bar{\nu}_i = \sum_{k=1}^{M} \nu_{ki} - 1. \tag{6.3-55}$$

(Note that, for the component species, equations 6.3-44 and 6.3-53 give trivially $n_i = n_i$). In equation 6.3-54 K_{pi} is the chemical equilibrium constant for the stoichiometric equation forming one mole of species i from the component species. Equation 6.3-53 is completely equivalent to equation 6.3-40. Equation 6.3-53, when used in place of equation 6.3-40, would make equations 6.3-41 and 6.3-42 quite different in *appearance* from their present form, although they would remain mathematically identical.

In terms of the classification schemes discussed at the beginning of the chapter, the Brinkley algorithm is essentially a nonlinear equation method. The equilibrium conditions are satisfied at each iteration, and the algorithm iterates to satisfy the element-abundance constraints. It is either an equilibrium-constant method or not, depending on one's point of view (i.e., depending on whether equation 6.3-53 is used). As in the case of the RAND variation, there are in general $(M + \pi)$ unknown variables to be determined. These are the total number of moles in each phase and the chemical potentials of M component species.

6.3.2.3 The NASA Variation

In our discussion of the RAND variation we have seen that in the original formulation (White et al., 1958) the mole numbers on each iteration satisfy the element-abundance constraints and that this restriction was relaxed in a later modification (Zeleznik and Gordon, 1962; Levine, 1962). In the Brinkley variation we have seen that the (logarithmic) mole-number variables on each iteration satisfy the equilibrium conditions. This restriction may also be relaxed, resulting in the NASA variation of the algorithm (Huff et al., 1951; Gordon et al., 1959).

The NASA algorithm was originally formulated to consider equilibrium calculations at specified pressure P and enthalpy H. However, we consider its derivation here for the usual case of specified temperature T and pressure P, to facilitate comparison with the RAND and Brinkley variations.

Again considering the case of an ideal solution and using logarithmic variables for the mole numbers (equation 6.3-32), we linearize equations 6.3-14 and 6.3-16 about estimates $(\mathbf{n}^{(m)}, \boldsymbol{\psi}^{(m)})$ to yield, respectively,

$$\delta(\ln n_k)^{(m)} - \delta(\ln n_t)^{(m)} = \sum_{j=1}^{M} a_{jk}\left(\psi_j^{(m)} + \delta\psi_j^{(m)}\right) - \frac{\mu_k^{(m)}}{RT};$$

$$k = 1, 2, \ldots, N' \quad (6.3\text{-}56)$$

and

$$\sum_{k=1}^{N'} a_{jk} n_k^{(m)} \delta(\ln n_k)^{(m)} = b_j - b_j^{(m)}; \quad j = 1, 2, \ldots, M. \quad (6.3\text{-}57)$$

The NASA variation always includes the atomic elements as species in the calculations. Numbering the species so that the first M are the elements, we have

$$\psi_j = \frac{\mu_j}{RT}; \quad j = 1, 2, \ldots, M. \tag{6.3-58}$$

Hence for the elemental species, from equation 6.3-56, we have

$$\psi_j^{(m)} + \delta\psi_j^{(m)} = \delta(\ln n_j)^{(m)} - \delta(\ln n_t)^{(m)} + \frac{\mu_j^{(m)}}{RT}; \quad j = 1, 2, \ldots, M. \tag{6.3-59}$$

When k does *not* denote an elemental species, equation 6.3-56 is written by using equation 6.3-59 as

$$\delta(\ln n_k)^{(m)} = \delta(\ln n_t)^{(m)}\left(1 - \sum_{i=1}^{M} a_{ik}\right) + \sum_{i=1}^{M} a_{ik}\delta(\ln n_i)^{(m)}$$

$$+ \sum_{i=1}^{M} a_{ik}\frac{\mu_i^{(m)}}{RT} - \frac{\mu_k^{(m)}}{RT}; \quad k = M+1, \ldots, N'. \tag{6.3-60}$$

Substitution of equation 6.3-60 in 6.3-57 yields a set of M linear equations involving $\{\delta(\ln n_j)^{(m)}; j = 1, 2, \ldots, M\}$:

$$n_j^{(m)}\delta(\ln n_j)^{(m)} + \sum_{i=1}^{M} \delta(\ln n_i)^{(m)}\left(\sum_{k=M+1}^{N'} a_{ik}a_{jk}n_k^{(m)}\right)$$

$$+ \delta(\ln n_t^{(m)}) \sum_{k=M+1}^{N'} a_{jk}n_k^{(m)}\left(1 - \sum_{i=1}^{M} a_{ik}\right)$$

$$= \sum_{k=M+1}^{N'} -a_{jk}n_k^{(m)}\left(\sum_{i=1}^{M} a_{ik}\frac{\mu_i^{(m)}}{RT} + \frac{\mu_k^{(m)}}{RT}\right) + b_j - b_j^{(m)};$$

$$j = 1, 2, \ldots, M. \tag{6.3-61}$$

A final equation is obtained by linearizing

$$\sum_{i=1}^{N'} \exp(\ln n_i) = \exp(\ln n_t) - n_z \tag{6.3-62}$$

about $(\mathbf{n}^{(m)}, n_t^{(m)})$. This gives

$$\sum_{i=1}^{M} \delta(\ln n_i)^{(m)} \left(n_i^{(m)} + \sum_{k=M+1}^{N'} a_{ik} n_k^{(m)} \right)$$

$$+ \delta(\ln n_t)^{(m)} \left\{ -n_t^{(m)} + \sum_{k=M+1}^{N'} n_k^{(m)} \left(1 - \sum_{i=1}^{M} a_{ik} \right) \right\}$$

$$= n_t^{(m)} - \sum_{k=1}^{N'} n_k^{(m)} - n_z - \sum_{k=M+1}^{N'} n_k^{(m)} \left(-\sum_{i=1}^{M} a_{ik} \frac{\mu_i^{(m)}}{RT} + \frac{\mu_k^{(m)}}{RT} \right). \quad (6.3\text{-}63)$$

Equations 6.3-61 and 6.3-63 are a set of $(M + 1)$ linear equations in the $(M + 1)$ unknowns $\{\delta(\ln n_i)^{(m)}; i = 1, 2, \ldots, M$ and $\delta(\ln n_t)^{(m)}\}$. The changes $(\delta \ln n_k)^{(m)}$ in the remaining species mole numbers are given by equation 6.3-60. These equations essentially comprise the NASA algorithm. The only minor difference between this presentation and the algorithm as it appears in the literature arises from the fact that the pressure P is used as a variable instead of the total number of moles n_t.

We can see the similarity to the Brinkley variation by noting that if the initial $(\mathbf{n}^{(m)}, n_t^{(m)})$ satisfied the equilibrium conditions, the first term on the right side of equation 6.3-61 would be absent. In addition, we see from equation 6.3-60 that the new mole numbers would also satisfy the equilibrium conditions. The resulting algorithm in this case would hence be exactly equivalent to the Brinkley variation, except for the minor fact that changes in the chemical potentials of the elemental species (or component species in the Brinkley variation) are given in the NASA algorithm by equation 6.3-59, whereas in the Brinkley variation these are determined by exponentiating equations 6.3-59 and combining this with the linear approximation

$$e^x \approx 1 + x \qquad (6.3\text{-}64)$$

to yield equation 6.3-49.

We thus see that the only essential difference between the Brinkley and NASA variations is the fact that successive iterations satisfy the equilibrium conditions in the former, but not in the latter. The inclusion of the elements as species in the NASA variation is not an essential part of the method. Thus the NASA variation complements the Brinkley variation similar to the way that the modification due to Zeleznik and Gordon (1962) and Levine (1962) complements the original RAND variation. The main difference between the three variations is between (1) the RAND variation and (2) the Brinkley and NASA variations.

This difference consists solely of the fact that the former variation uses the mole numbers n_i as variables and the latter variations use $\ln n_i$ as variables. Computationally, one would expect little difference between the performance of the three variations, and this has been confirmed by Zeleznik and Gordon (1960).

In terms of the classifications in Section 6.1, the NASA algorithm is a nonlinear-equation method. It satisfies neither the element-abundance nor the equilibrium conditions on each iteration. It can be, although it seldom is, formulated by using equilibrium constants. There are in general $(M + \pi)$ variables to be determined, which are the chemical potentials of the elemental species and the total number of moles in each phase.

6.3.3 Other Nonstoichiometric Algorithms

In this section we briefly discuss some other chemical equilibrium algorithms that have appeared in the literature and that are based directly on equations 6.3-14 and 6.3-16. They are motivated from the numerical viewpoint of either minimization or nonlinear equations, but it seems that none of these has any particular advantage over the BNR algorithm.

Several numerical schemes other than the BNR algorithm have been published for solving the set of nonlinear equations that result when the equilibrium conditions (equations 6.3-14) are substituted into the element-abundance constraints (equations 6.3-16). As we have seen, the Brinkley algorithm results from the application of the Newton-Raphson method to these nonlinear equations. Other ways of solving these equations have been described by Scully (1962), Storey and Van Zeggeren (1967), and Stadtherr and Scriven (1974).

Other methods based on minimization techniques have also been suggested in the literature. Madeley and Toguri (1973a) have developed an approach that uses the first-order algorithm due to Storey and Van Zeggeren (1964, 1970) in the initial stages and the RAND algorithm in the final stages. George et al. (1976) use Powell's method of minimization (Powell, 1970) on an unconstrained objective function that incorporates G and the element-abundance and nonnegativity constraints. Gautam and Seider (1979) have suggested a method based on the use of quadratic programming. Finally, Castillo and Grossman (1979) have used the variable-metric projection method due to Sargent and Murtagh (1973).

A somewhat different approach uses an optimization technique called *geometric programming* (Duffin et al., 1967). Minimization of any function is equivalent to maximization of the exponential of the negative of the function. Thus the chemical equilibrium problem may be formulated (for one phase) as

$$\max_{\mathbf{n}} \left[\exp\left(\frac{-G}{RT} \right) \right] = n_t^{n_t} \prod_{i=1}^{N} \left[\frac{\exp(-\mu_i^*/RT)}{n_i} \right]^{n_i}. \qquad (6.3\text{-}65)$$

When the element-abundance constraints are formulated appropriately, these equations together with equation 6.3-65 form a problem in geometric programming. Algorithms have been developed for solving such a problem, and these have been applied to the chemical equilibrium problem (Wilde and Beightler, 1967; Passey and Wilde, 1968; Dinkel and Lakshmanan, 1977).

6.3.4 Illustrative Example for the BNR Algorithm

Example 6.1 We illustrate the use of the RAND variation of the BNR algorithm by considering equilibrium in a system investigated by White et al. (1958). This involves determination of the composition of the gas resulting from the combustion of hydrazine (N_2H_4) with oxygen in a 1:1 ratio at 3500 K and 51.0 atm. Using the species listed by White et al., we represent the system by $\{(H_2O, N_2, H_2, OH, H, O_2, NO, O, N, NH), (H, N, O)\}$. We use the free-energy data provided by them and one mole of initial reacting system.

Solution We first construct an input data file for the BNR computer program given in Appendix C, in accordance with the *User's Guide* (see Figure 6.3). The first line indicates that there is one problem to be considered, and the second that there are 10 species and three components (in this case, equal to the number of elements). Each of the following 10 lines contains a species name, its formula vector, its phase designation (1 denotes a gaseous multi-species phase), and its standard chemical potential ($\mu°/RT$ in this case). The next two lines contain the initial equilibrium estimate (taken from the original paper), and these are followed by a line giving the element abundances. The final three lines show, respectively, the temperature and pressure, the names of the elements, and an arbitrary title.

Using this input file, we obtain the output shown in Figure 6.4. Convergence is achieved after eight iterations, and the results are given both as equilibrium mole numbers and as mole fractions. They essentially agree with

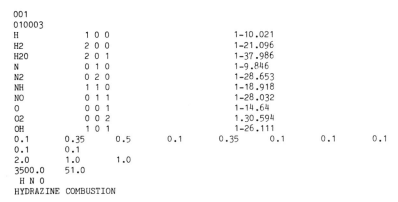

```
001
010003
H            1 0 0                    1-10.021
H2           2 0 0                    1-21.096
H2O          2 0 1                    1-37.986
N            0 1 0                    1-9.846
N2           0 2 0                    1-28.653
NH           1 1 0                    1-18.918
NO           0 1 1                    1-28.032
O            0 0 1                    1-14.64
O2           0 0 2                    1.30.594
OH           1 0 1                    1-26.111
0.1        0.35       0.5     0.1       0.35      0.1     0.1     0.1
0.1        0.1
2.0        1.0        1.0
3500.0     51.0
  H N O
HYDRAZINE COMBUSTION
```

Figure 6.3 Input data file for Examples 6.1 and 6.2.

RAND CALCULATION METHOD

HYDRAZINE COMBUSTION

```
  10 SPECIES         3 ELEMENTS              3 COMPONENTS
  10 PHASE1 SPECIES          0 PHASE2 SPECIES        0 SINGLE SPECIES PHASES
  PRESSURE                51.000 ATM
  TEMPERATURE           3500.000 K
  MOLES INERT GAS          0.0
```

ELEMENTAL ABUNDANCES   CORRECT   FROM ESTIMATE

```
                         H    2.000000000000D 00   1.999999821186D 00
                         N    1.000000000000D 00   9.999998211861D-01
                         O    1.000000000000D 00   9.999998211861D-01
```

STAN. CHEM. POT. IS MU/RT

SPECIES	FORMULA VECTOR				STAN. CHEM. POT.	EQUILIBRIUM EST.
	H	N	O	SI (I)		
H	1	0	0	1	-1.0021D 01	1.0000D-01
H2	2	0	0	1	-2.1096D 01	3.5000D-01
H2O	2	0	1	1	-3.7986D 01	5.0000D-01
N	0	1	0	1	-9.8460D 00	1.0000D-01
N2	0	2	0	1	-2.8653D 01	3.5000D-01
NH	1	1	0	1	-1.8918D 01	1.0000D-01
NO	0	1	1	1	-2.8032D 01	1.0000D-01
O	0	0	1	1	-1.4640D 01	1.0000D-01
O2	0	0	2	1	-3.0594D 01	1.0000D-01
OH	1	0	1	1	-2.6111D 01	1.0000D-01

8 ITERATIONS

SPECIES	EQUILIBRIUM MOLES	MOLE FRACTION	FINAL DELTA
H	4.0672827D-02	2.4824005D-02	2.8357D-11
H2	1.4773719D-01	9.0169014D-02	2.8383D-11
H2O	7.8314179D-01	4.7797799D-01	-6.2506D-11
N	1.4143462D-03	8.6323241D-04	3.8327D-09
N2	4.8524621D-01	2.9616221D-01	-1.8934D-09
NH	6.9318974D-04	4.2307720D-04	-1.2869D-12
NO	2.7400048D-02	1.6723178D-02	-4.4493D-11
O	1.7949416D-02	1.0955137D-02	1.6698D-11
O2	3.7316357D-02	2.2775438D-02	2.4562D-11
OH	9.6876036D-02	5.9126729D-02	4.1186D-11

G/RT = -4.7761368D 01
TOTAL PHASE1 MOLES = 1.6384D 00

ELEMENTAL ABUNDANCES
```
                         H    2.000000000000D 00
                         N    1.000000000000D 00
                         O    1.000000000000D 00
```

FINAL LAGRANGE MULTIPLIERS (LAMBDA/RT)

```
  -9.78511842D 00
  -1.29690111D 01
  -1.52221206D 01
```

Figure 6.4 Computer output for Example 6.1 from BNR algorithm in Appendix C.

those given by White et al. (1958) to within four significant figures. The entries under "FINAL DELTA" give the final mole-number corrections at convergence.

The dominant products of combustion are H_2O and N_2, as expected. If we had assumed that the system behaved as a simple system ($R = 1$) with complete combustion, the amount of H_2O would have been 1 mole (cf. 0.7831), and that of N_2 would have been 0.5 moles (cf. 0.4852); n_t would have been 1.5 moles (cf. 1.638). It has been assumed that N_2H_4 is completely consumed (note that it has been excluded from the list of species). Finally, the amount of O_2 remaining is 0.03732 mole, rather than zero.

6.4 STOICHIOMETRIC ALGORITHMS

6.4.1 Introduction

Stoichiometric algorithms eliminate the element-abundance constraints from the minimization problem, resulting in an unconstrained formulation. As discussed in Section 3.4, this is accomplished by transforming from the N unknown mole numbers $\mathbf{n}$, which are constrained by the M element-abundance equations, to a new set of "reaction-extent" variables ξ, equal in number to $R = (N' - M)$. Mole numbers on each iteration for these methods always satisfy the element-abundance constraints.

The changes in the mole numbers $\delta \mathbf{n}^{(m)}$ from any estimate $\mathbf{n}^{(m)}$ satisfying the element-abundance constraints are related to new ξ variables by

$$\delta n_i^{(m)} = \sum_{j=1}^{R} \nu_{ij} \delta \xi_j^{(m)}; \qquad i = 1, 2, \ldots, N'. \tag{6.4-1}$$

The matrix $\mathbf{N}$ has $R = (N' - M)$ linearly independent columns and is related to the formula matrix $\mathbf{A}$ by

$$\sum_{i=1}^{N'} a_{ki} \nu_{ij} = 0; \qquad \begin{array}{l} k = 1, 2, \ldots, M \\ j = 1, 2, \ldots, R \end{array} \tag{6.4-2}$$

Viewing the Gibbs function G as a function of the reaction-extent variables ξ, we see that the chemical equilibrium problem is that of minimizing $G(\xi)$. The necessary conditions for this are the nonlinear equations (cf. equation 3.4-2)

$$\frac{\partial G}{\partial \xi} = \mathbf{0}. \tag{6.4-3}$$

We have seen in Chapter 3 that equation 6.4-3 is equivalent to the classical chemical formulation of the equilibrium conditions (cf. equation 3.4-5)

$$\Delta \mathbf{G} \equiv \mathbf{N}^T \boldsymbol{\mu}(\xi) = \mathbf{0}. \tag{6.4-4}$$

Analogous to the discussion in Section 6.3, we may treat this formulation of the chemical equilibrium problem numerically from either the minimization or the nonlinear equation point of view.

One of the main differences between stoichiometric and nonstoichiometric algorithms concerns the total number of independent variables that must essentially be determined. Using equations 6.3-14 and 6.3-16 directly, nonstoichiometric algorithms incorporate the element-abundance constraints by the introduction of an additional set of M variables (the Lagrange multipliers). We have seen that in several such algorithms a new variable is also introduced for each phase in the system. This results in a total of $(N' + M + \pi)$ variables altogether. When the phases are ideal, this number is reduced to $(M + \pi)$. In the stoichiometric algorithms the number of variables is always $N' - M$, regardless of whether the phases are ideal. Thus, for nonideal systems, the stoichiometric algorithms always have fewer variables. For ideal systems with a small number of phases, the nonstoichiometric algorithms usually have fewer variables.

For problems involving single-species phases, stoichiometric algorithms have certain numerical advantages over nonstoichiometric algorithms, and these are discussed in Chapter 9.

We note that the mere appearance of stoichiometric coefficients in an algorithm does not justify classifying it as a stoichiometric algorithm in terms of the classification schemes presented in Section 6.1. For example, the Brinkley algorithm uses stoichiometric coefficients, but it does so only in an incidental way, and hence we do not classify it as a stoichiometric algorithm.

A number of general-purpose algorithms have appeared in the literature using the stoichiometric formulation. One of the first of these was that due to Naphtali (1959, 1960, 1961), who suggested using a first-order method to minimize $G(\xi)$. At about the same time Villars (1959, 1960) devised an algorithm for solving the set of nonlinear equations 6.4-4. Cruise (1964) subsequently made several improvements to this algorithm. Smith (1966) and Smith and Missen (1968) reformulated the Villars-Cruise algorithm as a minimization method, resulting in improved convergence properties. Hutchison (1962) suggested the use of the Newton-Raphson method in equations 6.4-4. This approach has also been suggested by Stone (1966) and by Bos and Meerschoek (1972). The coefficient matrix of the linear equations in the algorithm is usually so large, however, that the method is rather unwieldy and apparently has not been widely used. Finally, Meissner et al. (1969) have discussed an approach that is very similar to the Villars algorithm. We consider each of these in turn in more detail.

Other stoichiometric methods, not discussed in detail here, have also appeared in the literature. For example, Sanderson and Chien (1973) have used Marquardt's algorithm (Marquardt, 1963) to solve equations 6.4-4.

6.4.2 First-Order Algorithm

Naphtali (1959, 1960, 1961) suggested use of the first-order method, discussed in Section 5.2.1.1, for minimizing $G(\xi)$. The variables ξ at each iteration are

Stoichiometric Algorithms

adjusted by amounts $\delta\xi$, where

$$\delta\xi_j^{(m)} = -\left(\frac{\partial G}{\partial \xi_j}\right)^{(m)} = -\Delta G_j^{(m)}$$

$$= -\sum_{i=1}^{N'} \nu_{ij}\mu_i^{(m)}; \quad j = 1,2,\ldots,R. \qquad (6.4\text{-}5)$$

The mole numbers are adjusted by means of equation 6.4-1. This algorithm has been found to converge rather slowly, especially near the solution, as is characteristic of first-order optimization methods in general. It hence does not appear to be widely used.

6.4.3 Second-Order Algorithm

Hutchison (1962) and others (Stone, 1966; Bos and Meerschoek, 1972) have suggested applying the Newton-Raphson method to equations 6.4-4. This yields

$$\delta\xi^{(m)} = -\left(\frac{\partial^2 G}{\partial \xi^2}\right)^{-1}_{\mathbf{n}^{(m)}} \left(\frac{\partial G}{\partial \xi}\right)_{\mathbf{n}^{(m)}}. \qquad (6.4\text{-}6)$$

This approach requires the solution of a set of $R = (N' - M)$ linear equations on each iteration. Since N' is usually large compared with M, the numerical solution of these linear equations can be a very time-consuming segment of the algorithm. Thus this approach does not appear to have been widely used as a general-purpose method, but we have used it in Chapter 4 for relatively simple systems. Ma and Shipman (1972) have developed a method that uses the first-order algorithm in the initial stages and the second-order algorithm in the final stages. The next approach to be discussed is related to equations 6.4-6 and is essentially a way of reducing the labor involved in the solution of the linear equations.

6.4.4 Optimized Stoichiometry—The Villars-Cruise-Smith (VCS) Algorithm

The Villars-Cruise-Smith (VCS) algorithm is intermediate between a first- and second-order method. The algorithm begins with equation 6.4-6. In the case of a single ideal phase, the Hessian matrix $(\partial^2 G/\partial \xi^2)$ is given by

$$\frac{\partial^2 G}{\partial \xi_i \partial \xi_j} = \frac{\partial}{\partial \xi_j}\left(\sum_{k=1}^{N'} \nu_{ki}\mu_k\right)$$

$$= RT \sum_{k=1}^{N'} \sum_{l=1}^{N'} \nu_{ki}\nu_{lj}\left(\frac{\delta_{kl}}{n_k} - \frac{1}{n_t}\right); \quad i,j = 1,2,\ldots,R, \qquad (6.4\text{-}7)$$

where δ_{kl} is the Kronecker delta. We may rewrite equation 6.4-7 as

$$\frac{1}{RT}\frac{\partial^2 G}{\partial \xi_i \partial \xi_j} = \sum_{k=1}^{N'} \frac{\nu_{ki}\nu_{kj}}{n_k} - \frac{\bar{\nu}_i \bar{\nu}_j}{n_t}; \quad i,j = 1,2,\ldots,R, \quad (6.4\text{-}8)$$

where

$$\bar{\nu}_i = \sum_{k=1}^{N'} \nu_{ki}. \quad (6.4\text{-}9)$$

We recall from Chapter 2 that we have considerable freedom in choosing the stoichiometric matrix **N**. If we can make use of this freedom to choose **N** so that the Hessian matrix in equation 6.4-7 is easily inverted, the use of equation 6.4-6 is very attractive. For example, if we can make the first terms in equation 6.4-8 vanish for $i \neq j$, the Hessian is easily inverted in closed form.

We can, in fact, choose **N** in this way by choosing $\{\nu_j\}$ so that

$$\sum_{k=1}^{N'} \frac{\nu_{ki}\nu_{kj}}{n_k} = \delta_{ij}; \quad (6.4\text{-}10)$$

that is, $\{\nu_j\}$ is orthonormal with respect to the inner product and vector norms,

$$\nu_i \cdot \nu_j = \sum_{k=1}^{N'} \frac{\nu_{ki}\nu_{kj}}{n_k} \quad (6.4\text{-}11)$$

and

$$\|\nu_i\| = \left(\sum_{k=1}^{N'} \frac{\nu_{ki}^2}{n_k} \right)^{1/2}, \quad (6.4\text{-}12)$$

respectively. Although it is possible in principle to compute $\{\nu\}$ in this way, it is probably not very useful since we would have to recalculate the $N \times R$ matrix **N** on each iteration, corresponding to each new composition $\mathbf{n}^{(m)}$.

The VCS algorithm essentially makes a compromise between computing **N** in this way on each iteration and computing **N** only once at the beginning of the procedure. We note that if our stoichiometric matrix is in canonical form, the product $\nu_{ki}\nu_{kj}$ for $i \neq j$ is zero when k refers to a noncomponent species ($k > M$) since each noncomponent species has a nonzero stoichiometric coefficient only in one stoichiometric vector. When $i = j$, $\nu_{ki}\nu_{kj} = 1$ for such k values. The entries of the Hessian matrix are thus, numbering the component species from 1 to M and the noncomponent species from $(M+1)$ to N',

$$\frac{1}{RT}\frac{\partial^2 G}{\partial \xi_i \partial \xi_j} = \frac{\delta_{ij}}{n_{j+M}} + \sum_{k=1}^{M} \frac{\nu_{ki}\nu_{kj}}{n_k} - \frac{\bar{\nu}_i \bar{\nu}_j}{n_t}; \quad i,j = 1,2,\ldots,R. \quad (6.4\text{-}13)$$

If we choose the component species to be those with the largest mole numbers, this tends to make the second term on the right side of equation 6.4-13 small and the first term large. The last term vanishes if either $\bar{\nu}_i$ or $\bar{\nu}_j$ vanishes, and in any event it is often small compared with the first term because of the presence of n_t in the denominator.

If we form the **N** matrix in this way, we may make the reasonable approximation that the Hessian matrix may be considered to be diagonal, and we invert it directly to give

$$RT\left(\frac{\partial^2 G}{\partial \xi_i \partial \xi_j}\right)^{-1} \approx \left(\frac{1}{n_{i+M}} + \sum_{k=1}^{M} \frac{\nu_{ki}^2}{n_k} - \frac{\bar{\nu}_i^2}{n_t}\right)^{-1} \delta_{ij}. \quad (6.4\text{-}14)$$

[In the literature (Villars, 1959, 1960; Cruise, 1964; Smith, 1966; Smith and Missen, 1968), a further approximation is usually made by neglecting the term involving $\bar{\nu}_i$.] The VCS algorithm for a single ideal phase thus consists of using equation 6.4-6 with 6.4-14 and iteratively adjusts each stoichiometric equation by an amount

$$\delta \xi_j^{(m)} = -\left(\frac{1}{n_{j+M}^{(m)}} + \sum_{k=1}^{M} \frac{\nu_{kj}^2}{n_k^{(m)}} - \frac{\bar{\nu}_j^2}{n_t}\right)^{-1} \frac{\Delta G_j^{(m)}}{RT}; \quad j = 1, 2, \ldots, R.$$

$$(6.4\text{-}15)$$

On each iteration the species mole numbers are examined to ensure that the component species are those with the largest mole numbers. If this is not the case, a new stoichiometric matrix is calculated.

In the case of an ideal multiphase system, equation 6.4-15 becomes

$$\delta \xi_j^{(m)} = \begin{cases} -\left[\frac{\delta_{j+M,\alpha}^*}{n_{j+M}^{(m)}} + \sum_{k=1}^{M} \frac{\nu_{kj}^2 \delta_{k\alpha}^*}{n_k^{(m)}} \right. \\ \left. - \sum_{\alpha=1}^{\pi_m} \sum_{k=1}^{N'} \frac{(\nu_{kj} \delta_{k\alpha})^2}{n_{t\alpha}}\right]^{-1} \frac{\Delta G_j^{(m)}}{RT} \\ \text{(provided that at least one species for which} \\ \nu_{kj} \neq 0 \text{ is in a multispecies phase)} \\ -\frac{\Delta G_j^{(m)}}{RT} \quad \text{(otherwise)} \end{cases} \quad (6.4\text{-}16)$$

Here α denotes a phase. The value of $\delta_{k\alpha}^*$ is unity if species k is in *any* multispecies phase α and is zero otherwise, and $\delta_{k\alpha}$ is unity if species k is in the particular multispecies phase α and is zero otherwise. We remark that the VCS algorithm is well suited to handle multiphase problems, especially those involving single-species phases, such as arise in metallurgical applications. This is due to the fact that the nonnegativity constraints on the species mole numbers are easily handled in this algorithm (as opposed, e.g., to the BNR algorithm discussed in Section 6.3). We discuss the treatment of the nonnegativity constraints in detail in Chapter 9.

The foregoing description of the VCS algorithm is essentially that due to Smith (1966). However, historically this algorithm was not originally viewed as a free-energy-minimization method, but as a method for solving the nonlinear equations represented by the classical equilibrium conditions of equation 6.4-4. Villars (1959, 1960) originally proposed the use of equation 6.4-15 using an arbitrarily chosen $\mathbf{N}$ matrix. He also adjusted each individual stoichiometric equation in turn and recomputed the system composition before adjusting the next equation. He viewed this approach as a way of using the Newton-Raphson method on the equilibrium conditions, adjusting the stoichiometric equations one at a time. The analogous method for a single stoichiometric equation had been proposed by Deming (1930). In the approach due to Meissner et al. (1969) each main iteration consists of bringing the reactions one at a time exactly, rather than approximately, to equilibrium. Cruise (1964) incorporated the optimized choice of $\mathbf{N}$ described previously, based on earlier work of Browne et al. (1960). Cruise also advocated the simultaneous adjustment of all stoichiometric equations by means of equation 6.4-15 on each iteration before recomputing the system composition. He found that these two modifications to Villars' method resulted in substantial improvements in computing speed and convergence. Finally, Smith (1966) and Smith and Missen (1968) reformulated the method as a minimization algorithm and incorporated the step-size parameter ω. This minimization point of view resulted in an algorithm that is both rapid and free of convergence problems.

We remark in conclusion that the VCS algorithm is a descent method of minimization since, for a single ideal phase,

$$\frac{1}{RT}\frac{\partial^2 G}{\partial \xi_j^2} = \sum_{k=1}^{N'} \frac{\nu_{kj}^2}{n_k} - \frac{\bar{\nu}_j^2}{n_t} = \sum_{k=1}^{N'} n_k \left(\frac{\nu_{kj}}{n_k} - \frac{\bar{\nu}_j}{n_t}\right)^2 > 0, \quad (6.4\text{-}17)$$

and equation 6.4-15 thus yields on each iteration

$$\left(\frac{dG}{d\omega^{(m)}}\right)_{\omega^{(m)}=0} = -\sum_{j=1}^{R} \left(\frac{\partial G}{\partial \xi_j}\right)^2_{\mathbf{n}^{(m)}} \left(\frac{\partial^2 G}{\partial \xi_j^2}\right)^{-1}_{\mathbf{n}^{(m)}} \leq 0. \quad (6.4\text{-}18)$$

One significant computational advantage of this algorithm is the fact that there are no linear equations to solve on each iteration. We recall that the BNR algorithm for an ideal system requires the solution of $(M + \pi)$ linear equations

Stoichiometric Algorithms

on each iteration. These equations can have a singular or nearly singular coefficient matrix for some problems, and this can cause practical difficulties. The VCS algorithm avoids these. We discuss some examples of this type of difficulty in Chapter 9.

In Figure 6.5 a flow chart is displayed for the VCS algorithm, as developed here. A FORTRAN computer program that implements this algorithm is given in Appendix D.

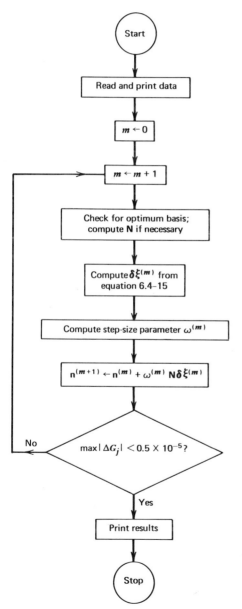

Figure 6.5 Flow chart for the VCS algorithm.

6.4.5 Illustrative Example for the VCS Algorithm

Example 6.2 We illustrate the use of the VCS algorithm by means of the system described in Example 6.1 (White et al., 1958). The input data file is the same as for the BNR algorithm (Figure 6.3)—see *User's Guide* in Appendix D. The output is shown in Figure 6.6. Convergence is achieved after 17 iterations, during which the stoichiometric matrix is calculated twice. Although the number of iterations (17) is greater than in Example 6.1 (8), the total

```
VCS CALCULATION METHOD

HYDRAZINE COMBUSTION

   10 SPECIES        3 ELEMENTS              3 COMPONENTS
   10 PHASE1 SPECIES          0 PHASE2 SPECIES     0 SINGLE SPECIES PHASES

   PRESSURE                51.000 ATM
   TEMPERATURE           3500.000 K
   PHASE1 INERTS              0.0

   ELEMENTAL ABUNDANCES         CORRECT              FROM ESTIMATE

                           H  2.000000000000D 00   2.000000000000D 00
                           N  1.000000000000D 00   1.000000000000D 00
                           O  1.000000000000D 00   1.000000000000D 00

   USER ESTIMATE OF EQUILIBRIUM
   STAN. CHEM. POT. IS MU/RT

   SPECIES        FORMULA VECTOR            STAN. CHEM. POT.    EQUILIBRIUM EST.

                  H  N  O  SI(I)
   H2O            2  0  1   1                 -3.79860D 01       5.00000D-01
   N2             0  2  0   1                 -2.86530D 01       3.50000D-01
   H2             2  0  0   1                 -2.10960D 01       3.50000D-01
   N              0  1  0   1                 -9.84600D 00       1.00000D-01
   H              1  0  0   1                 -1.00210D 01       1.00000D-01
   NH             1  1  0   1                 -1.89180D 01       1.00000D-01
   NO             0  1  1   1                 -2.80320D 01       1.00000D-01
   O              0  0  1   1                 -1.46400D 01       1.00000D-01
   O2             0  0  2   1                 -3.05940D 01       1.00000D-01
   OH             1  0  1   1                 -2.61110D 01       1.00000D-01

   ITERATIONS =     17
   EVALUATIONS OF STOICHIOMETRY =     2

   SPECIES             EQUILIBRIUM MOLES    MOLE FRACTION      DG/RT REACTION

   H2O                   7.8314153D-01      4.7797781D-01
   H2                    1.4773739D-01      9.0169136D-02
   N2                    4.8524622D-01      2.9616221D-01
   OH                    9.6876244D-02      5.9126854D-02      -3.3228D-07
   H                     4.0672719D-02      2.4823939D-02       2.9316D-08
   O2                    3.7316404D-02      2.2775466D-02      -2.3448D-12
   NO                    2.7400034D-02      1.6723169D-02      -4.4711D-07
   O                     1.7949382D-02      1.0955116D-02      -2.0911D-07
   N                     1.4143465D-03      8.6322362D-04       2.4821D-08
   NH                    6.9318773D-04      4.2307596D-04      -7.9993D-07

   G/RT =    -4.7761377D 01
   TOTAL PHASE 1 MOLES =    1.6384D 00

   ELEMENTAL ABUNDANCES        H       2.00000000D 00
                               N       1.00000000D 00
                               O       1.00000000D 00
```

Figure 6.6 Computer output for Example 6.2 from VCS algorithm in Appendix D.

Composition Variables Other Than Mole Fraction

computation time is about the same in both cases. The results calculated in this example agree with those in Example 6.1 to within five significant figures. Finally, each number below "DG/RT REACTION" gives $\Delta G/RT$ for the stoichiometric equation in which one mole of the indicated species is formed from the first three species (H_2O, H_2, and N_2) as components.

6.5 COMPOSITION VARIABLES OTHER THAN MOLE FRACTION

For the algorithms in Chapters 4 and 6, we have expressed the composition in terms of mole fraction. The computer programs in the appendixes also use mole fraction as the composition variable. This is appropriate for gaseous systems and solutions of nonelectrolytes, but for solutions of electrolytes (e.g., aqueous solutions of acids, bases, and salts), the composition is usually expressed in molality or molarity, as described in Chapter 3. In this section we describe how the algorithms must be modified to consider problems involving such systems.

For the RAND algorithm, the equations corresponding to equations 6.3-24 to 6.3-27 must be rederived. We leave this as an exercise in Problem 6.9. For the VCS algorithm, the only change that must be made in the computer program in Appendix D is to calculate the chemical potential in the appropriate way in the subroutine DFE. We illustrate how this is done by means of an example from Denbigh (1981, p. 328).

Example 6.3 Consider the system {($Cl_2(g)$, $Cl_2(\ell)$, $H^+(\ell)$, $Cl^-(\ell)$, $HClO(\ell)$, $ClO^-(\ell)$, $H_2O(\ell)$), (Cl, H, O, p)} resulting from bubbling $Cl_2(g)$ at a partial pressure of 0.5 atm through water at 25°C. Calculate the concentrations of the species in the liquid (aqueous) phase, if the solution is ideal, and the standard free energies of formation, in kJ mole^{-1}, are $\Delta \mathbf{G}_f^\circ = (0, 6.90, 0, -131.25, -79.58, -27.20, -236.65)^T$.

Solution The chemical potential of H_2O is given by $\mu[H_2O(\ell)] = \Delta G_f^\circ[H_2O(\ell)] + RT \ln x_{H_2O}$ and of each of the other species in the liquid phase by $\mu_i = \Delta G_{fi}^\circ + RT \ln m_i$; for $Cl_2(g)$, $\mu[Cl_2(g)] = \Delta G_f^\circ[Cl_2(g)] + RT \ln p_{Cl_2}$. In subroutine DFE in Appendix D three FORTRAN statements are modified as follows. Statement number 11 is replaced by

```
11 FE(I) = FF(I) + ALOG(Z(I)) − ALOG(Z(1)*0.018016D0)
   IF(I.EQ.1)FE(I) = FF(I) + ALOG(Z(I)) − Y
```

Statement number 21 is replaced by

```
21 FE(L) = FF(L) + ALOG(Z(L)) − ALOG(Z(1)*0.018016D0)
```

Statement number 31 is replaced by

31 FE(L) = FF(L) + ALOG(Z(L)) − ALOG(Z(1)*0.018016D0)

The vector **b** is defined by $n°[H_2O(\ell)] = 1000/18.016$ by choosing an arbitrarily large initial amount of $Cl_2(g)$ so that all of it does not dissolve {here we choose $n°[Cl_2(g)] = 1$} and, finally, by the electroneutrality requirement. Thus **b** = (2.0, 2000/18.016, 1000/18.016, 0)^T. The computer output from the VCS algorithm is shown in Figure 6.7. For the species in the liquid phase, the equilibrium mole numbers are virtually the molalities because of the choice of $n°[H_2O(\ell)]$.

```
VCS CALCULATION METHOD

CHLORINE-SOLUTION PROBLEM
THIS SPECIES:CL2 (G)    IS THE ONLY GAS.  IT WILL THEREFORE BE TREATED AS A SOLID.

    7 SPECIES         4 ELEMENTS               4 COMPONENTS
    0 PHASE1 SPECIES              6 PHASE2 SPECIES         1 SINGLE SPECIES PHASES

    PRESSURE                 0.500 ATM
    TEMPERATURE              298.150 K
    PHASE2 INERTS            0.0

    ELEMENTAL ABUNDANCES           CORRECT                  FROM ESTIMATE

                            CL   2.000000000000D 00    2.000000000000D 00
                            H    1.110124333925D 02    1.110120000000D 02
                            O    5.550621669627D 01    5.550600000000D 01
                            P    0.0                   1.387778780781D-17

    USER ESTIMATE OF EQUILIBRIUM
    STAN. CHEM. POT. IN KJ./MOLE

    SPECIES        FORMULA VECTOR             STAN. CHEM. POT.     EQUILIBRIUM EST.

              CL  H  O  P SI(I)
    H2O (L)    0  2  1  0   2                  -2.36650D 02         5.52060D 01
    CL2 (G)    2  0  0  0   0                  -1.71825D 00         6.00000D-01
    H+ (L)     0  1  0  1   2                   0.0                 5.00000D-01
    CL- (L)    1  0  0 -1   2                  -1.31250D 02         3.00000D-01
    HCLO (L)   1  1  1  0   2                  -7.95800D 01         1.00000D-01
    CLO- (L)   1  0  1 -1   2                  -2.72000D 01         2.00000D-01
    CL2 (L)    2  0  0  0   2                   6.90000D 00         1.00000D-01

    ITERATIONS =    10
    EVALUATIONS OF STOICHIOMETRY =      2

    SPECIES                  EQUILIBRIUM MOLES       MOLE FRACTION        DG/RT REACTION

    H2O (L)                   5.5481594D 01          9.9811525D-01
    CL2 (G)                   9.4447957D-01          1.0000000D 00
    H+ (L)                    2.4622867D-02          4.4296600D-04
    CL- (L)                   2.4622867D-02          4.4296598D-04
    CL2 (L)                   3.0897566D-03          5.5584798D-04        -3.9043D-06
    HCLO (L)                  2.4622866D-02          4.4296597D-04         8.0456D-09
    CLO- (L)                  6.6534481D-10          1.1969570D-11        -4.7070D-06

    G/RT =   -5.2997211D 03
    TOTAL PHASE2 MOLES =    5.5586D 01

    ELEMENTAL ABUNDANCES      CL    2.00000000D 00
                              H     1.11012433D 02
                              O     5.55062167D 01
                              P     7.18250180D-19
```

Figure 6.7 Computer output for Example 6.3 from VCS algorithm in Appendix D.

PROBLEMS

6.1 Derive equations 6.3-28 to 6.3-30, the RAND algorithm for a multiphase ideal system.

6.2 Show that, in the case of a multiphase ideal system, the working equations of the Brinkley algorithm, corresponding to equations 6.3-45 and 6.3-46, are

$$\sum_{i=1}^{M} \sum_{k=1}^{N'} \nu_{ik}\nu_{jk} n_k^{(m)}\delta\psi_i^{(m)} + \sum_{\alpha=1}^{\pi} q_{j\alpha}^{(m)} v_\alpha = q_j - q_j^{(m)}; \quad j=1,2,\ldots,M$$

and

$$\sum_{i=1}^{M} q_i^{(m)}\delta\psi_i^{(m)} - n_{z\alpha}v_\alpha = n_{t\alpha}^{(m)} - \sum_{k=1}^{N'} n_{k\alpha}^{(m)} - n_{z\alpha}; \quad \alpha=1,2,\ldots,\pi_s \mid \pi_m.$$

6.3 Prove equation 6.4-17.

6.4 Determine the composition at equilibrium at 4000 K and 1.5 atm of the product stream resulting from the reaction of 1 mole of CH_4 and 1 mole of N_2, based on the following standard free energies of formation at 4000 K (in kJ mole^{-1}) (JANAF, 1971):

C(gr):	0	CH_3:	236.98	H:	−15.32
C(g):	90.06	CH_4:	352.08	H_2:	0
C_2(g):	80.41	CHN:	9.991	HN:	258.66
C_3(g):	21.46	CN:	41.56	H_2N:	331.16
C_4(g):	142.13	C_2H_2:	14.46	H_3N:	411.06
C_5(g):	169.55	C_2H_4:	367.25	N:	210.77
CH:	154.41	C_2N_2:	133.62	N_2:	0
CH_2:	238.27	C_4N_2:	187.44		

6.5 Extend Problem 4.5 by considering equilibrium involving the *additional* species: C(gr), C(g), C_2(g), C_3(g), C_4(g), C_5(g), CH(g), CH_2(g), CH_3(g), C_2H(g), C_2H_2(g), C_2H_4(g), O(g), H(g), OH(g), and HO_2(g). Additional standard free energies of formation (JANAF, 1971), in the order cited, are (0, 479.87, 546.98, 497.52, 658.98, 653.79, 426.11, 325.68, 171.55, 282.87, 143.00, 160.10, 154.93, 136.61, 16.90, 92.03)^T, in kJ mole^{-1}.

6.6 Ethylene can be made in a tubular reactor by the dehydrogenation of ethane, with outlet conditions of about 1100 K and 2.0 atm. Suppose that the feed consists of steam (assume it to be inert) and ethane in the ratio 0.4 mole of steam per mole of ethane, and that the composition of the product stream on a steam-free basis is 36.0 mole % H_2, 11.7% CH_4,

0.4% C_2H_2, 31.7% C_2H_4, 16.8% C_2H_6, 1.1% C_3H_6, 0.3% C_3H_8, 1.1% C_4 fraction (assume that it is 1-C_4H_8), and 0.9% C_{5+} fraction (assume that it is C_6H_6). Examine the approach to equilibrium with respect to the entire set of species or selected subsets of them. Examine also whether the formation of carbon (gr) is possible thermodynamically. Standard free energies of formation at 1100 K in kJ mole^{-1} are as follows (Zwolinski et al., 1974):

H_2:	0	C_3H_6:	199.91
CH_4:	30.21	C_3H_8:	223.13
C_2H_2:	164.59	1-C_4H_8:	292.46
C_2H_4:	126.19	C_6H_6:	280.41
C_2H_6:	130.88		

6.7 Martin and Yachter (1951) have considered equilibrium in the system $\{(H, HCl, HF, H_2, Cl, ClF, Cl_2, F, F_2, N, N_2), (H, Cl, F, N)\}$ at 3000 K and 20.42 atm, in connection with performance calculations for a rocket fuel. Determine the composition at equilibrium at these conditions if $\mathbf{b} = (1, 0.2381, 0.7143, 0.3333)^T$ and standard free energies of formation, in kJ mole^{-1} (JANAF, 1971), are $\Delta G_f^\circ = (46.170, -112.02, -282.16, 0, -56.16, -64.23, 0, -109.58, 0, 279.00, 0)^T$.

6.8 Vonka and Holub (1975) have studied the effect of using different models (ideal-gas solution, ideal solution, and nonideal solution) on the calculation of equilibrium in the reaction of C_2H_5OH, H_2, and S_2 (initial ratio $1:1:0.5$) to form ethanethiol and diethylsulfide at 600 K and 80 atm. The first of these is used in Appendix B to illustrate sample input and output for the BASIC stoichiometric algorithm, and the third is used as a problem in Chapter 7. Here we investigate the use of the ideal solution model in conjunction with an equation of state to calculate the fugacity (actually the fugacity coefficient f/P) of each pure species at (T, P).

The Redlich-Kwong equation of state (Chapter 7) provides the following expression for the fugacity coefficient (Vonka and Holub, 1975):

$$\ln \phi = \ln\left(\frac{RT}{P(v-b)}\right) + \frac{b}{v-b} - \frac{a}{RT^{1.5}}\left[\frac{1}{v+b} + \frac{1}{b}\ln\left(\frac{v+b}{v}\right)\right], \quad (A)$$

where the two parameters a and b are given by

$$a = 0.4278(RT_C)^2 \quad (B)$$

and

$$b = 0.0867\, RT_C/P_C. \quad (C)$$

Problems

The system considered by Vonka and Holub is $\{(C_2H_5OH, H_2, S_2, C_2H_5SH, (C_2H_5)_2S, H_2S, (C_2H_5)_2O, CH_3CHO, C_2H_6, C_2H_4, H_2O), (C, H, O, S)\}$. Calculate the equilibrium mole numbers at these conditions, using the ideal-solution model and free-energy (Stull et al., 1969 and Stull and Sinke, 1956, p. 193) and critical-constant (Kobe and Lynn, 1953) data tabulated in the following list, and compare with the results given in Appendix B for the ideal-gas-solution model:

Species	ΔG_f°, kJ mole^{-1}	T_C, K	P_C, atm
C_2H_5OH	−95.52	516	63.0
H_2	0	33.3	12.8
S_2	35.26	1313	116
C_2H_5SH	44.77	499	54.2
$(C_2H_5)_2S$	130.83	557	39.1
H_2S	−12.22	373.6	88.9
$(C_2H_5)_2O$	17.53	467	35.6
CH_3CHO	−96.11	461	54.7*
C_2H_6	24.94	305.5	48.2
C_2H_4	87.53	282.4	50.5
H_2O	−214.05	647.4	218.3

*Weast, 1979–1980, p. F-88.

6.9 Derive the form of the RAND algorithm corresponding to equations 6.3-24, 6.3-26, and 6.3-27, when the chemical potentials of the solutes are given by $\mu_i = \mu_{mi}^* + RT \ln m_i$ (equation 3.7-19) and the chemical potential of the solvent, by $\mu_i = \mu_i^* + RT \ln x_i$.

6.10 Repeat Example 6.3 with the inclusion of OH$^-$ (ℓ) in the species list. For OH$^-$ (ℓ), ΔG_f° is −37.594 kJ mole^{-1} (Wagman et al., 1965–1973). Is the exclusion of OH$^-$ in Example 6.3 justified?

6.11 Calculate the species distribution (in molality or molarity) and the solubility in water at 25°C in each of the following cases (cf. Chaston, 1975):

(a) Solubility of $Mg(OH)_2$; system is $\{(Mg(OH)_2(s), Mg^{2+}(\ell), MgOH^+(\ell), H_2O(\ell), H^+(\ell), OH^-(\ell)), (Mg, H, O)\}$; $\Delta G_f^\circ = (-833.58, -454.8, -626.8, -237.18, 0, -157.29)^T$.

(b) Solubility of $Fe(OH)_2$; system is $\{(Fe(OH)_2(s), Fe^{2+}(\ell), FeOH^+(\ell), H_2O(\ell), H^+(\ell), \Delta H^-(\ell)), (Fe, H, O)\}$; $\Delta G_f^\circ = (-486.6, -78.87, -277.4, -237.18, 0, -157.29)^T$.

(c) Solubility of $Ca_3(PO_4)_2$; system is $\{(Ca_3(PO_4)_2(s, \beta), Ca^{2+}(\ell), H_3PO_4(\ell), H_2PO_4^-(\ell), HPO_4^{2-}(\ell), PO_4^{3-}(\ell), H_2O(\ell), H^+(\ell), OH^-(\ell)), (Ca, P, H, O)\}$; $\Delta G_f^\circ = (-3884.8, -553.54, -1119.2, -1088.6, -1089.3, -1018.8, -237.18, 0, -157.29)^T$.

(d) Solubility of $CaCO_3$; system is {($CaCO_3$(s, calcite), $CaCO_3(\ell)$, $H_2CO_3(\ell)$, $HCO_3^-(\ell)$, $CO_3^{2-}(\ell)$, $Ca^{2+}(\ell)$, $CO_2(\ell)$, $H_2O(\ell)$, $H^+(\ell)$, $OH^-(\ell)$), (Ca, C, H, O)}; $\Delta G_f^\circ = (-1128.8, -1081.4, -623.2, -586.8, -527.9, -553.54, -385.0, -237.18, 0, -157.29)^T$.

Note: The standard free energies of formation are in kJ mole^{-1}, and for dissolved species indicated by (ℓ), other than $H_2O(\ell)$, refer to the infinitely dilute standard state usually denoted by (aq). Data are from Wagman et al. (1965–1973).

6.12 Consider the system described in Problem 4.10 with the additional species $C_2H_5OH(g)$, $CH_3COOH(g)$, $CH_3COOC_2H_5(g)$, and $H_2O(g)$. Calculate the equilibrium composition at 358 K and 0.9 atm with the assumption that both phases are ideal (vapor phase is an ideal-gas solution, and liquid phase is an ideal solution). (We note that the assumption is not a good one for the liquid phase, as indicated by the existence of a ternary azeotrope involving ethyl alcohol, ethyl acetate, and water.) At 358 K the vapor pressures of the four substances are 1.286, 0.327, 1.299, and 0.567 atm, respectively.

6.13 *Suppose that the product from a crude styrene unit consists of 2 mole % benzene (C_6H_6), 3% toluene (C_7H_8), 45% styrene (C_8H_8), and 50% ethylbenzene (C_8H_{10}) and enters a vacuum distillation column for separation between toluene and ethylbenzene. If the stream is at 30°C and 0.015 atm, what is the composition of each of the two phases (liquid and vapor) present? At 30°C the vapor pressures are 0.1570, 0.0482, 0.0166, and 0.0109 atm, respectively. Assume that the vapor phase is an ideal-gas solution, that the liquid phase is an ideal solution, and that only phase equilibrium is involved. (In solving this problem, consider the implications of the restriction to phase equilibrium with regard to free-energy data for the individual species and an appropriate formula matrix for the system, as discussed in Section 2.4.5.)

---

*Because of the assumptions made, this problem can be reduced to the solution of one nonlinear equation in one unknown. Thus it does not require an elaborate algorithm for its solution. However, it illustrates how such a problem can be solved by a general procedure, and if the phases were nonideal (see Chapter 7), the reduction could not be achieved.

CHAPTER SEVEN

Chemical Equilibrium Algorithms for Nonideal Systems

In Chapters 4 and 6 we presented algorithms for systems involving phases that are either pure species or ideal solutions, including the special case for the latter of ideal-gas solutions. In this chapter we see how the general-purpose algorithms presented in Chapter 6 may be adapted for use when the assumption of ideal-solution behavior is not appropriate. We first discuss in general terms the conditions and types of system for which nonideal behavior must be taken into account. We then present further comments on the determination and representation of the chemical potential of a species in a nonideal solution, as a continuation of Section 3.7; finally, we consider the basic structure of appropriate algorithms, presenting three approaches to the problem.

7.1 THE TRANSITION FROM IDEALITY TO NONIDEALITY

As has been emphasized in previous chapters, to solve the equations expressing the conditions for equilibrium, we must have an appropriate expression for the chemical potential of each species that relates it to composition, in addition to temperature and pressure.

The chemical potential for a species in an ideal solution given, for example, by

$$\mu_i(T, P, x_i) = \mu_i^*(T, P) + RT \ln x_i, \qquad (3.7\text{-}15a)$$

depends only on (the measure of) its own composition (x_i in equation 3.7-15a) and not on the composition of other species in the solution. This applies regardless of whether ideality is based on the Raoult convention or the Henry convention and regardless of the particular variable used to express composition. This makes possible the construction of algorithms for the calculation of equilibrium whose relatively simple forms are due to the fact that $\partial \mu_i / \partial n_j$ can be written as a simple analytical expression.

The chemical potential for a species in a nonideal solution given, for example, by

$$\mu_i(T, P, \mathbf{x}) = \mu_i^*(T, P) + RT \ln \gamma_i(T, P, \mathbf{x}) x_i, \qquad (3.7\text{-}29)$$

depends on composition in general, as reflected in the dependence of the activity coefficient γ_i. This dependence may be complex and difficult to represent even when considerable experimental information is available [see Prausnitz (1969) for an extensive discussion of the phenomenological behavior and treatment of activity coefficients]. In principle, the accurate prediction of the compositional dependence of the chemical potential of a species is a problem in statistical thermodynamics. It is only in relatively recent years that progress has been made in the statistical mechanics of fluids, for example, and such approaches are just beginning to be used in the treatment of real fluids (Rowlinson, 1969; Reed and Gubbins, 1973).

Although we do not distinguish between phase equilibrium and reaction equilibrium, as the terms are commonly used, we note that much of the work devoted to the treatment of nonideal behavior has been done in the context of single phases and phase equilibrium, without the consequences of "chemical reaction" being taken into account. Relatively little attention has been paid to the *general* problem of determining chemical equilibrium (both intra- and interphase) in systems made up of nonideal solutions.

In considering the breakdown of ideal behavior as an appropriate assumption, we should distinguish between the transitions (1) from ideal-gas to non-ideal-gas behavior and (2) from ideal-solution to non-ideal-solution behavior. The former occurs as the density of the gas increases from a relatively low value, as a result of either increasing pressure, decreasing temperature or both. Even at relatively high density, however, a non-ideal-gas mixture may be essentially an ideal solution. It is in liquid and solid solutions that we must be most conscious of the likelihood of nonideal, rather than ideal, solution behavior. In qualitative terms, the key to this likelihood lies in the loosely defined term "chemical similarity." For example, a solution of chemically similar pentane and hexane, which are adjacent members of an homologous series of hydrocarbons, may be considered to be virtually ideal, but if one of the two is replaced by the dissimilar species methyl alcohol, the resulting solution is very nonideal (Tenn and Missen, 1963).

For nonideal solutions, since μ is often a very complex function of composition, this results, in turn, in complex expressions for $\partial \mu_i / \partial n_j$. This complexity destroys the relatively simple forms of the algorithms obtained for ideal systems in Chapters 4 and 6.

Before examining the structure of algorithms for nonideal systems, we consider further, following Section 3.7, the representation of the chemical potential for nonideal systems.

7.2 FURTHER DISCUSSION OF CHEMICAL POTENTIALS IN NONIDEAL SYSTEMS

In this section we amplify the very brief comments given in Section 3.7.2. The chemical potential of a species in a solution is determined ultimately by the nature of the intermolecular forces among the molecules. All thermodynamic properties may be calculated in principle from these forces by the methods of statistical mechanics (Reed and Gubbins, 1973). The difficulties are formidable, however, in the present state of knowledge. Not only is the precise nature of these forces usually unknown, but also, even given such knowledge, the exact numerical calculation of the properties is often impossible. Any reasonably accurate solutions to this problem must involve approximations in terms of both these aspects.

In face of these difficulties, most chemical potential information has been obtained from macroscopic experimental data, guided, in the sense of correlation and prediction, where possible, by the more fundamental approach, which attempts to solve the statistical mechanical problem approximately for approximate intermolecular potential models. We outline three approaches: use of excess free-energy expressions, equations of state, and corresponding states theory. We then consider separately the case of electrolytes.

7.2.1 Use of Excess Free-Energy Expressions

For liquid solutions of nonelectrolytes, chemical-potential information is commonly given in terms of the molar excess free energy (g^E) of the solution or the activity coefficient of each species (see Section 3.7.2 for the definition of an excess function). The former provides a convenient summary for all species, and the interrelationships are as follows:

$$\mu_i^E = RT \ln \gamma_i, \tag{7.2-1}$$

where μ_i^E is the excess chemical potential of species i and

$$g^E = \sum_{i=1}^{N} x_i \mu_i^E. \tag{7.2-2}$$

The activity coefficient may be calculated from g^E by means of an equation analogous to equation 3.7-34:

$$RT \ln \gamma_i = \mu_i^E = g^E - \sum_{j \neq 1} x_j \left(\frac{\partial g^E}{\partial x_j} \right)_{T, P, x_{k \neq j}}. \tag{7.2-3}$$

The compositional dependence of g^E or γ_i is often given by means of an empirical or semiempirical correlation of experimental data. The temperature and pressure dependence and the Gibbs-Duhem relation are given by equations analogous to equations 3.2-10 to 3.2-17. We consider some of the commonly used correlations for g^E for binary systems; the extension to multispecies systems may have to be done on an ad hoc basis. More elaborate methods, not described here, are used by Prausnitz et al. (1980) in computing vapor-liquid and liquid-liquid equilibria; see also Skjold-Jørgensen et al. (1982).

7.2.1.1 Power-Series Expansion of $g^E/x_1 x_2$

An example of the power-series expansion of $g^E/x_1 x_2$ is given by the equation of Redlich and Kister (1948):

$$\frac{g^E}{x_1 x_2} = \sum_{k \geq 0} a_k(T, P)(x_1 - x_2)^k, \qquad (7.2\text{-}4)$$

where the a_k's are parameters determined from experimental data and x_1 and x_2 are the mole fractions of species 1 and 2, respectively. Application of equation 7.2-3 to equation 7.2-4 results in the power-series expansions of $\ln \gamma_1$ and $\ln \gamma_2$ that are due to Margules (1895).

7.2.1.2 Power-Series Expansion of $(g^E/x_1 x_2)^{-1}$

The reciprocal of $g^E/x_1 x_2$ may also be represented by a power-series expansion (Van Ness, 1959; Otterstedt and Missen, 1962):

$$\left(\frac{g^E}{x_1 x_2}\right)^{-1} = \sum_{k \geq 0} b_k(T, P)(x_1 - x_2)^k, \qquad (7.2\text{-}5)$$

where the b_k's are parameters determined from experimental data. The first-order form of this leads to the van Laar equations for activity coefficients (van Laar, 1910) on application of equation 7.2-3.

7.2.1.3 The Wohl Expansion

The equation of Wohl (1946) is

$$\frac{g^E}{RT(x_1 q_1 + x_2 q_2)} = 2a_{12} z_1 z_2 + 3a_{112} z_1^2 z_2 + 3a_{122} z_1 z_2^2$$

$$+ 4a_{1112} z_1^3 z_2 + 4a_{1222} z_1 z_2^3 + 6a_{1122} z_1^2 z_2^2 + \cdots, \qquad (7.2\text{-}6)$$

Further Discussion of Chemical Potentials in Nonideal Systems

where q is an effective volume parameter,

$$z_1 = \frac{x_1 q_1}{x_1 q_1 + x_2 q_2}, \tag{7.2-7}$$

$$z_2 = \frac{x_2 q_2}{x_1 q_1 + x_2 q_2}, \tag{7.2-8}$$

and the a's are interaction parameters, the subscripts to which indicate the nature and number of molecules involved in a particular interaction. Both the Margules and van Laar equations can be obtained as special cases of the Wohl equation. The equation can also be extended to multispecies systems.

7.2.1.4 The Wilson Equation

The equation given by Wilson (1964) is

$$\frac{g^E}{RT} = -x_1 \ln(x_1 + A_{12} x_2) - x_2 \ln(x_2 + A_{21} x_1), \tag{7.2-9}$$

where

$$A_{12} = \frac{v_2}{v_1} \exp\left[-\frac{(\lambda_{12} - \lambda_{11})}{RT}\right], \tag{7.2-10}$$

$$A_{21} = \frac{v_1}{v_2} \exp\left[-\frac{(\lambda_{12} - \lambda_{22})}{RT}\right], \tag{7.2-11}$$

and v_1 and v_2 are the molar volumes of pure (liquid) species 1 and 2, respectively, and the λ's are interaction energies. This equation can also be extended to multispecies systems.

7.2.1.5 The Regular-Solution Equation

The concept of a regular solution (Hildebrand et al., 1970) provides the following expression for g^E:

$$g^E = v \phi_1 \phi_2 A_{12}, \tag{7.2-12}$$

where

$$v = x_1 v_1 + x_2 v_2, \tag{7.2-13}$$

ϕ_1 and ϕ_2 are volume fractions of species 1 and 2, respectively, with, for example,

$$\phi_1 = \frac{x_1 v_1}{v} \quad (7.2\text{-}14)$$

and

$$A_{12} = (\delta_1 - \delta_2)^2. \quad (7.2\text{-}15)$$

δ is the solubility parameter and is formally defined as the square root of the cohesive-energy density of the species; thus, for species 1,

$$\delta_1 = \left(\frac{\Delta u_1}{v_1}\right)^{1/2} \simeq \left(\frac{\Delta H_1^{\text{vap}} - RT}{v_1}\right)^{1/2}, \quad (7.2\text{-}16)$$

where Δu is the cohesive energy, which is approximately equal to the energy of vaporization ($\Delta H^{\text{vap}} - RT$).

7.2.2. Use of Equations of State

The fugacity f_i of a species in a solution, whether gas, liquid, or solid, can be determined, in principle, from $PvT\mathbf{x}$ (volumetric) information by means of equation 3.7-32 (Prausnitz, 1969, p. 30),

$$RT \ln \frac{f_i}{x_i P} = \int_0^P \left(\bar{v}_i - \frac{RT}{P}\right) dP. \quad (3.7\text{-}22)$$

The activity coefficient γ_i of the species is related to the fugacity by (Van Ness, 1964, p. 31)

$$\gamma_i = \frac{f_i}{x_i f_i^*}, \quad (7.2\text{-}17)$$

where f_i^*, the fugacity of species i in the standard state, is similarly determined, for example, from equation 3.7-7. It follows that γ_i can be determined from volumetric data by, in the case of a gaseous system,

$$\gamma_i = \exp\left[\frac{1}{RT} \int_0^P (\bar{v}_i - v_i^*) \, dP\right], \quad (7.2\text{-}18)$$

where v_i^* is the molar volume of species i in the standard state [for the Raoult convention, it is the pure-component molar volume at (T, P), and for the Henry convention, it is the partial molar volume at infinite dilution at (T, P)]. Thus, in principle, it is possible to determine the chemical potential of a species

Further Discussion of Chemical Potentials in Nonideal Systems 159

in solution from volumetric data for the solution since

$$\mu_i = \mu_i^* + RT \ln \gamma_i x_i. \tag{3.7-29}$$

This procedure is limited by the requirement that the volumetric data be available as a function of T, P, and $\mathbf{x}$. This, in turn, requires either a very large amount of experimental data or an appropriate equation of state,

$$\phi(P, v, T, \mathbf{x}) = 0, \tag{7.2-19}$$

the form of which must be obtained from experimental data or a theoretical model.

If ϕ is volume explicit, it is convenient to use equation 3.7-32. If ϕ is pressure explicit, it is more convenient to use the equivalent form (Prausnitz, 1969, p. 41)

$$RT \ln \frac{f_i}{x_i P} = \int_v^\infty \left[\left(\frac{\partial P}{\partial n_i} \right)_{T, V, n_{j \neq i}} - \frac{RT}{V} \right] dV - RT \ln z, \tag{7.2-20}$$

where z is the compressibility factor defined by

$$z = \frac{PV}{n_t RT}. \tag{7.2-21}$$

In spite of the limitations noted, we briefly outline three of the most widely used equations of state.

7.2.2.1 The Virial Equation

The pressure-explicit (Leiden) form of the virial equation is

$$z = \frac{Pv}{RT} = 1 + \frac{B(T, \mathbf{x})}{v} + \frac{C(T, \mathbf{x})}{v^2} + \cdots, \tag{7.2-22}$$

where $B, C, \ldots$ are called the second, third, $\ldots$ virial coefficients. The compositional dependence of the nth virial coefficient is a generalized nth-degree linear form in the mole fractions (Mason and Spurling, 1969, p. 57). That is,

$$B = \sum_{i,j} x_i x_j B_{ij} \tag{7.2-23}$$

and

$$C = \sum_{i,j,k} x_i x_j x_k C_{ijk}. \tag{7.2-24}$$

Although $B_{ij}, C_{ijk}, \ldots$ are determinable in principle from intermolecular potential functions, these are rarely known accurately except for the simplest of molecules. The virial coefficients for pure species can be obtained experimentally from volumetric data, but those for species in solutions are rarely available. It is thus usually necessary to postulate "mixing rules" relating the virial coefficients of the solution to those of the pure species (Mason and Spurling, 1969, pp. 257–265).

Equation 7.2-22 may be inverted to give the volume-explicit (Berlin) form of the virial equation (Putnam and Kilpatrick, 1953).

The main disadvantage of the virial equation is its inapplicability to high densities (and hence to liquids).

7.2.2.2 The Redlich-Kwong Equation

The equation of Redlich and Kwong (1949) is a two-parameter, pressure-explicit form:

$$z = \frac{Pv}{RT} = \frac{v}{v-b} - \frac{a}{RT^{3/2}(v+b)}, \qquad (7.2\text{-}25)$$

where a and b are the two parameters. There are no rigorous expressions relating these parameters to x corresponding to equations 7.2-23 and 7.2-24, and arbitrary mixing rules must be used (Redlich and Kwong, 1949).

7.2.2.3 The Benedict-Webb-Rubin Equation

The equation of Benedict, Webb, and Rubin (1940, 1942) is also in pressure-explicit form and requires arbitrary mixing rules; it contains eight parameters:

$$P = \frac{RT}{v} + \frac{B_0 RT - A_0 + C_0/T^2}{v^2} + \frac{bRT - a}{v^3}$$

$$+ \frac{a\alpha}{v^6} + \frac{c}{T^2 v^3} \frac{1+\gamma}{v^2} \exp\left(\frac{-\gamma}{v^2}\right). \qquad (7.2\text{-}26)$$

This equation has been widely used for liquid-vapor equilibrium in light hydrocarbon systems (Benedict et al., 1951), and values of the parameters are available for a number of species (Holub and Vonka, 1976, Appendix 11).

7.2.3 Use of Corresponding States Theory

The theory of corresponding states provides an alternative, generally less precise, approach to determining information about chemical potentials from volumetric data [for a review, see Mentzer et al. (1980)]. This approach is often useful when insufficient data are available to use the methods of the previous two sections. The theory has a rigorous statistical mechanical basis for simple species with spherically symmetrical molecular force fields, such as argon and krypton (Pitzer, 1939). For these, in its simplest (two-parameter) macroscopic form, the theory may be written as

$$z_i(T_R, P_R) = z_0(T_R, P_R), \qquad (7.2\text{-}27)$$

where z_i is the compressibility factor of any species i, and z_0 is that of a reference species; T_R and P_R are respectively the reduced temperature and pressure defined by

$$T_R = \frac{T}{T_C}, \qquad P_R = \frac{P}{P_C}, \qquad (7.2\text{-}27\text{a})$$

where T_C and P_C are the critical temperature and pressure, respectively, of the species. Equation 7.2-27 expresses the idea that any two species behave the same volumetrically at the same reduced conditions of temperature and pressure. This is only approximate for most species, however, but it can be improved by the addition of a third parameter. Thus Pitzer (1955) and Pitzer et al. (1955) have introduced the *acentric factor* ω as a measure of the departure from spherical symmetry and defined empirically by

$$\omega = -\log_{10}\left(\frac{p^*}{P_C}\right)_{T_R = 0.7}. \qquad (7.2\text{-}28)$$

Tables of values of ω and derived thermodynamic quantities for pure species ("normal" fluids) are given by Lewis and Randall (1961, pp. 605–629). Normal fluids are essentially nonpolar, and for polar species, an additional parameter must be introduced.

The theory of corresponding states may be extended to solutions in several ways. One way is to assume that the properties of the solution are those of a hypothetical fluid characterized by critical constants that depend in some way on the critical constants of the species in the solution. This may be called a *one-fluid model* of the solution. Two- and three-fluid models may also be used in which the properties of the solution are determined from averages of the properties of two and three hypothetical species, respectively (Scott, 1956).

The implementation of a one-fluid model, to which we confine attention, requires equations expressing the properties of the hypothetical fluid in terms of those of the (pure) species of the solution. The first and simplest of these was suggested by Kay (1936) for the critical constants:

$$T_C = \sum_i x_i T_{Ci} \qquad (7.2\text{-}29)$$

and

$$P_C = \sum_i x_i P_{Ci}, \qquad (7.2\text{-}30)$$

to which may be added

$$\omega = \sum_i x_i \omega_i. \qquad (7.2\text{-}31)$$

Equations 7.2-29 and 7.2-30 are known as *Kay's rule for pseudocritical constants*. It is not necessary that the equations be of this form, and other expressions have been postulated. For example, Leland et al. (1962) have proposed, in terms of T_C and v_C, the critical volume, rather than T_C and P_C,

$$T_C v_C = \sum_{i,j} x_i x_j T_{Cij} v_{Cij} \qquad (7.2\text{-}32)$$

and

$$v_C = \sum_{i,j} x_i x_j v_{Cij}, \qquad (7.2\text{-}33)$$

where

$$T_{Cii} = T_{Ci}, \qquad (7.2\text{-}34)$$

$$v_{Cii} = v_{Ci}, \qquad (7.2\text{-}35)$$

$$v_{Cij}^{1/3} = \tfrac{1}{2}\left(v_{Cii}^{1/3} + v_{Cjj}^{1/3}\right), \qquad (7.2\text{-}36)$$

and

$$T_{Cij} = \xi_{ij}\left(T_{Cii} T_{Cjj}\right)^{1/2} (\xi_{ij} \leq 1). \qquad (7.2\text{-}37)$$

Once the critical properties of the hypothetical fluid are determined, the compressibility factor is determined in the usual way from equation 7.2-27 or its equivalent, augmented by the acentric factor. The equations due to Leland et al. (1962) are not completely arbitrary since they have some justification from statistical mechanics (Leland et al., 1968).

7.2.4 Electrolyte Solutions

The correlation and prediction of activity coefficients of electrolytes in solution is perhaps an even more difficult task than that for nonelectrolytes, to which the previous three sections have primarily been directed. Caution (cf. Nordstrom et al., 1979) is required in selecting, interpreting, and applying appropriate equations for single ions and electrolytes, let alone mixed electrolytes. The concept of excess free energy has been applied to solutions of electrolytes, but

Further Discussion of Chemical Potentials in Nonideal Systems

it is not necessarily defined (Harned and Robinson, 1968, pp. 10, 33) in the same way as for solutions of nonelectrolytes (Section 3.7.2). These remarks notwithstanding, we attempt a brief description of methods for estimating activity coefficients for single electrolytes, but original sources should be consulted for greater detail and for mixed electrolytes.

As an empirical extension of the Debye-Hückel limiting law, the following equation has been provided by Davies (1962) for the mean-ion activity coefficient of a single electrolyte in dilute aqueous solution at 25°C:

$$-\log_{10}\gamma_\pm = 0.5\,|z_+ z_-|\left(\frac{I^{0.5}}{1+I^{0.5}} - 0.30I\right), \qquad (7.2\text{-}38)$$

where z_+ and z_- are the cation and anion charges, respectively, and I is the ionic strength of the solution, defined by

$$I = 0.5 \sum_i m_i z_i^2. \qquad (7.2\text{-}39)$$

Equation 7.2-38 is intended to provide a mean-ion activity coefficient that takes ion association into account. For uni-univalent electrolytes, equation 7.2-38 predicts a value of $\gamma_\pm = 0.785$ at $m = 0.1$; experimental values for 50 electrolytes were shown to agree with a mean deviation of less than 2%. For 40 uni-bivalent and bi-univalent electrolytes, the mean deviation from the calculated value of 0.545 was about 4%. For several bi-bivalent electrolytes, the agreement is less satisfactory, and the equation may be limited in its use for these to concentrations of less than about 0.05 m. The relation between mean-ion activity coefficients that do and do not take association into account is given by Davies (1962). Since the concentrations at which the equation is valid are relatively low, activity coefficients calculated from equation 7.2-38 may be used either on a molality or on a molarity basis.

For higher concentrations of a single electrolyte, Bromley (1973) has presented a correlation, which for 25°C becomes

$$\log_{10}\gamma_\pm = -\frac{0.511\,|z_+ z_-|\,I^{0.5}}{1+I^{0.5}} + \frac{(0.06 + 0.6B)\,|z_+ z_-|\,I}{(1 + 1.5I/|z_+ z_-|)^2} + BI,$$

$$(7.2\text{-}40)$$

where B is a parameter, values of which are given by Bromley for many electrolytes at 25°C.

Representation of activity coefficients for mixed electrolytes has been considered, for example, by Meissner and Kusik (1972), Bromley (1973), and Pitzer and Kim (1973); and has been reviewed by Gautam and Seider (1979).

7.3 ALGORITHMS FOR NONIDEAL SYSTEMS

We describe three classes of method for performing equilibrium calculations in nonideal systems. The first of these consists of "indirect" methods based on algorithms for ideal systems, which are well developed and hence serve as points of departure for algorithms for nonideal systems. The second class consists of "direct" methods, which consider the nonideality explicitly from the outset and whose algorithms are derived in a manner similar to the one that has been used for ideal systems in Chapter 6. The third class is intermediate between the first two and consists of approaches that use the same working equations of the ideal-system algorithms but use the appropriate nonideal values of the chemical potentials. We consider each of these three classes in turn.

7.3.1 Indirect Methods Based on Algorithms for Ideal Systems

An indirect method, first suggested by Brinkley (1947), has been used by Fickett (1963, 1976) and Cowperthwaite and Zwisler (1973) in calculating the detonation properties of explosives. Vonka and Holub (1975) have also used it in computing equilibrium compositions of real gaseous systems.

The approach is based on the fact that equation 3.7-29 may be written as

$$\mu_i = \mu_i^* + RT \ln \gamma_i(T, P, \mathbf{n}) + RT \ln x_i. \tag{7.3-1}$$

The first two terms are combined, and the equation is formally rewritten as

$$\mu_i = \mu_i^*[T, P, \mathbf{n}^*(T, P)] + RT \ln x_i, \tag{7.3-2}$$

where μ_i^* is now a function of T and P through the (unknown) equilibrium solution $\mathbf{n}^*$. Equation 7.3-2 is written in the ideal-solution form for the chemical potential (equation 3.7-15a). The calculation procedure is an iterative one, in which the first step is to compute the equilibrium composition assuming ideality ($\gamma_i = 1$), yielding a first approximation to the system mole numbers $\mathbf{n}^{(1)}$. Then the activity coefficients γ for the nonideal system are computed from a known chemical potential expression at this composition $\mathbf{n}^{(1)}$. In the next step the equilibrium composition in the "ideal" system is computed from equation 7.3-2, with μ_i^* replaced by

$$\mu_i^{*(1)} = \mu_i^* + RT \ln \gamma_i(T, P, \mathbf{n}^{(1)}). \tag{7.3-3}$$

That is, we assume that γ_i remains fixed at $\gamma_i^{(1)}$. This yields a second approximation $\mathbf{n}^{(2)}$. The procedure is repeated, and equation 7.3-3 is replaced in

Algorithms for Nonideal Systems

general by

$$\mu_i^{*(m)} = \mu_i^* + RT\ln\gamma_i(T, P, \mathbf{n}^{(m)}); \quad m = 1, 2, 3, \ldots, \quad (7.3\text{-}3a)$$

until the composition on successive iterations remains constant to within some specified tolerance. The procedure is illustrated schematically in the flow chart shown in Figure 7.1.

This procedure can be used in conjunction with any ideal-system calculation method, such as the BNR or VCS algorithm in Chapter 6. Folkman and Shapiro (1968) have given a set of sufficient conditions on the γ_i for such a scheme to produce a decreasing sequence of free-energy values that converges

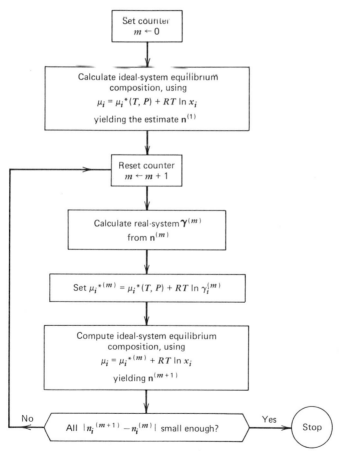

Figure 7.1 Flow chart for calculating equilibrium composition in nonideal system by using ideal-system algorithm in an iterative procedure.

on the unique solution to the problem (provided that this exists). The iteration procedure may become subject to convergence difficulties, however, if the deviations from ideality are large.

A simple approach to alleviate convergence difficulties is the parameter-variation technique discussed in Chapter 5. We write

$$\mu_i = \mu_i^* + \alpha RT \ln \gamma_i + RT \ln x_i, \quad (7.3\text{-}4)$$

in which we regard α as a parameter that is zero in the ideal system and unity in the nonideal system. The technique is to calculate equilibrium compositions by the method illustrated in Figure 7.1 for a sequence of values $\{\alpha^{(m)}\}$ corresponding to a sequence of hypothetical nonideal systems with

$$\mu_i = \mu_i^* + \alpha^{(m)} RT \ln \gamma_i + RT \ln x_i. \quad (7.3\text{-}5)$$

The value of $\alpha^{(m)}$ is changed gradually from zero to unity. At each step the equilibrium composition for $\alpha = \alpha^{(m)}$ is used as the initial estimate of the solution for the calculation at $\alpha = \alpha^{(m+1)}$. The important practical problem in the implementation of this approach is, of course, a wise choice of the sequence $\{\alpha^{(m)}\}$.

Example 7.1 Calculate the composition (molalities) at equilibrium at 25°C for the aqueous system $\{(H_3PO_4, H_2PO_4^-, HPO_4^{2-}, PO_4^{3-}, CaHPO_4, CaH_2PO_4^+, CaPO_4^-, Ca^{2+}, CaOH^+, H_2O, H^+, OH^-), (H, P, O, Ca, p)\}$ (cf. Feenstra, 1979). The value of **b** is based on 0.0009 mole of H_3PO_4 and 0.0015 mole of $CaHPO_4$ dissolved in 1 kg of water. The standard free energies of formation of the species are $\Delta G_f^\circ / RT = (-451.49, -439.12, -439.40, -410.98, -662.70, -665.66, -649.16, -223.30, -289.80, -95.677, 0, -63.452)^T$. The data are from Wagman et al. (1965–1973), with the exception of the values for $CaH_2PO_4^+$ and $CaPO_4^-$, which are calculated from information given by Feenstra.

Solution In this system, as for many electrolyte systems, $C \neq M$. We discuss such problems in general in Section 9.3, and we note here that, in solving this problem, we may ignore any row of **A**, provided that $C = \text{rank}\,(\mathbf{A}, \mathbf{b})$.

The solution follows the flow chart in Figure 7.1. For the activity coefficients, it is assumed that the Davies equation (equation 7.2-38) is valid for the individual ionic species and that neutral species are ideal. The ideal-solution values and results for the first three iterations for the nonideal solution are shown in the following tabular list (the fourth iteration gives the same values as the third). The number of moles of H_2O and the ionic strength I, calculated from equation 7.2-39, are given below the molalities. The charge balance (not shown in the list) is satisfied to within 10^{-14} on a molality basis.

Algorithms for Nonideal Systems

Species	Molality			
	Ideal Solution	Iteration 1	Iteration 2	Iteration 3
$H_3PO_4 \times 10^4$	3.817	3.055	3.045	3.044
$H_2PO_4^- \times 10^6$	1.564	1.304	1.301	1.300
$HPO_4^{2-} \times 10^3$	2.014	2.092	2.093	2.093
$PO_4^{3-} \times 10^{12}$	0.881	1.364	1.372	1.372
$CaHPO_4 \times 10^6$	3.014	1.533	1.519	1.519
$CaH_2PO_4^+ \times 10^8$	5.996	3.500	3.474	3.473
$CaPO_4^- \times 10^9$	3.825	2.027	2.010	2.010
$Ca^{2+} \times 10^3$	1.497	1.498	1.498	1.498
$CaOH^+ \times 10^{13}$	3.082	2.248	2.239	2.239
$H^+ \times 10^3$	1.035	1.188	1.190	1.190
$OH^- \times 10^{12}$	9.771	1.018	1.019	1.019
H_2O, moles	55.5062	55.5062	55.5062	55.5062
$I \times 10^3$	7.5393	7.7746	7.7778	7.7778

7.3.2 Direct Methods

We consider here the structure of algorithms that attack the problem taking into account nonideality from the outset. There are three types of approach: (1) first-order methods; (2) second-order methods; and (3) quasi-Newton or variable metric methods (Powell, 1980). As for ideal-system algorithms, basic differences also result from whether they are constructed as nonstoichiometric algorithms or as stoichiometric algorithms. In the literature these have usually been developed in the context of a specific form of chemical potential. For this reason, the general features of the algorithms are somewhat obscured, and we elaborate each of these types in the following.

7.3.2.1 First-Order Methods

In these methods only μ_i itself is used. Any of the ideal-system algorithms discussed in Chapter 6 that do not use compositional derivatives of μ_i remain relatively unchanged for nonideal systems. The only difference is that the appropriate nonideal model for μ_i is used instead of the ideal-solution form. The stoichiometric algorithm given by Naphtali (1959, 1960, and 1961), the nonstoichiometric algorithm due to Storey and van Zeggeren (1964), and the gradient-projection algorithm given in Section 6.3.1.1 are examples of this type.

7.3.2.2 Second-Order Methods

In methods of this type $\partial \mu_i / \partial n_j$ is used explicitly. We consider both the nonideal versions of the second-order nonstoichiometric and stoichiometric algorithms presented in Chapter 6.

The nonideal versions of the nonstoichiometric Brinkley-NASA-RAND (BNR) algorithm result from direct consideration of the equilibrium conditions (equations 3.5-3 and 3.5-4) and follow the derivation of Chapter 6 up to the point where the ideal-solution model for the chemical potentials is invoked. The nonideal version of the RAND variation is obtained by employing the Newton-Raphson method in equations 3.5-3 and 3.5-4. This gives (cf. Boynton, 1963; Michels and Schneiderman, 1963; Zeleznik and Gordon, 1966; and Gautam and Seider, 1979)

$$-\frac{1}{RT}\sum_{j=1}^{N'}\left(\frac{\partial\mu_i}{\partial n_j}\right)_{\mathbf{n}^{(m)}}\delta n_j^{(m)} + \sum_{k=1}^{M}a_{ki}\delta\psi_k^{(m)} = \frac{\mu_i^{(m)}}{RT} - \sum_{k=1}^{M}a_{ki}\psi_k^{(m)};$$

$$i = 1, 2, \ldots, N' \quad (6.3\text{-}16)$$

and

$$\sum_{j=1}^{N'} a_{kj}\delta n_j^{(m)} = b_k - b_k^{(m)}; \quad k = 1, 2, \ldots, M, \quad (6.3\text{-}19)$$

where

$$b_k^{(m)} = \sum_{j=1}^{N'} a_{kj} n_j^{(m)}; \quad k = 1, 2, \ldots, M. \quad (6.3\text{-}20)$$

The quantities $\mathbf{n}^{(m)}$ and $\boldsymbol{\psi}^{(m)}$ are estimates of the solution at iteration m. Equations 6.3-16 and 6.3-19 are a set of $(N' + M)$ linear algebraic equations in as many unknowns, which must be solved on each iteration of the method. For an ideal solution, this number may be reduced to M, as in Chapter 4, or to $(M + 1)$, as in Chapter 6, but this is not possible in general.

The nonideal version of the stoichiometric algorithm in Section 6.4.3 is based on consideration of equations 3.4-5. On each iteration, the linear equations resulting from the Newton-Raphson method,

$$\sum_{l=1}^{R}\delta\xi_l^{(m)}\sum_{i=1}^{N'}\sum_{k=1}^{N'}\nu_{ij}\nu_{kl}\left(\frac{\partial\mu_i}{\partial n_k}\right)_{\mathbf{n}^{(m)}} = -\sum_{i=1}^{N'}\nu_{ij}\mu_i^{(m)}; \quad j = 1, 2, \ldots, R,$$

$$(7.3\text{-}6)$$

(cf. equation 4.3-2) must be solved for the reaction-adjustment parameters $\delta\xi^{(m)}$. Equation 7.3-6 consists of a total of $R = (N' - M)$ equations, as opposed to $(N' + M)$ in the case of equations 6.3-16 and 6.3-19. Thus equation 7.3-6 is to be preferred. This should be contrasted with the situation discussed in Chapter 6, where the ideal-solution form of equations 6.3-16 and 6.3-19 is preferable for systems with relatively small numbers of elements and phases.

A number of stoichiometric algorithms have been developed since about 1970 for computing equilibrium in aqueous systems. These have been reviewed by Nordstrom et al. (1979), who compare a number of them and provide an extensive bibliography, mostly from the field of geochemistry. These algorithms generally use the Newton-Raphson or a related method to solve the nonlinear equations 3.4-5. The computer programs are rather specialized for this particular application and usually have thermodynamic data files as internal components.

7.3.2.3 Quasi-Newton or Variable Metric Methods

Essentially, numerical information about μ_i is used on successive iterations to construct approximations to $\partial \mu_i / \partial n_j$ in these methods. This type of approach has been used to solve chemical equilibrium problems by George et al. (1976) and Castillo and Grossman (1979).

7.3.3 Intermediate Methods Based on Algorithms for Ideal Systems

A third class of algorithm, intermediate between the first two, consists of using the working equations of an ideal-system algorithm, except that the chemical potential is replaced by its nonideal value in the calculation procedure (e.g., Eriksson and Rosen, 1973). This amounts to using the ideal-solution values for the compositional derivatives of μ and the nonideal values for μ itself.

7.3.4 Discussion

Computational experience with the three classes of method described in this section is rather limited. Given this, we believe that the indirect and intermediate classes appear to be most useful since they can be employed in conjunction with any available ideal-system algorithm. Of the direct class of methods, the second-order stoichiometric algorithm is considerably simpler than the nonstoichiometric version. The quasi-Newton methods also appear promising.

PROBLEMS

7.1 Continue Problem 6.8 by calculating the equilibrium mole numbers, using the third model studied by Vonka and Holub (1975)—a nonideal solution. Use the Redlich-Kwong equation of state to calculate fugacity coefficients (and hence activity coefficients). For a species in a nonideal solution, the fugacity coefficient is then given by (Vonka and Holub, 1975)

$$\ln \phi_i = \ln\left(\frac{RT}{P(v-b)}\right) + \frac{b_i}{v-b} + \frac{a}{RT^{1.5}}$$
$$\times \left\{\frac{b_i}{b(v+b)} + \left[\frac{2(a_i/a)^{0.5} - (b_i/b)}{b}\right]\ln\left(\frac{v+b}{v}\right)\right\}, \quad (7.2\text{-}25a)$$

where a and b for the solution are obtained by the mixing rules:

$$a = \left(\sum x_i a_i^{0.5}\right)^2 \qquad (7.2\text{-}25b)$$

and

$$b = \sum x_i b_i, \qquad (7.2\text{-}25c)$$

and v is the molar volume of the mixture at (T, P). This can be converted into a relationship for the activity coefficient γ_i for use with the algorithm shown in Figure 7.1 since, from equations 3.7-33 and 7.2-17,

$$\gamma_i = \frac{\phi_i}{\phi_i^*}, \qquad (7.2\text{-}25d)$$

where ϕ_i^* is the fugacity coefficient of pure species i at (T, P), obtained as shown in Problem 6.8.

7.2 Methanol (CH_3OH) is synthesized commercially from hydrogen and carbon oxides (CO and CO_2) in one process at about 600 K and 340 atm by use of a catalyst based on zinc and chromic oxides. The main products are CH_3OH and H_2O, and secondary products are CH_4 and $(CH_3)_2O$. Assuming that the feed consists only of H_2 and CO in the stoichiometric ratio for the synthesis, compare the equilibrium yield of CH_3OH (moles of H_2 reacted to form CH_3OH per mole of H_2 in the feed) for the two systems

(a) $\{(H_2, CO, CH_3OH), (C, H, O)\}$,
(b) $\{(H_2, CO, CH_3OH, CO_2, H_2O, CH_4, (CH_3)_2O), (C, H, O)\}$.

Use the three models and the approach described in Problems 6.8 and 7.1. Data (sources as in Problem 6.8) are as follows:

Species	ΔG_f°, kJ mole^{-1}	T_C, K	P_C, atm
H_2	0	33.3	12.8
CO	−164.68	133	34.5
CH_3OH	−119.33	513.2	78.5
CO_2	−395.18	304.2	72.9
H_2O	−214.05	647.4	218.3
CH_4	−23.05	191.1	45.8
$(CH_3)_2O$	−35.31	400.1	53

Problems

7.3 The feed to an ammonia synthesis converter consists of N_2 and H_2 in the stoichiometric ratio. Calculate the maximum possible yield of NH_3 at 450°C and 300 atm. Use the approach described in Problem 7.1. Data (JANAF, 1971; Kobe and Lynn, 1953) are as follows:

Species	ΔG_f°, kJ mole^{-1}	T_C, K	P_C, atm
H_2	0	33.3	12.8
N_2	0	126.2	33.5
NH_3, 600 K	15.86		
700 K	27.17	405.5	111.3
800 K	38.64		
900 K	50.23		

The observed value (Dodge, 1944, p. 495) is 0.526.

7.4 Repeat Problem 4.10 without the assumption that the liquid phase is an ideal solution. Assume that liquid-phase activity coefficients are given by the multispecies equivalent of the Wilson equation (equation 7.2-9), with the parameters provided by Suzuki et al. (1970) for the six binary systems as follows:

Binary System	Wilson Parameters	
	A_{12}	A_{21}
$CH_3COOH + C_2H_5OH$	0.27558	2.28180
$CH_3COOH + H_2O$	0.26838	1.22642
$CH_3COOH + CH_3COOC_2H_5$	0.61790	0.89277
$C_2H_5OH + H_2O$	0.15347	0.92038
$CH_3COOC_2H_5 + C_2H_5OH$	0.55046	0.76670
$CH_3COOC_2H_5 + H_2O$	0.12353	0.14907

Activity coefficients for the quaternary system are given from these parameters by

$$\ln \gamma_k = 1 - \ln\left(\sum_{j=1}^{4} x_j A_{kj}\right) - \sum_{i=1}^{4} \left(\frac{x_i A_{ik}}{\sum_{j=1}^{4} x_j A_{ij}}\right).$$

7.5 For the system described in Problem 6.13, calculate the composition of each phase with the assumption that the liquid phase is a regular solution. For this purpose, the solubility parameters are 595, 575, 570, and 600 J$^{1/2}$ liter$^{-1/2}$, respectively, and the molar volumes of the pure liquids are 0.089, 0.107, 0.116, and 0.123 liters mole^{-1}, respectively. For each species in a multispecies regular solution, the activity coefficient is given by (Prausnitz, 1969, p. 279)

$$RT \ln \gamma_i = v_i (\delta_i - \bar{\delta})^2,$$

where

$$\bar{\delta} = \sum \phi_i \delta_i,$$

and the volume fraction ϕ_i is

$$\phi_i = \frac{x_i v_i}{\sum x_i v_i}.$$

7.6 Repeat Example 7.1 on the assumption that the solution is also 0.15 m in KNO_3 (Feenstra, 1979).

7.7 Repeat Example 6.3 by treating the aqueous phase as a nonideal solution.

7.8 Repeat Problem 6.11 by treating the aqueous phase as a nonideal solution.

CHAPTER EIGHT

The Effects of Problem Parameter Changes on Chemical Equilibria (Sensitivity Analysis)

In this chapter we discuss the effects on chemical equilibria of changing one, or more than one, of the parameters of the problem from a given set of initial values, for which the solution is known, to another set. We consider these parameters to be temperature T, pressure P, elemental abundances $\mathbf{b}$, standard chemical potentials of the species $\boldsymbol{\mu}^*$, and the amounts of inert species in the various phases $\mathbf{n}_z$.

We consider both qualitative and quantitative effects of these parameter changes on the equilibrium composition. For the former, we discuss some inequalities relating equilibrium compositions corresponding to different sets of parameters. For the latter, we discuss the calculation of the marginal rates of change of the equilibrium amounts n_i with respect to the parameters p_j. These rates determine the sensitivity matrix of the solution with respect to the parameters, the Jacobian matrix with elements $\partial n_i / \partial p_j$. This type of calculation, in which the derivatives of the solution of a problem are determined with respect to the problem parameters, is called "sensitivity analysis" (McKeown, 1980).

To set the stage for the general treatment of sensitivity analysis, we begin by reviewing the overall effects of changes in some of the parameters for a simple system ($R = 1$) in stoichiometric terms. Then after discussing the effects of problem parameter changes on optimization problems and sets of nonlinear equations in a general mathematical setting, we discuss the qualitative aspects of these changes on equilibria. We next derive the sensitivity matrix and results of its use for the various parameters in both stoichiometric and nonstoichiometric formulations. Finally, we consider the effects of errors in free-energy data and the calculation of thermodynamic derivatives.

8.1 OVERALL EFFECTS OF PARAMETER CHANGES

The overall effects we consider here are those caused by changes in the parameters temperature, pressure, and moles of inert species. We confine attention to the stoichiometric formulation for a simple system ($R = 1$) based on calculations at different values of the parameters. These calculations can be done by the methods described in previous chapters. The free-energy data required may be available at the stipulated values of T and P (see Table 3.2 for sources of data) or may have to be calculated from data at other conditions. Illustrating the main features of the dependence of equilibrium on each of T, P, and n_z in turn, we first show how the free-energy data may be determined at various values of T and P. The results generally illustrate quantitative aspects of Le Chatelier's principle (Section 1.6).

8.1.1 Effect of Temperature

Integration of the van't Hoff equation (equation 3.10-4), together with the use of the Kirchhoff equation for the temperature dependence of the standard enthalpy of reaction (Denbigh, 1981, p. 145),

$$\left(\frac{\partial \Delta H^*}{\partial T}\right)_P = \Delta C_P^* \equiv \sum_{i=1}^{N} \nu_i C_{Pi}^*, \tag{8.1-1}$$

results in

$$\ln K_a = I + \frac{1}{R} \int \left[\frac{\Delta H_0^* + \int \Delta C_P^*(T, P) \, dT}{T^2}\right] dT. \tag{8.1-2}$$

Here I and ΔH_0^* are two integration constants, the evaluation of which requires knowledge of each of K_a and ΔH^* at one temperature, together with knowledge of $\Delta C_P^*(T, P)$. The effect of pressure is often insignificant, and the quantities ΔH^* and ΔC_P^* are usually replaced by $\Delta H^o(T)$ and $\Delta C_P^o(T)$. Alternatively, the standard free energy of reaction may be determined from equation 3.10-1 as a function of temperature by

$$\frac{\Delta G^*}{T} = -IR - \int \left[\frac{\Delta H_0^* + \int \Delta C_P^*(T, P) \, dT}{T^2}\right] dT. \tag{8.1-3}$$

In principle, the value of K_a or ΔG^* can be calculated at any temperature, given the minimum amount of information specified for the determination of the integration constants. The extent of reaction ξ may then be calculated at any temperature. If standard free-energy data are already available at various temperatures, ξ can be calculated at these temperatures without the use of equations 8.1-2 or 8.1-3.

Overall Effects of Parameter Changes

From the van't Hoff equation (equation 3.10-4), K_a increases with increasing T for an endothermic reaction ($\Delta H^* > 0$) and decreases with increasing T for an exothermic reaction ($\Delta H^* < 0$). Similar conclusions hold for ξ since ξ increases as K increases. These two different results are illustrated in Figures 8.1 and 8.2 for the dehydrogenation of ethane and the oxidation of sulfur dioxide, respectively.

In both Figures 8.1 and 8.2 the quantity plotted against temperature is the fractional conversion of (limiting) reactant f at equilibrium. Regardless of whether the system is at equilibrium, f_i for species i is defined by

$$f_i = \frac{n_i^\circ - n_i}{n_i^\circ}, \tag{8.1-4}$$

where n_i° is the initial amount and is related to ξ by

$$f_i = \left(\frac{|\nu_i|}{n_i^\circ}\right)\xi. \tag{8.1-5}$$

Thus f_i may be regarded as a normalized ξ.

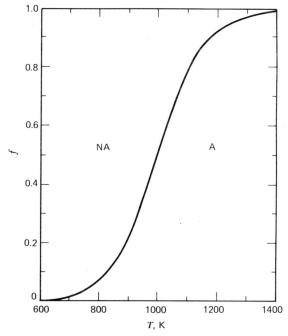

Figure 8.1 Dependence of equilibrium conversion f of C_2H_6 on temperature for

$$C_2H_6 = C_2H_4 + H_2$$

where $P = 1$ atm, feed is pure ethane, and A = accessible and NA = nonaccessible (regions).

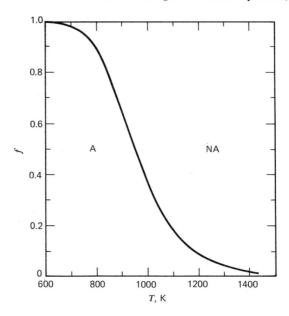

Figure 8.2 Dependence of equilibrium conversion f of SO_2 on temperature for

$$SO_2 + \tfrac{1}{2}O_2 = SO_3$$

where $P = 1$ atm; feed is 9.5 mole %SO_2, 11.5% O_2, 79% N_2; and A = accessible and NA = nonaccessible (regions).

Figure 8.1 shows the typical sigmoidal increase in equilibrium conversion for an endothermic reaction, from virtually no reaction at relatively low temperature to virtually complete reaction at relatively high temperature. For the dehydrogenation of ethane, the range shown is from 600 to 1400 K. It also illustrates the equilibrium conversion line as a boundary between states that are thermodynamically accessible (region A) and nonaccessible (region NA) from the initial state at the given conditions.

Figure 8.2 shows the converse behavior for a typical exothermic reaction. The temperature range shown is coincidentally about the same as that in Figure 8.1.

For reaction in a chemical cell, the standard emf $E°$ may be used in place of K_a or $\Delta G°$. Since $E° = -\Delta G°/zF$, the result corresponding to the van't Hoff equation is

$$\frac{d(E°/T)}{dT} = \frac{\Delta H°}{zFT^2}, \qquad (8.1\text{-}6)$$

which may be integrated to give a result comparable to equation 8.1-2 or 8.1-3.

8.1.2 Effect of Pressure

The effect of pressure on ξ (or f) at equilibrium may be considerable for a gas-phase reaction for which $\bar{\nu} \neq 0$ (see equation 4.3-6); otherwise, it is usually small and may be insignificant, unless large pressure changes occur. The principal conclusion obtained from equilibrium calculations is that ξ increases with increasing P if $\bar{\nu} < 0$ and conversely decreases with increasing P if $\bar{\nu} > 0$. The classic illustration of the former case for the synthesis of ammonia at 723 K is shown in Figure 8.3, based on data given by Dodge (1944, p. 494). The equilibrium conversion, which is based on either N_2 or H_2, since the feed is in the stoichiometric ratio, increases to over 80% as P increases to 1000 atm.

The effect of pressure on ξ should be distinguished from the effect of pressure on K_a. If the standard state chosen is independent of pressure [e.g., as indicated by $\Delta G°(T)$], K_a is independent of P, regardless of the value of $\bar{\nu}$. If the standard state chosen depends on P, the dependence of K_a on P is given by equation 3.10-5, which involves ΔV^*. In turn, ΔV^* may be zero (or relatively small), positive, or negative depending primarily on $\bar{\nu}$. In any case the dependence of ξ (or f) on P does not necessarily reflect directly the dependence of K_a on P.

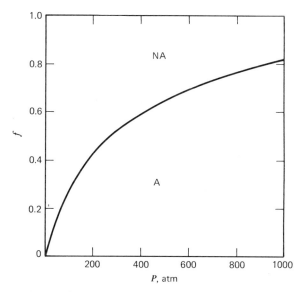

Figure 8.3 Dependence of equilibrium conversion f of reactants on pressure for

$$N_2 + 3H_2 = 2NH_3$$

where $T = 723$ K, $n°_{H_2}/n°_{N_2} = 3$, and A = accessible and NA = nonaccessible (regions).

8.1.3 Effect of Inert Species

The effect of the relative amount of inert species present in the system may similarly be calculated on a point-by-point basis for various values of n_z, with T and P fixed. The dependence of ξ (or f) on n_z is similar, *but opposite*, to that on P, and similar cases can be distinguished.

The main conclusions are that for a gas-phase reaction for which $\bar{\nu} \neq 0$, ξ (or f) increases as n_z increases if $\bar{\nu} > 0$ and decreases as n_z increases if $\bar{\nu} < 0$. The former case is illustrated in Figure 8.4 for the dehydrogenation of n-butene, with the effect of isomerization of n-butene taken into account (Section 9.6). In this case, as in some other dehydrogenation reactions, steam is used as the inert species.

8.2 EFFECTS OF PROBLEM PARAMETER CHANGES ON OPTIMIZATION PROBLEMS AND NONLINEAR EQUATIONS

We consider the optimization problem

$$\min_{\mathbf{x} \in \Omega} f(\mathbf{x}, \mathbf{p}), \qquad (8.2\text{-}1)$$

where f is a real-valued function, $\mathbf{x} \in R^N$ (i.e., $\mathbf{x}$ is an N-vector), where the x_i are variables, $\mathbf{p} \in R^L$, where the p_k are parameters, and Ω is the constraint set

$$\Omega = \{\mathbf{x}\colon \mathbf{g}(\mathbf{x}, \mathbf{p}) = \mathbf{0}\}, \qquad (8.2\text{-}2)$$

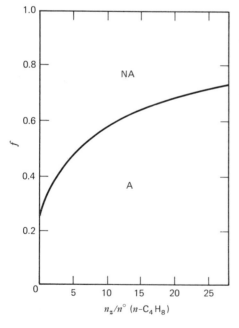

Figure 8.4 Dependence of equilibrium conversion f of n-C_4H_8 on molar ratio of inert species to n-C_4H_8 in feed for

$$n\text{-}C_4H_8 = 1{,}3\text{-}C_4H_6 + H_2$$

where $T = 900$ K, $P = 1.25$ atm, and A = accessible and NA = nonaccessible (regions).

where $g(x, p) \in R^M$ (M is the number of constraints). We assume that f and g are twice differentiable over Ω.

The necessary conditions for x to be a local solution of this minimization problem are the existence of a Lagrange multiplier vector $\lambda \in R^M$ such that (x, λ) satisfies the nonlinear equations

$$\frac{\partial f}{\partial x_i} - \sum_{j=1}^{M} \lambda_j \left(\frac{\partial g_j}{\partial x_i} \right) = 0; \quad i = 1, 2, \ldots, N \quad (8.2\text{-}3)$$

and

$$g_j(x, p) = 0; \quad j = 1, 2, \ldots, M. \quad (8.2\text{-}4)$$

We also consider a general set of N nonlinear equations

$$h(x, p) = 0, \quad (8.2\text{-}5)$$

where $h(x, p) \in R^N$ and is twice differentiable. The set of equations 8.2-3 and 8.2-4, resulting from the minimization problem 8.2 1, is of the form of equation 8.2-5.

In both these cases we consider the solution x to the problem to be a function of the parameters p. We write $x(p)$ to denote this dependence. Sensitivity analysis is a way of studying certain aspects of this functional relationship.

8.2.1 Qualitative Effects of Parameter Changes

We consider two distinct parameter vectors p^0 and p^1 with corresponding solution vectors x^0 and x^1, the latter pair of which we wish to relate in some way.

In the case of the minimization problem (statement 8.2-1), because each x is the solution of the minimization problem for its corresponding parameter vector, we must have

$$f(x^0, p^0) \leq f(x^1, p^0)$$

and

$$f(x^1, p^1) \leq f(x^0, p^1). \quad (8.2\text{-}6)$$

Adding the preceding inequalities, we obtain

$$f(x^0, p^0) + f(x^1, p^1) \leq f(x^1, p^0) + f(x^0, p^1). \quad (8.2\text{-}7)$$

We discuss the use of equation 8.2-7 in Section 8.3. If each solution x corresponds to a unique parameter vector p, the weak inequalities in relations 8.2-6 and 8.2-7 become strong inequalities.

8.2.2 Quantitative Effects of Parameter Changes and the Sensitivity Matrix

When the solution **x** to some problem depends on a set of parameters **p**, the Jacobian matrix $(\partial \mathbf{x}^T/\partial \mathbf{p})^T$ describes the marginal change in the solution as the parameters change. This matrix is called the *first-order sensitivity matrix* for the problem. One way of utilizing this matrix is to compute solution changes approximately through the use of

$$\mathbf{x} = \mathbf{x}^0 + \left(\frac{\partial \mathbf{x}^T}{\partial \mathbf{p}}\right)^T_0 \delta \mathbf{p}, \qquad (8.2\text{-}8)$$

or

$$x_i = x_i^0 + \sum_{j=1}^{L} \left(\frac{\partial x_i}{\partial p_j}\right)_0 \delta p_j, \qquad (8.2\text{-}8\text{a})$$

as the solution and parameters change respectively from $\mathbf{x}^0$ and $\mathbf{p}^0$ to final values **x** and $\mathbf{p}^0 + \delta \mathbf{p}$. The subscript 0 denotes evaluation at $(\mathbf{x}^0, \mathbf{p}^0)$.

The calculation of the sensitivity matrix in a given case depends on the details of the particular problem involved. For example, in the case when **x** is a solution of a set of N nonlinear equations 8.2-5, we differentiate with respect to the $L \times 1$ parameter vector to obtain

$$\sum_{k=1}^{N} \left(\frac{\partial h_i}{\partial x_k}\right)_0 \left(\frac{\partial x_k}{\partial p_j}\right)_0 + \left(\frac{\partial h_i}{\partial p_j}\right)_0 = 0; \qquad \begin{array}{l} i = 1, 2, \ldots, N; \\ j = 1, 2, \ldots, L. \end{array} \qquad (8.2\text{-}9)$$

Thus the sensitivity matrix at $(\mathbf{x}^0, \mathbf{p}^0)$ is obtained by solving the matrix system of L sets of N linear equations given by equation 8.2-9. Provided that the coefficient matrix $(\partial \mathbf{h}^T/\partial \mathbf{x})_0$ is nonsingular, we may formally write the solution of these equations as

$$\left(\frac{\partial \mathbf{x}^T}{\partial \mathbf{p}}\right)_0 = -\left(\frac{\partial \mathbf{h}^T}{\partial \mathbf{p}}\right)_0 \left(\frac{\partial \mathbf{h}^T}{\partial \mathbf{x}}\right)_0^{-1}. \qquad (8.2\text{-}10)$$

Applying this analysis to equations 8.2-3 and 8.2-4 and differentiating with respect to any parameter p_k, we obtain the following set of $(N + M)$ linear equations in $\partial \mathbf{x}/\partial p_k$ and $\partial \boldsymbol{\lambda}/\partial p_k$:

$$\sum_{j=1}^{N} \left(\frac{\partial x_j}{\partial p_k}\right) \left\{ \left(\frac{\partial^2 f}{\partial x_i \partial x_j}\right)_0 - \sum_{l=1}^{M} \lambda_l \left(\frac{\partial^2 g_l}{\partial x_i \partial x_j}\right)_0 \right\} - \sum_{j=1}^{M} \left(\frac{\partial \lambda_j}{\partial p_k}\right) \left(\frac{\partial g_j}{\partial x_i}\right)$$

$$= \sum_{j=1}^{M} \lambda_j \left(\frac{\partial^2 g_j}{\partial x_i \partial p_k}\right) - \frac{\partial^2 f}{\partial x_i \partial p_k}; \qquad i = 1, 2, \ldots, N \qquad (8.2\text{-}11\text{a})$$

and

$$\sum_{l=1}^{N} \left(\frac{\partial x_l}{\partial p_k}\right)\left(\frac{\partial g_j}{\partial x_l}\right)_0 = -\left(\frac{\partial g_j}{\partial p_k}\right)_0; \quad j=1,2,\ldots,M. \quad (8.2\text{-}11\text{b})$$

Equations 8.2-11 are L sets (one set for each parameter) of $(N+M)$ linear equations. Provided that the coefficient matrices of the preceding linear equations are nonsingular, the sensitivity matrix is readily calculated.

We discuss the use of equations 8.2-9 and 8.2-11 in the context of chemical equilibrium problems in Section 8.5.

8.3 QUALITATIVE EFFECTS OF PARAMETER CHANGES ON CHEMICAL EQUILIBRIA

In this section we consider several inequalities that relate equilibrium compositions corresponding to the same value of **b** but possibly different values of T, P, μ^*, and $\mathbf{n}_z$. The first inequality is applicable to general nonideal systems and results directly from equation 8.2-7. The remaining inequalities are then given as corollaries.

Denoting by $\mathbf{n}^0$ the equilibrium solution corresponding to the parameter set $(T^0, P^0, \mu^{*0}, \mathbf{n}_z^0)$, and by $\mathbf{n}^1$ that corresponding to $(T^1, P^1, \mu^{*1}, \mathbf{n}_z^1)$, we have the following

THEOREM

$$\sum_{i=1}^{N} \left(n_i^0 - n_i^1\right)\left(\mu_i^{*0} - \mu_i^{*1}\right)$$

$$+ R \sum_{i=1}^{N} \left\{ n_i^0 \left[T^0 \ln a_i(\mathbf{n}^0, T^0, P^0, \mathbf{n}_z^0) - T^1 \ln a_i(\mathbf{n}^0, T^1, P^1, \mathbf{n}_z^1) \right] \right.$$

$$\left. - n_i^1 \left[T^0 \ln a_i(\mathbf{n}^1, T^0, P^0, \mathbf{n}_z^0) - T^1 \ln a_i(\mathbf{n}^1, T^1, P^1, \mathbf{n}_z^1) \right] \right\} \leq 0. \quad (8.3\text{-}1)$$

PROOF

We write the Gibbs function, following equation 3.2-16, as

$$G(\mathbf{n}, T, P, \mu^*, \mathbf{n}_z) = \sum_{i=1}^{N} n_i \left[\mu_i^* + RT \ln a_i(\mathbf{n}, T, P, \mathbf{n}_z) \right]. \quad (8.3\text{-}2)$$

From the inequalities 8.2-6, we have

$$\sum_{i=1}^{N} n_i^0 \left[\mu_i^{*0} + RT^0 \ln a_i(\mathbf{n}^0, T^0, P^0, \mathbf{n}_z^0) \right]$$

$$\leq \sum_{i=1}^{N} n_i^1 \left[\mu_i^{*0} + RT^0 \ln a_i(\mathbf{n}^1, T^0, P^0, \mathbf{n}_z^0) \right] \quad (8.3\text{-}3)$$

and

$$\sum_{i=1}^{N} n_i^1 \left[\mu_i^{*1} + RT^1 \ln a_i(\mathbf{n}^1, T^1, P^1, \mathbf{n}_z^1) \right]$$

$$\leq \sum_{i=1}^{N} n_i^0 \left[\mu^{*1} + RT^1 \ln a_i(\mathbf{n}^0, T^1, P^1, \mathbf{n}_z^1) \right]. \quad (8.3\text{-}4)$$

We obtain the statement of the theorem by adding inequalities 8.3-3 and 8.3-4.

COROLLARY 1 For an ideal solution, relation 8.3-1 becomes

$$\sum_{i=1}^{N} (n_i^0 - n_i^1)(\mu_i^{*0} - \mu_i^{*1}) + R \sum_{i=1}^{N} \left(n^i \left\{ T^0 \ln \frac{n_i^0}{n_t^0} - T^1 \ln \frac{n_i^0}{n_t^0 + \delta n_z} \right\} \right.$$

$$\left. - n_i^1 \left\{ T^0 \ln \frac{n_i^1}{n_t^1 - \delta n_z} - T^1 \ln \frac{n_i^1}{n_t^1} \right\} \right) \leq 0, \quad (8.3\text{-}5)$$

where $\delta n_z = n_z^1 - n_z^0$.

COROLLARY 2 Provided that $T^0 = T^1$, $\delta \mathbf{n}_z = 0$, and a_i is independent of pressure, relation 8.3-1 becomes

$$\sum_{i=1}^{N} (n_i^0 - n_i^1)(\mu_i^{*0} - \mu_i^{*1}) \leq 0. \quad (8.3\text{-}6)$$

This result was originally proved by Shapiro (1964, 1969) for the case of a system consisting of ideal solutions.

Example 8.1 (Shapiro, 1964, 1969). If an ideal-system chemical-equilibrium problem with solution $\mathbf{n}^0$ is altered by an increase in one of the standard chemical potentials μ_i^*, and if $\mathbf{n}^1$ is any solution of the resulting problem, then

$$n_i^1 \leq n_i^0. \quad (8.3\text{-}7)$$

This result follows directly from relation 8.3-6 and essentially states that the equilibrium amount of any species is a monotonically decreasing function of its standard free-energy parameter, all other conditions being fixed. Smith and Missen (1974) have made use of this result to show that, for systems containing a substance that is capable of isomerization, the relative equilibrium amount of the substance is greater than it would be if only a strict subset of the isomeric forms were accessible. We discuss the general treatment of isomers in equilibrium calculations in Chapter 9. We show in Section 8.4.2 that the equality sign in relation 8.3-7 may be removed, provided that the species i is not inert.

Example 8.2 For an ideal-gas solution, with $T^0 = T^1$ and $\delta n_z = 0$, relation 8.3-6 yields

$$\left(n_t^0 - n_t^1\right)\ln\frac{P^0}{P^1} \leq 0. \tag{8.3-8}$$

8.4 SENSITIVITY MATRICES FOR (T, P) PROBLEMS

The first-order sensitivity matrix contains the elements $\partial n_i/\partial T$, $\partial n_i/\partial P$, $\partial n_i/\partial b_j$, $\partial n_i/\partial \mu_j^*$, and $\partial n_i/\partial n_{z_j}$, under the constraint of chemical equilibrium. The second-order sensitivity matrix contains the corresponding second derivatives. In this section we are primarily concerned with the computation of the first-order matrix. After reviewing the literature dealing with this problem, we describe the computation first using the stoichiometric formulation and then the nonstoichiometric formulation. In the former case we consider in detail the simple case of $R = 1$, as well as the general treatment. Finally, we briefly describe the determination of the second-order sensitivity matrix.

8.4.1 Review of Literature

All papers in the literature consider only the first derivatives of the equilibrium composition, and most are restricted to ideal systems. Hochstim and Adams (1962) considered the effects of T and P. By using a stoichiometric formulation and differentiating the R equilibrium-constant expressions and the M element-abundance constraints, they obtained in general two sets of N linear equations involving, respectively, $\partial n_i/\partial T$ and $\partial n_i/\partial P$. For the special case of air with 30 species, they effected a reduction to three linear equations in three unknowns. They suggested similar techniques to reduce the size of the linear system in other cases. This reduction technique is, of course, different for each problem, and is thus not a general approach.

Neumann (1962) discussed the effects of errors in μ^* on the equilibrium composition of ideal gaseous mixtures. He examined some special cases in which he formulated the equilibrium condition in terms of a real zero of a 10th-degree polynomial whose coefficients are functions of μ^*. He then analyzed the effects of an inaccurate μ^* on these coefficients, on the resulting zero of the polynomial, and finally on the equilibrium composition itself. The method is not a general one, and the reduction to a polynomial becomes quite tedious for some problems. Subsequently, Neumann (1963) derived a method that yielded a set of N simultaneous linear equations in $\partial n_i/\partial \mu_k^*$, using an approach similar to that used by Hochstim and Adams (1962) for the T and P derivatives.

Storey (1965) devised a method to calculate the effects of changes in **b** on the equilibrium composition of solid-gas systems. He obtained a set of $(M + 2 - \pi)$ linear equations (where π is the total number of phases). An extra step must be taken at the outset of the calculations involving the solid species. If this step were removed from the procedure, a set of ($M + \pi$) linear equations would result.

Zeleznik and Gordon (1962) considered the effects of variations in T and P in connection with their version of the BNR chemical equilibrium algorithm. They obtained sets of ($N + M$) linear equations that yield the appropriate derivatives of the equilibrium mole numbers with respect to either T or P. In the case of an ideal system they demonstrated a reduction to a set of ($M + \pi$) equations.

Smith (1969) developed a technique to enable the simultaneous calculation of the first derivatives of the equilibrium composition of ideal systems with respect to any or all of the data parameters by means of the solution of a set of ($M + \pi$) linear equations. In the case of T and P, the equations are essentially identical to those due to Zeleznik and Gordon (1962). In the case of **b** the equations are similar but not identical to those of Storey (1965). Van Zeggeren and Storey (1970, pp. 49–51, 149) later derived equations for considering changes in $\mu°$ that, when corrected for a minor error, are identical to those of Smith (1969) for this case. The effects of changes in $\mathbf{n}_z$ have apparently not been considered elsewhere in the literature. Since Smith's technique is the most general one, we discuss it in detail in Section 8.4.3.

Bigelow and Shapiro (1971) discussed the sensitivity matrix for the parameters **b** and μ^* (and also with respect to the elements of the formula matrix **A**) in a manner similar to that of Smith (1969), as did Shapley et al. (1968). Dinkel and Lakshmanan (1977) have considered sensitivity analysis with respect to T, P, and **b** using the viewpoint of geometric programming. They considered gaseous systems, and a matrix of order ($N - M - 1$) must be inverted since they focus on the stoichiometric formulation of the problem.

For the simple case of $R = 1$ in the stoichiometric formulation, a considerable body of results exists, dating back to the work of van't Hoff (1884, p. 161) and Le Chatelier (1884). More recently, Pings (1961, 1963, 1964, 1966) has calculated the first derivatives of ξ with respect to a number of variables under various constraints. Lu (1967) and Larsen et al. (1968) have extended some of this work to the case of more than one stoichiometric equation. Samuels and Eliassen (1972) have shown that there may be an optimal feed composition for maximum conversion at equilibrium. Holub and Vonka (1976, Chapter 4) have also discussed these various aspects.

8.4.2 Stoichiometric Formulation

8.4.2.1 *System Involving One Stoichiometric Equation (R = 1)*

We illustrate the general equations given in Section 8.2 for a simple system, since the results are generally familiar and can be stated in definite terms. They also illustrate the basis for the main features of Figures 8.1 to 8.4.

We write the equilibrium condition in the form of equation 8.2-5 as

$$\frac{\partial G}{\partial \xi} \equiv \Delta G(\xi, \mathbf{p}) \equiv \sum_{i=1}^{N} \nu_i \mu_i(\xi, \mathbf{p}) = 0, \qquad (8.4\text{-}1)$$

Sensitivity Matrices for (T, P) Problems

in which the extent of reaction ξ is the single variable, and the vector of parameters is

$$\mathbf{p} = (T, P, \mathbf{b}, \boldsymbol{\mu}^*, \mathbf{n}_z)^T. \tag{8.4-2}$$

Since, in the stoichiometric formulation,

$$\mathbf{n} = \mathbf{n}^\circ(\mathbf{b}) + \boldsymbol{\nu}\xi, \tag{8.4-3}$$

we rewrite equation 8.4-1 more explicitly as

$$\Delta G\{\mathbf{n}[\xi, \mathbf{n}^\circ(\mathbf{b})], T, P, \boldsymbol{\mu}^*, \mathbf{n}_z\} = 0. \tag{8.4-4}$$

Following the procedure leading to equation 8.2-9, we differentiate equation 8.4-4 with respect to p_k and obtain

$$\left(\frac{\partial \Delta G}{\partial \xi}\right)_0 \left(\frac{\partial \xi}{\partial p_k}\right) + \sum_{i=1}^{N}\sum_{j=1}^{N} \left(\frac{\partial \Delta G}{\partial n_j}\right)_0 \left(\frac{\partial n_j}{\partial n_i^\circ}\right)_0 \left(\frac{\partial n_i^\circ}{\partial p_k}\right)_0 + \left(\frac{\partial \Delta G}{\partial p_k}\right)_0 = 0, \tag{8.4-5}$$

where subscript 0 denotes evaluation at $(\xi^0, \mathbf{p}^0)$ satisfying equation 8.4-1. From this we obtain (dropping the subscripts 0 for ease of notation),

$$\frac{\partial \xi}{\partial p_k} = -\frac{(\partial \Delta G/\partial p_k) + \sum_{i=1}^{N}\sum_{j=1}^{N}(\partial \Delta G/\partial n_j)(\partial n_j/\partial n_i^\circ)(\partial n_i^\circ/\partial p_k)}{\partial^2 G/\partial \xi^2}, \tag{8.4-6}$$

where we have used

$$\frac{\partial \Delta G}{\partial \xi} = \frac{\partial^2 G}{\partial \xi^2}. \tag{8.4-7}$$

Based on the discussion in Section 3.9, we take the denominator of equation 8.4-6 to be positive, and hence the sign of $\partial \xi/\partial p_k$ is opposite to that of the numerator. We consider the effect of each of the parameters T, P, $\boldsymbol{\mu}^*$, $\mathbf{n}_z$, and $\mathbf{b}$ in turn.

8.4.2.1.1 Effect of Temperature

From equation 8.4-6, since $\partial n_i^\circ/\partial T = 0$,

$$\frac{\partial \xi}{\partial T} = -\frac{\partial \Delta G/\partial T}{\partial^2 G/\partial \xi^2} = -\frac{\sum_{i=1}^{N} \nu_i(\partial \mu_i/\partial T)}{\partial^2 G/\partial \xi^2}$$

$$= \frac{\sum_{i=1}^{N} \nu_i \bar{s}_i}{\partial^2 G/\partial \xi^2} = \frac{\Delta H/T}{\partial^2 G/\partial \xi^2}, \tag{8.4-8}$$

since

$$\sum_{i=1}^{N} \nu_i \bar{s}_i = \Delta S = \frac{\Delta H - \Delta G}{T} = \frac{\Delta H}{T} \equiv \sum_{i=1}^{N} \nu_i \bar{h}_i, \qquad (8.4\text{-}9)$$

where $\bar{s}_i$ is the partial molar entropy of species i. Hence $\partial \xi / \partial T$ has the sign of ΔH, the enthalpy of reaction. This has already been illustrated in Figures 8.1 and 8.2.

From equations 8.4-3 and 8.4-8,

$$\frac{\partial n_i}{\partial T} = \nu_i \left(\frac{\partial \xi}{\partial T} \right) = \frac{\nu_i \Delta H / T}{\partial^2 G / \partial \xi^2}. \qquad (8.4\text{-}10)$$

For an endothermic reaction, the amount of a product ($\nu_i > 0$) increases as T increases, and the amount of a reactant ($\nu_i < 0$) decreases. For an exothermic reaction, the opposite is true.

To calculate numerical values of $\partial \xi / \partial T$, a chemical potential expression is required. For the case of a single ideal solution,

$$\mu_i = \mu_i^* + RT \ln \frac{n_i}{n_t} \qquad (4.3\text{-}3)$$

and

$$\frac{\partial^2 G}{\partial \xi^2} = RT \left(\sum_{i=1}^{N} \frac{\nu_i^2}{n_i} - \frac{\bar{\nu}^2}{n_t} \right) = RT \sum_{i=1}^{N} n_i \left(\frac{\nu_i}{n_i} - \frac{\bar{\nu}}{n_t} \right)^2, \qquad (8.4\text{-}11)$$

where

$$\bar{\nu} = \sum_{i=1}^{N} \nu_i. \qquad (8.4\text{-}12)$$

From this and also because ΔH becomes ΔH^* for an ideal solution, equation 8.4-8 becomes

$$\frac{\partial \xi}{\partial T} = \frac{\Delta H^*}{RT^2} \left[\sum_{i=1}^{N} n_i \left(\frac{\nu_i}{n_i} - \frac{\bar{\nu}}{n_t} \right)^2 \right]^{-1}, \qquad (8.4\text{-}13)$$

and $\partial n_i / \partial T$ is obtained similarly from equation 8.4-10.

8.4.2.1.2 Effect of Pressure

The equations corresponding to 8.4-8 and 8.4-10 are, respectively,

$$\frac{\partial \xi}{\partial P} = -\frac{\Delta V}{\partial^2 G / \partial \xi^2} \qquad (8.4\text{-}14)$$

Sensitivity Matrices for (T, P) Problems

and

$$\frac{\partial n_i}{\partial P} = -\frac{\nu_i \Delta V}{\partial^2 G/\partial \xi^2}, \qquad (8.4\text{-}15)$$

where

$$\Delta V = \sum_{i=1}^{N} \nu_i \bar{v}_i. \qquad (8.4\text{-}16)$$

The sign of $\partial \xi/\partial P$ is opposite to that of ΔV. For a system not involving a gas phase, ΔV and, hence, $\partial \xi/\partial P$ are generally very small. For a system involving a gas phase that is an ideal solution, we write

$$\Delta V \simeq \sum_{(g)} \nu_i v_i^* = \frac{RT}{P} \sum_{(g)} \nu_i z_i^*, \qquad (8.4\text{-}17)$$

where (g) denotes that the summation is over the gaseous species. If we write

$$z_i^* = \bar{z} + \delta z_i^*, \qquad (8.4\text{-}18)$$

equation 8.4-17 becomes

$$\Delta V \simeq \frac{RT}{P} \left(\bar{\nu}_g \bar{z} + \sum_{(g)} \nu_i \delta z_i^* \right). \qquad (8.4\text{-}19)$$

If z_i^* is the same for all gaseous species, or if δz_i^* is small, then ΔV is approximately proportional to $\bar{\nu}_g$ (the sum of the stoichiometric coefficients for the gaseous species) and vanishes if $\bar{\nu}_g$ vanishes. As a result, we can say that the equilibrium value of ξ is essentially independent of P if $\bar{\nu}_g$ is zero, increases with increasing P if $\bar{\nu}_g < 0$, and decreases with increasing P if $\bar{\nu}_g > 0$. This has been illustrated in Figure 8.3.

From equation 8.4-15 and the definition of $n_{t\alpha}$, the total number of moles in phase α,

$$\frac{\partial n_{t\alpha}}{\partial P} = -\frac{\sum_{(\alpha)} \nu_i \Delta V}{\partial^2 G/\partial \xi^2} \equiv -\frac{\bar{\nu}_\alpha \Delta V}{\partial^2 G/\partial \xi^2}, \qquad (8.4\text{-}20)$$

where (α) denotes that the summation is over all species in phase α. Thus if $\bar{\nu}_\alpha$ is zero, pressure has no effect on $n_{t\alpha}$.

For an ideal solution, the result corresponding to equation 8.4-13 is

$$\frac{\partial \xi}{\partial P} = -\frac{\Delta V^*}{RT} \left[\sum_{i=1}^{N} n_i \left(\frac{\nu_i}{n_i} - \frac{\bar{\nu}}{n_t} \right)^2 \right]^{-1}. \qquad (8.4\text{-}21)$$

8.4.2.1.3 Effect of μ^*

The equations corresponding to 8.4-8 and 8.4-10 are, respectively,

$$\frac{\partial \xi}{\partial \mu_k^*} = -\frac{\nu_k}{\partial^2 G/\partial \xi^2} \qquad (8.4\text{-}22)$$

and

$$\frac{\partial n_i}{\partial \mu_k^*} = -\frac{\nu_i \nu_k}{\partial^2 G/\partial \xi^2}. \qquad (8.4\text{-}23)$$

For a product, ξ decreases with increasing μ_k^*; for a reactant, the opposite is true. According to equation 8.4-23, $\partial n_i/\partial \mu_i^* < 0$ (provided that ν_i is not zero; i.e., species i is not inert).

The chemical potential μ_k^* can change as a result of a change in temperature or in pressure or an error in its value. The first two of these cases are treated in Sections 8.4.2.1.1 and 8.4.2.1.2, and equations 8.4-22 and 8.4-23 are useful for the third case, as discussed in Section 8.5.

8.4.2.1.4 Effect of Inert Species

The equations corresponding to 8.4-8 and 8.4-10 are, respectively,

$$\frac{\partial \xi}{\partial n_z} = -\frac{\sum_{i=1}^{N} \nu_i (\partial \mu_i/\partial n_z)}{\partial^2 G/\partial \xi^2} \qquad (8.4\text{-}24)$$

and

$$\frac{\partial n_j}{\partial n_z} = -\frac{\nu_j \sum_{i=1}^{N} \nu_i (\partial \mu_i/\partial n_z)}{\partial^2 G/\partial \xi^2}. \qquad (8.4\text{-}25)$$

Unlike the previous three cases, it is not possible to interpret these further without introducing a chemical potential expression. For an ideal solution, equation 8.4-24 becomes

$$\frac{\partial \xi}{\partial n_z} = \frac{\bar{\nu}}{n_t} \left[\sum_{i=1}^{N} n_i \left(\frac{\nu_i}{n_i} - \frac{\bar{\nu}}{n_t} \right)^2 \right]^{-1}. \qquad (8.4\text{-}26)$$

According to this result, the sign of $\partial \xi/\partial n_z$ is that of $\bar{\nu}$. Thus the equilibrium value of ξ is independent of n_z if $\bar{\nu}$ is zero, increases with increasing n_z if $\bar{\nu} > 0$, and decreases with increasing n_z if $\bar{\nu} < 0$. The second case has been illustrated in Figure 8.4.

8.4.2.1.5 Effect of b

We examine the effect of **b** through the effect of the initial composition $\mathbf{n}^\circ$ in equation 8.4-3. Unlike the previous four cases, the effect of $\mathbf{n}^\circ$ is obtained from the second term in the numerator of equation 8.4-6, rather than the first term. Since, in this case, $\partial n_j^\circ/\partial n_i^\circ = \delta_{ij}$ and $\partial n_i^\circ/\partial n_k^\circ = \delta_{ik}$, it follows that

$$\frac{\partial \xi}{\partial n_k^\circ} = -\frac{\sum_{i=1}^N \sum_{j=1}^N (\partial \Delta G/\partial n_j)\delta_{ij}\delta_{ik}}{\partial^2 G/\partial \xi^2} = -\frac{\partial \Delta G/\partial n_k}{\partial^2 G/\partial \xi^2}. \qquad (8.4\text{-}27)$$

For an ideal solution,

$$\frac{\partial \xi}{\partial n_k^\circ} = \frac{\bar{\nu}/n_t - \nu_k/n_k}{\sum_{i=1}^N n_i(\nu_i/n_i - \bar{\nu}/n_t)^2}, \qquad (8.4\text{-}28)$$

and since

$$n_k = n_k^\circ + \nu_k \xi \qquad (8.4\text{-}29)$$

and

$$n_t = n_t^\circ + \bar{\nu}\xi, \qquad (8.4\text{-}30)$$

equation 8.4-28 becomes

$$\frac{\partial \xi}{\partial n_k^\circ} = \frac{n_k^\circ \bar{\nu} - \nu_k n_t^\circ}{n_t n_k \sum_{i=1}^N n_i(\nu_i/n_i - \bar{\nu}/n_t)^2}. \qquad (8.4\text{-}31)$$

Since the sign of the denominator is positive, the sign of $\partial \xi/\partial n_k^\circ$ depends on the sign of the numerator, which we rewrite as

$$n_k^\circ \bar{\nu} - \nu_k n_t^\circ = n_k^\circ(\bar{\nu} - \nu_k) - \nu_k n_{t'}^\circ, \qquad (8.4\text{-}32)$$

where

$$n_{t'}^\circ = n_t^\circ - n_k^\circ. \qquad (8.4\text{-}33)$$

The sign of $\partial \xi/\partial n_k^\circ$ may be examined through the quantity $\bar{\nu} - \nu_k$ of equation 8.4-32. If k is a reactant and $\bar{\nu} - \nu_k \geq 0$, then $\partial \xi/\partial n_k^\circ > 0$; if $\bar{\nu} - \nu_k < 0$, $\partial \xi/\partial n_k^\circ$ may change sign, depending on the values of $n_{t'}^\circ$ and n_k°, and ξ has an optimal value that can be shown to be a maximum. The significance of this, as originally shown by Samuels and Eliassen (1972), is that normally the addition of relatively more of a reactant to the feed increases the amount of reaction [i.e., amount of product(s)] with respect to another reactant, but in some cases there is an optimal ratio. If k is a product and $\bar{\nu} - \nu_k \leq 0$, then $\partial \xi/\partial n_k^\circ < 0$; if $\bar{\nu} - \nu_k > 0$, ξ may go through a minimum.

In either case the optimal value occurs when, from equations 8.4-31 and 8.4-32,

$$n_k^o(\bar{\nu} - \nu_k) - \nu_k n_{i'}^o = 0, \qquad (8.4\text{-}34)$$

or when the ratio of the amount of species k to the total amount of the other species in the feed is

$$\frac{n_k^o}{n_{i'}^o} = \frac{\nu_k}{\bar{\nu} - \nu_k}. \qquad (8.4\text{-}35)$$

In the special case in which there are two reactants, of which one is k, and no product(s) in the feed, the ratio on the left side of equation 8.4-35 is the ratio of the amount of reactant k to the other reactant in the feed.

Example 8.3 For the production of cyclohexane by the hydrogenation of benzene, determine whether there is an optimal molar ratio of (1) benzene to hydrogen (i.e., an optimal amount of benzene for a fixed amount of hydrogen) to maximize the amount of cyclohexane at equilibrium and (2) hydrogen to benzene to achieve a similar maximum.

Solution The system is $\{(C_6H_6, H_2, C_6H_{12}), (C, H)\}$, and the stoichiometric equation may be written as

$$C_6H_6 + 3H_2 = C_6H_{12}.$$

1 Here C_6H_6 is species k (i.e., $k = 1$). Then $\nu_k = -1$ and $\bar{\nu} - \nu_k = -2$. There is an optimal value of n_1^o, which, from equation 8.4-35, is $n_2^o/2$.
2 Here $k = 2$, $\nu_k = -3$, and $\bar{\nu} - \nu_k = 0$. There is no optimal value of n_1^o in this case.

These two apparently different results are due to the fact that ξ, which is a measure of the amount of cyclohexane, is a function of both n_1^o and n_2^o (for fixed values of the other parameters). Hence $\xi(n_1^o, n_2^o)$ is a two-dimensional surface bounded by the constraint planes $0 \leq \xi \leq \min(-n_1^o/\nu_1, -n_2^o/\nu_2)$, as dictated by whichever reactant is limiting in the stoichiometric sense. Evidently, sections of this surface at fixed n_2^o are curves that have maxima (part 1), and sections at fixed n_1^o are monotonically increasing curves.

8.4.2.2 System Involving $R > 1$

For $R > 1$, we write the equilibrium conditions corresponding to equation 8.4-4 as

$$\sum_{i=1}^{N} \Delta G_j\{n[\xi, n^o(b)], T, P, \mu^*, n_z\} = 0; \qquad j = 1, 2, \ldots, R. \qquad (8.4\text{-}36)$$

Sensitivity Matrices for (T, P) Problems

Equation 8.4-5 becomes the system of linear equations in $\partial \xi_i / \partial p_k$; $i = 1, 2, \ldots, R$:

$$\sum_{i=1}^{R} \left(\frac{\partial \xi_i}{\partial p_k} \right) \left(\frac{\partial^2 G}{\partial \xi_i \, \partial \xi_j} \right) = -\left(\frac{\partial \Delta G_j}{\partial p_k} \right) - \sum_{l=1}^{N} \sum_{m=1}^{N} \left(\frac{\partial \Delta G_j}{\partial n_l} \right) \left(\frac{\partial n_l}{\partial n_m^\circ} \right) \left(\frac{\partial n_m^\circ}{\partial p_k} \right);$$

$$j = 1, 2, \ldots, R, \quad (8.4\text{-}37)$$

where all quantities are evaluated at the (equilibrium) solution of equation 8.4-36.

The sensitivity coefficients $\partial n_i / \partial p_k$ are obtained by substituting the solution of equation 8.4-37 into the derivative of equations 2.3-1a to give

$$\frac{\partial n_i}{\partial p_k} = \sum_{l=1}^{N} \left(\frac{\partial n_i}{\partial n_l^\circ} \right) \left(\frac{\partial n_l^\circ}{\partial p_k} \right) + \sum_{j=1}^{R} \nu_{ij} \left(\frac{\partial \xi_j}{\partial p_k} \right); \quad i = 1, 2, \ldots, N.$$

$$(8.4\text{-}38)$$

The right side of equation 8.4-37 must be evaluated for this purpose, and we summarize the results for the various parameters in Table 8.1. The entries for the parameters T, P, and n_k° have been derived by Larsen et al. (1968). The effects of T and P have been obtained by Hochstim and Adams (1962) in a different manner, as have the effects of μ_k^* by Neumann (1963).

In contrast to the case of $R = 1$, when $R > 1$, it is difficult to make general statements about the sensitivity coefficients, and for a particular case, it is necessary to obtain $\partial n_i / \partial p_k$ from equations 8.4-38, using the solution to equations 8.4-37. We note that the coefficient matrix of equations 8.4-37 is generally positive definite, and the equations thus have a unique solution. It *is* possible, however, to make some general statements about the effects of P, μ_k^*, and n_{z_k}.

Table 8.1 Right Side of Equations 8.4-37 for the Various Parameters

Parameter	General Case	Ideal Solution
T	$\dfrac{\Delta H_j}{T}$	$\dfrac{\Delta H_j^*}{T}$
P	$-\Delta V_j$	$-\Delta V_j^*$
μ_k^*	$-\nu_{kj}$	$-\nu_{kj}$
n_{z_k}	$-\sum\limits_{i=1}^{N} \nu_{ij} \dfrac{\partial \mu_i}{\partial n_{z_k}}$	$RT \dfrac{\bar{\nu}_j}{n_t}$
n_k°	$-\sum\limits_{i=1}^{N} \nu_{ij} \dfrac{\partial \mu_i}{\partial n_k}$	$RT \left(\dfrac{\bar{\nu}_j}{n_t} - \dfrac{\nu_{kj}}{n_k} \right)$

In the case of P, if ΔV_j is zero for each stoichiometric equation, the solution at equilibrium, **n**, is independent of pressure. It can also be shown that, for a single phase, if all $\bar{v}_i$'s are the same (e.g., as in an ideal-gas solution), then $\partial n_t/\partial P < 0$, unless all $\bar{v}_j$'s are zero, in which case $\partial n_t/\partial P = 0$. The proof is as follows. From the definition of n_t and equation 8.4-38, with $p_k = P$,

$$\frac{\partial n_t}{\partial P} = \sum_{k=1}^{N}\sum_{j=1}^{R} \nu_{kj}\left(\frac{\partial \xi_j}{\partial P}\right) = \sum_{j=1}^{R} \bar{\nu}_j\left(\frac{\partial \xi_j}{\partial P}\right). \qquad (8.4\text{-}39)$$

From equation 8.4-37, it follows that

$$\sum_{i=1}^{R}\left(\frac{\partial \xi_i}{\partial P}\right)\left(\frac{\partial^2 G}{\partial \xi_i \partial \xi_j}\right) = -\sum_{i=1}^{N} \nu_{ij}\bar{v}_i = -v\bar{\nu}_j, \qquad (8.4\text{-}40)$$

where v is the common value of $\bar{v}_i$. Multiplying equation 8.4-40 by $\partial \xi_j/\partial P$, summing over j, and using equation 8.4-39, we obtain

$$\frac{\partial n_t}{\partial P} = -\frac{1}{v}\sum_{i=1}^{R}\sum_{j=1}^{R}\left(\frac{\partial \xi_i}{\partial P}\right)\left(\frac{\partial \xi_j}{\partial P}\right)\left(\frac{\partial^2 G}{\partial \xi_i \partial \xi_j}\right). \qquad (8.4\text{-}41)$$

Since $\partial^2 G/\partial \xi_i \partial \xi_j$ is assumed to be positive definite, the right side of equation 8.4-41 is negative, unless $\partial \boldsymbol{\xi}/\partial P = \mathbf{0}$. This is the case only if every $\bar{\nu}_j$ is zero in equation 8.4-40.

In the case of μ_k^*, in similar fashion, but assuming only that $\partial^2 G/\partial \xi_i \partial \xi_j$ is positive definite, we can show that

$$\frac{\partial n_k}{\partial \mu_k^*} < 0, \qquad (8.4\text{-}41\text{a})$$

provided that at least one ν_{kj} is not zero (i.e., species k is not inert).

In the case of n_z, for an ideal solution, from Table 8.1, $\partial \boldsymbol{\xi}/\partial n_z = \mathbf{0}$ if and only if $\bar{\nu}_j = 0$ for all j. Furthermore, $\partial n_t/\partial n_z > 0$ for the same conditions.

8.4.3 Nonstoichiometric Formulation

For simplicity, we consider only the case of a single phase. The generalization to the multiphase case is straightforward but tedious and is given by Smith (1969) for ideal solutions. We begin by writing equations 8.2-5 in the form of the equilibrium conditions and the element-abundance constraints:

$$\frac{\mu_i}{RT} = \sum_{i=1}^{M} a_{ki}\psi_k; \qquad i = 1, 2, \ldots, N'; \qquad (3.6\text{-}1)$$

$$\sum_{i=1}^{N'} a_{ji}n_i - b_j = 0; \qquad j = 1, 2, \ldots, M. \qquad (2.2\text{-}1)$$

Sensitivity Matrices for (T, P) Problems

The sensitivity coefficients of $\mathbf{n}$ and $\boldsymbol{\psi}$ are obtained by using equation 8.2-9, which becomes

$$\sum_{l=1}^{N'} \left(\frac{\partial n_l}{\partial p_k}\right) \frac{\partial (\mu_i/RT)}{\partial n_l} - \sum_{l=1}^{M} \left(\frac{\partial \psi_l}{\partial p_k}\right) a_{li} = -\frac{\partial (\mu_i/RT)}{\partial p_k}; \quad i = 1, 2, \ldots, N' \tag{8.4-42a}$$

and

$$\sum_{l=1}^{N'} \left(\frac{\partial n_l}{\partial p_k}\right) a_{jl} = \frac{\partial b_j}{\partial p_k}; \quad j = 1, 2, \ldots, M. \tag{8.4-42b}$$

Equations 8.4-42 are a set of $(N' + M)$ linear equations in the unknowns $\partial \mathbf{n}/\partial p_k$ and $\partial \boldsymbol{\psi}/\partial p_k$, which can be solved when a chemical potential expression is specified.

In the special case of ideal solutions we may reduce equations 8.4-42 in number to $(M + \pi)$. This is analogous to the situation for the ideal-system equilibrium algorithms discussed in Chapter 6. We give here the derivation for a single ideal solution. Then $\partial \mu_i/\partial n_j$ is given by equation 4.3-4, and equation 8.4-42a yields

$$\frac{\partial n_j}{\partial p_k} = n_j \left[\sum_{l=1}^{M} a_{lj} \left(\frac{\partial \psi_l}{\partial p_k}\right) + \frac{\partial \ln n_t}{\partial p_k} - \frac{\partial (\mu_j^*/RT)}{\partial p_k} \right]. \tag{8.4-43}$$

Substitution of equation 8.4-43 in equation 8.4-42b then yields the M linear equations

$$\sum_{l=1}^{M} \frac{\partial \psi_l}{\partial p_k} \sum_{i=1}^{N'} a_{ji} a_{li} n_i + b_j \left(\frac{\partial \ln n_t}{\partial p_k}\right) = \frac{\partial b_j}{\partial p_k} + \sum_{i=1}^{N'} a_{ji} n_i \frac{\partial (\mu_i^*/RT)}{\partial p_k};$$

$$j = 1, 2, \ldots, M. \tag{8.4-44a}$$

We obtain a final equation by combining equation 8.4-43 with the derivative of the definition of n_t,

$$\frac{\partial n_t}{\partial p_k} = \sum_{l=1}^{N'} \frac{\partial n_l}{\partial p_k} + \frac{\partial n_z}{\partial p_k}, \tag{8.4-45}$$

to yield

$$\sum_{l=1}^{M} b_l \left(\frac{\partial \psi_l}{\partial p_k}\right) - n_z \frac{\partial \ln n_t}{\partial p_k} = \sum_{i=1}^{N'} n_i \frac{\partial (\mu_i^*/RT)}{\partial p_k} - \frac{\partial n_z}{\partial p_k}. \tag{8.4-44b}$$

Equations 8.4-44 are the required set of $(M+1)$ linear equations in the unknowns $\partial\psi/\partial p_k$ and $\partial \ln n_t/\partial p_k$. The sensitivity coefficients of $\mathbf{n}$ are obtained by substituting these quantities in equation 8.4-43. For all the parameters $\{T, P, \mathbf{b}, \boldsymbol{\mu}^*, n_z\}$, the coefficient matrices of equations 8.4-44 are identical, and only the right sides differ. The right sides for the different parameters are given in Table 8.2, together with $\partial(\mu_j^*/RT)/\partial p_k$ of equation 8.4-43.

The preceding results show qualitatively that the most important species in determining the sensitivity coefficients are the most abundant ones since they contribute the most to the coefficient matrix and the right sides of equations 8.4-44. Species present in only minor amounts affect the sensitivity coefficients insignificantly. For such species, an estimate of the sensitivity coefficient $\partial n_k/\partial \mu_k^*$ is given by

$$\frac{\partial n_k}{\partial \mu_k^*} = -\frac{n_k}{RT}. \qquad (8.4\text{-}46)$$

This is derived from equation 8.4-43 by setting $\partial\psi_l/\partial\mu_k^* = \partial \ln n_t/\partial\mu_k^* = 0$.

Example 8.4 To illustrate the use of sensitivity coefficients to calculate the approximate effect of parameter changes on the equilibrium solution, we consider the system described in Examples 6.1 and 6.2 (White et al., 1958) for a change of pressure (Smith, 1969). For this purpose, we use equation 8.2-8 in the form

$$n_i = n_i^\circ + \left(\frac{\partial n_i}{\partial P}\right)_0 \delta P, \qquad (A)$$

where $\mathbf{n}^\circ$ is the equilibrium solution at the original conditions of the problem and δP is arbitrarily chosen to be 5 atm, an increase of about 10% of the original pressure.

Table 8.2 Evaluation of Quantities in Equations 8.4-43 and 8.4-44 for the Various Parameters (Ideal Solution)

Parameter	$\partial(\mu_j^*/RT)/\partial p_k$ (equation 8.4-43)	Right Side of Equation 8.4-44a	Right Side of Equation 8.4-44b
T	$\dfrac{-h_j^*}{RT^2}$	$-\left(\dfrac{1}{RT^2}\right)\sum_{i=1}^{N'} a_{ji} n_i h_i^*$	$-\left(\dfrac{1}{RT^2}\right)\sum_{i=1}^{N'} n_i h_i^*$
P	$\dfrac{v_j^*}{RT}$	$\dfrac{1}{RT}\sum_{i=1}^{N'} a_{ji} n_i v_i^*$	$\left(\dfrac{1}{RT}\right)\sum_{i=1}^{N'} n_i v_i^*$
b_k	0	δ_{jk}	0
μ_k^*	$\dfrac{\delta_{jk}}{RT}$	$\dfrac{a_{jk} n_k}{RT}$	$\dfrac{n_k}{RT}$
n_z	0	0	-1

Sensitivity Matrices for (T, P) Problems

Solution The value of $\partial n_i/\partial P$ is calculated by using equations 8.4-43 and 8.4-44 with the aid of Table 8.2. For this problem, these equations are

$$\sum_{l=1}^{3}\left(\frac{\partial \psi_l}{\partial P}\right)\sum_{i=1}^{10} a_{ji}a_{li}n_i^\circ + b_j\left(\frac{\partial \ln n_t}{\partial P}\right) = \frac{b_j}{P}; \quad j=1,2,3 \quad \text{(from 8.4-44a)},$$

$$\sum_{l=1}^{3} b_l\left(\frac{\partial \psi_l}{\partial P}\right) = \frac{n_t^\circ}{P} \quad \text{(from 8.4-44b)},$$

and

$$\frac{\partial n_j}{\partial P} = n_j^\circ\left[\sum_{l=1}^{3} a_{lj}\left(\frac{\partial \psi_l}{\partial P}\right) + \frac{\partial \ln n_t}{\partial P} - \frac{1}{P}\right] \quad \text{(from 8.4-43)},$$

where $P = 51$ atm, and $\mathbf{n}^\circ$ and $\mathbf{b}$ are given in Figure 6.4. The results are shown in Table 8.3. The approximate values at 56 atm agree with the correct values to within about 0.5%.

8.4.4 Second-Order Sensitivity Coefficients

We now present an extension of the preceding technique that enables us to calculate the second derivatives of the equilibrium mole numbers with respect to the problem parameters.

Table 8.3 Comparison of Correct[a] and Approximate[b] Equilibrium Mole Numbers for Example 8.4

Species	Pressure, atm		
	51	56	
	Original	Correct[a]	Approximate[b]
H_2O	0.7831	0.7894	0.7898
N_2	0.4852	0.4855	0.4855
H_2	0.1477	0.1440	0.1438
OH	0.09688	0.09425	0.09409
H	0.04067	0.03826	0.03807
O_2	0.03732	0.03624	0.03617
NO	0.02740	0.02701	0.02699
O	0.01795	0.01686	0.01677
N	0.001414	0.001348	0.001343
NH	0.0006932	0.0006845	0.0006841

[a] Calculated from VCS algorithm in Appendix D.
[b] Calculated from equation A in Example 8.4.

To obtain the second derivatives, we can differentiate twice either equations 3.4-5 or 3.6-1 and 2.2-1. To illustrate the approach, we state the result for the nonstoichiometric formulation obtained from the latter. The following set of $(M+1)$ linear equations in the unknowns $\partial^2 \psi / \partial p_k \partial p_m$ and $\partial^2 \ln n_t / \partial p_k \partial p_m$ is obtained in the case of an ideal solution:

$$\sum_{l=1}^{M} \left(\frac{\partial^2 \psi_l}{\partial p_k \partial p_m} \right) \sum_{i=1}^{N'} a_{ji} a_{li} n_i + b_j \left(\frac{\partial^2 \ln n_t}{\partial p_k \partial p_m} \right) = \left(\frac{\partial^2 b_j}{\partial p_k \partial p_m} \right)$$

$$+ \sum_{i=1}^{N'} a_{ji} n_i \frac{\partial^2 (\mu_i^*/RT)}{\partial p_k \partial p_m} - \sum_{i=1}^{N'} a_{ji} \frac{(\partial n_i/\partial p_k)(\partial n_i/\partial p_m)}{n_i} ;$$

$$j = 1, 2, \ldots, M; \quad (8.4\text{-}47a)$$

$$\sum_{l=1}^{M} b_l \left(\frac{\partial^2 \psi_l}{\partial p_k \partial p_m} \right) - n_z \left(\frac{\partial^2 \ln n_t}{\partial p_k \partial p_m} \right) = \sum_{i=1}^{N'} n_i \frac{\partial^2 (\mu_i^*/RT)}{\partial p_k \partial p_m}$$

$$- \sum_{i=1}^{N'} \frac{(\partial n_i/\partial p_k)(\partial n_i/\partial p_m)}{n_i}$$

$$+ n_t \left(\frac{\partial \ln n_t}{\partial p_k} \right) \left(\frac{\partial \ln n_t}{\partial p_m} \right) - \frac{\partial^2 n_z}{\partial p_k \partial p_m}. \quad (8.4\text{-}47b)$$

The first-order sensitivity coefficients appearing in equations 8.4-47 are obtained from equations 8.4-43 and 8.4-44. The second-order sensitivity coefficients are then given by

$$\frac{\partial^2 n_j}{\partial p_k \partial p_m} = n_j \frac{\partial^2 \ln n_t}{\partial p_k \partial p_m} + \sum_{l=1}^{M} a_{lj} \frac{\partial^2 \psi_l}{\partial p_k \partial p_m}$$

$$- \frac{\partial^2 (\mu_j^*/RT)}{\partial p_k \partial p_m} + \frac{(\partial n_j/\partial p_k)(\partial n_j/\partial p_m)}{n_j}. \quad (8.4\text{-}48)$$

8.5 EFFECT OF ERRORS IN μ^* ON EQUILIBRIUM COMPOSITION

We consider here the problem of estimating the precision of the computed equilibrium composition in terms of the precision of the standard chemical potentials (μ^*). For example, these may be given as $\mu^* \pm \delta\mu^*$, where $\delta\mu_j^*$ represents the standard deviation of the uncertainty in μ_j^* and $(\delta\mu_j^*)^2$ is the variance. In such a case it is useful to be able to estimate the corresponding measure of the uncertainty in the equilibrium composition $\delta\mathbf{n}$.

Effect of Errors in μ^* on Equilibrium Composition

This problem may be solved exactly by means of a Monte Carlo technique (Bard, 1974, pp. 46–47) that consists of calculating equilibrium compositions at a large number of sets of μ^* values within the given uncertainty range and then computing the corresponding uncertainty range of the composition. This, however, is time consuming and rarely justified. A more convenient alternative is to use equation 8.2-8 in an approximate way to calculate the variance of the uncertainty δn_i^2. If we assume that the uncertainties in the parameters are statistically independent, δn_i^2 can be estimated by

$$\delta n_i^2 = \sum_{j=1}^{N} \left(\frac{\partial n_i}{\partial \mu_j^*}\right)^2 (\delta \mu_j^*)^2. \tag{8.5-1}$$

If $\delta \mu_j^*$ is not known, calculation of the sensitivity coefficients $\partial n_i/\partial \mu_j^*$ is a guide as to whether this lack of knowledge is significant. For example, if $|\partial n_i/\partial \mu_j^*|$ is small, the effect of uncertainties in μ_j^* on n_i is also small, as when n_j is present at equilibrium in only a minor amount. In general, examination of the entries $\partial n_i/\partial \mu_j^*$ in the sensitivity matrix provides a guide as to which $\delta \mu_j^*$ are the most important.

Example 8.5 For the system $\{(C_6H_5CH_3, H_2, C_6H_6, CH_4), (C, H)\}$ at 980 K and 43 atm with $\mathbf{b} = (7, 16)^T$ (Björnbom, 1975), suppose that the dimensionless standard free energies of formation (μ_j°/RT) together with the standard deviations of their uncertainties are $(39.54 \pm 0.50, 0, 32.35 \pm 0.50, 2.054 \pm 0.075)^T$. Estimate the standard deviations of the uncertainties in the computed equilibrium mole numbers, using equation 8.5-1 (Smith, 1978).

Solution From equations 8.4-37 and 8.4-38 and with the aid of Table 8.1, or from 8.4-43 and 8.4-44 with the aid of Table 8.2, and $\mathbf{n}^\circ = (0.1183, 0.0787, 0.5439, 2.9081)^T$ (Smith, 1976), we obtain the sensitivity coefficients relative to n_i°, $[\partial n_i/\partial(\mu_j^\circ/RT)]/n_i^\circ$, given in the matrix

$$\mathbf{J} = \begin{pmatrix} -0.790 & -0.006 & 0.877 & 0.267 \\ -0.009 & -0.991 & -0.100 & 0.663 \\ 0.191 & -0.014 & -0.213 & -0.054 \\ 0.011 & 0.018 & -0.010 & -0.016 \end{pmatrix}.$$

Entry i, j of the matrix is the rate of change of the equilibrium mole number of species i, n_i, with respect to μ_j°/RT of species j relative to n_i. The second column is not important since μ°/RT for H_2 is exactly zero. Apart from this column, the entries largest in magnitude are -0.790 (J_{11}) and 0.877 (J_{13}), indicating that uncertainties in μ_1°/RT and μ_3°/RT can significantly affect n_1°.

From equations 8.5-1, the standard deviations of the uncertainties in the computed mole numbers relative to n_i°, $\delta n_i^\circ/n_i^\circ$ are $(0.591, 0.071, 0.143, 0.008)^T$.

This indicates that the uncertainty of the equilibrium mole number of $C_6H_5CH_3$ is relatively large and that the mole numbers of the other three species are much more precise.

8.6 CALCULATION OF THERMODYNAMIC DERIVATIVES FOR A SYSTEM AT CHEMICAL EQUILIBRIUM

It is often required to determine the value of a first partial derivative involving any three of the functions P, V, T, S, U, H, A, and G. In most cases this is done by relating the derivative in question to other quantities, including other derivatives, which are assumed to be measurable or calculable. For example, the heat capacity at constant volume C_V may be determined from C_P and volumetric quantities by

$$C_V \equiv \left(\frac{\partial U}{\partial T}\right)_V = C_P + T\frac{(\partial V/\partial T)_P^2}{(\partial V/\partial P)_T}, \qquad (8.6\text{-}1)$$

where $C_P = (\partial H/\partial T)_P$. Hence C_V is related to three thermodynamic first partial derivatives.

We may wish to carry out such a determination for a system under the constraint of chemical equilibrium. Since the three derivatives on the right side of equation 8.6-1 frequently occur in these determinations, we need to be able to evaluate them, and sensitivity analysis is important for this. Thus for a system at equilibrium, the three derivatives are given by

$$C_P = \sum_{i=1}^{N} n_i \left(\frac{\partial \bar{h}_i}{\partial T}\right) + \sum_{i=1}^{N} \bar{h}_i \left(\frac{\partial n_i}{\partial T}\right), \qquad (8.6\text{-}2)$$

$$\left(\frac{\partial V}{\partial T}\right)_P = \sum_{i=1}^{N} n_i \left(\frac{\partial \bar{v}_i}{\partial T}\right) + \sum_{i=1}^{N} \bar{v}_i \left(\frac{\partial n_i}{\partial T}\right), \qquad (8.6\text{-}3)$$

and

$$\left(\frac{\partial V}{\partial P}\right)_T = \sum_{i=1}^{N} n_i \left(\frac{\partial \bar{v}_i}{\partial P}\right) + \sum_{i=1}^{N} \bar{v}_i \left(\frac{\partial n_i}{\partial P}\right), \qquad (8.6\text{-}4)$$

where $\partial n_i/\partial T$ and $\partial n_i/\partial P$ are the sensitivity coefficients of n_i with respect to T and P, respectively. The calculation of these has been discussed in Section 8.4.

PROBLEMS

8.1 The standard free energy of formation of acetaldehyde [C_2H_4O (g)] is -133.72 kJ mole^{-1} at 25°C. Calculate the value at 750 K, if the standard enthalpy of formation is -166.36 kJ mole^{-1} at 25°C and molar heat capacities are as follows, in J mole^{-1} K^{-1} with T in kelvins (data from Smith and Van Ness, 1975, pp. 106, 120, 393):

C(gr): $C_P = 16.86 + 4.77 \times 10^{-3}T - 8.54 \times 10^5 T^{-2}$
H$_2$: $C_P = 27.28 + 3.26 \times 10^{-3}T + 0.50 \times 10^5 T^{-2}$
O$_2$: $C_P = 29.96 + 4.18 \times 10^{-3}T - 1.67 \times 10^5 T^{-2}$
C$_2$H$_4$O: $C_P = 14.07 + 149.46 \times 10^{-3}T - 5.12 \times 10^{-5} T^2$

8.2 Consider the stoichiometric equation $\nu_1 A_1 + \nu_2 A_2 = A_3$, and suppose that the *total* feed input (no A_3 in feed) is fixed. Determine whether there is an optimal ratio of reactants in the feed, n_2°/n_1°, to maximize the production of A_3 at equilibrium. Assume that the system is a single-phase ideal solution (cf. Denbigh, 1981, p. 176; Prigogine and Defay, 1954, p. 134).

8.3 Show, by considering the stoichiometric formulation in the form

$$\sum_{i=1}^{N} \nu_{ij}\mu_i(\mathbf{n}, T, P, \mathbf{b}, \boldsymbol{\mu}^w, \mathbf{n}_z) = 0; \quad j = 1, 2, \ldots, R$$

$$\sum_{i=1}^{N} a_{li} n_i = b_l; \quad l = 1, 2, \ldots, M$$

and differentiating with respect to a parameter p_k, that $\partial n_i/\partial p_k$ is determined by the solution of the N linear equations

$$\sum_{i=1}^{N} \sum_{m=1}^{N} \nu_{ij}\left(\frac{\partial \mu_i}{\partial n_m}\right)\left(\frac{\partial n_m}{\partial p_k}\right) = -\sum_{i=1}^{N} \nu_{ij}\left(\frac{\partial \mu_i}{\partial p_k}\right); \quad j = 1, 2, \ldots, R$$

$$\sum_{i=1}^{N} a_{li}\left(\frac{\partial n_i}{\partial p_k}\right) = \frac{\partial b_l}{\partial p_k}; \quad l = 1, 2, \ldots, M$$

(cf. Hochstim and Adams, 1962; Neumann, 1963). Show that these latter equations are essentially equivalent to equations 8.4-37 and 8.4-38.

8.4 Show that for an ideal solution and $R = 1$,

$$\frac{\partial n_{t'}}{\partial n_z} = \frac{\bar{\nu}^2}{n_t}\left[\sum n_i\left(\frac{\nu_i}{n_i} - \frac{\bar{\nu}}{n_t}\right)^2\right]^{-1}$$

and for an ideal-gas solution,

$$\frac{\partial n_t}{\partial P} = -\frac{RT\bar{\nu}}{P}\left[\sum n_i\left(\frac{\nu_i}{n_i} - \frac{\bar{\nu}}{n_t}\right)^2\right]^{-1}.$$

8.5 Show that, where x is mole fraction,

$$\frac{\partial x_i}{\partial p_k} = \frac{n_t(\partial n_i/\partial p_k) - n_i(\partial n_t/\partial p_k)}{n_t^2}.$$

8.6 Verify the entries in Table 8.2 from the three equations indicated.

8.7 Derive equations 8.4-47 and 8.4-48.

8.8 Derive equations 8.4-43 and 8.4-44 for the case of several ideal multi-species and single-species phases.

8.9 (a) For two ideal-solution systems at the same T, P, $\mathbf{b}$, and $\boldsymbol{\mu}^*$ but with differing amounts of inert species n_z^0 and n_z^1, show that

$$n_t^0 \ln\left(1 + \frac{\delta n_z}{n_t^0}\right) + n_t^1 \ln\left(1 - \frac{\delta n_z}{n_t^1}\right) \leq 0,$$

where

$$\delta n_z = n_z^1 - n_z^0.$$

(b) Show that the result of part a follows directly from the identity $\ln(1 + x) \leq x$.

8.10 When μ_i^* for a species present in a small amount at equilibrium is in error by an amount $\delta\mu_i^*$, show that

$$\frac{\partial n_i}{\partial \mu_i^*} \simeq \frac{-n_i}{RT},$$

and hence that the change in the equilibrium amount of species i is given approximately by

$$\delta n_i = n_i^0\left[\exp\left(\frac{-\delta\mu_i^*}{RT}\right) - 1\right].$$

8.11 Calculate $\partial n_i/\partial \mu_j^0$ and $\partial n_i/\partial b_j$ for the system in Example 6.1.

8.12 Ratajczykowa (1972) has considered various feed ratios of H_2O/C_2H_6 for the system $\{(H_2, O_2, CO, CO_2, H_2O, CH_4, C_2H_2, C_2H_4, C_2H_6, C_6H_6), (C, H, O)\}$ at 1400 K and 0.0263 atm. The standard free energies of formation (JANAF, 1971; Zwolinski et al., 1974, for C_2H_6 and C_6H_6), in kJ mole^{-1}, are $\Delta G_f^\circ = (0, 0, 0, -235.09, -396.29, -170.13, 63.58, 148.25, 151.80, 195.77, 340.33)^T$. Use the result of Problem 8.5 and the sensitivity coefficients $\partial x_i/\partial \mathbf{b}$ to examine whether there is an optimal feed ratio of H_2O/C_2H_6 for the formation of H_2 and/or CO.

CHAPTER NINE

Practical Considerations and Special Topics

In this chapter we discuss several practical considerations and special topics that are important in the use or adaptation of chemical equilibrium algorithms in general. The practical considerations involve making initial estimates, possible numerical singularities, the situation when $C \neq M$, utilization of free-energy data when given in the form of equilibrium constants, and the treatment of minor species. The special topics are equilibrium problems in which there are compositional restrictions in addition to the usual element-abundance constraints, treatment of isomers and isotopes, and aspects of equilibrium computations when the thermodynamic constraints are other than (T, P).

9.1 INITIAL ESTIMATES FOR EQUILIBRIUM ALGORITHMS

Making initial estimates for the equilibrium algorithms given in the appendixes involves determining either $\mathbf{n}^{(0)}$ satisfying the element-abundance and nonnegativity constraints or Lagrange multipliers $\boldsymbol{\lambda}^{(0)}$. The former is required for all stoichiometric algorithms and is desirable for the BNR algorithm presented in Appendix C, and the latter is required for the nonstoichiometric algorithms in Appendix B. In this section we discuss each of these two cases in turn.

9.1.1 Estimate of Mole Numbers

An initial estimate of the mole numbers $\mathbf{n}^{(0)}$ may be generated either by a hand calculation procedure or by a linear programming procedure. The latter is incorporated as an option in the BNR and VCS algorithms in Appendixes C and D, and the former must be used for the other two programs. Because of the logarithmic term in the chemical potential expression, $n_i^{(0)}$ in a multispecies phase must be chosen to be positive, but for a single-species phase, $n_i^{(0)}$ may be taken to be zero.

9.1.1.1 Hand Calculation Procedure

A relatively straightforward way of determining $\mathbf{n}^{(0)}$ by hand calculation is to proceed as follows:

1. Calculate $\mathbf{b}$ from the given data, if it is not specified directly.
2. Choose a set of $C \equiv$ rank $(\mathbf{A})$ component species with linearly independent formula vectors either by inspection or by the method discussed in Section 2.3.3; if elemental species are present in the system, they may be conveniently chosen as components; it may also be useful to use μ^* as a guide in choosing the components.
3. Set $\{n_i^{(0)}; i = C + 1, C + 2, \ldots, N'\}$ to an arbitrary small positive value, and solve the linear equations

$$\sum_{i=1}^{C} a_{ki} n_i^{(0)} = b_k - \sum_{i=C+1}^{N'} a_{ki} n_i^{(0)} \qquad (9.1\text{-}1)$$

for $\{n_i^{(0)}; i = 1, 2, \ldots, C\}$.

9.1.1.2 Linear Programming Procedure

The linear programming procedure incorporated as an option in the VCS algorithm (Smith and Missen, 1968) is based on the use of the chemical potential expression without the logarithmic term. A simplified version is incorporated in the BNR algorithm. It involves solving a certain linear programming problem and then proceeding according to the nature of its solution.

The linear programming problem is

$$\left.\begin{array}{c} \min \mathbf{n}^T \boldsymbol{\mu}^*, \\ \text{such that} \quad \mathbf{An} = \mathbf{b}, \quad n_i \geqslant 0. \end{array}\right\} \qquad (9.1\text{-}2)$$

Linear programming algorithms are readily available, and the algorithms in Appendixes C and D allow the use of any such suitable program. The solution is one of two types.

The first type is characterized by C positive values, denoted by $\{n_i^{\text{LP}}; i = 1, 2, \ldots, C\}$, with the rest being zero (superscript LP denotes a linear programming value). In this case the total number of moles $n_{t\alpha}$ in each multispecies phase is determined, and the stoichiometric matrix $\mathbf{N}$ is calculated in canonical form, using these C species as component species. Then, making the approximations that $n_{t\alpha}$ and $\{n_i^{\text{LP}}; i = 1, 2, \ldots, C\}$ remain constant, we compute μ_k for these component species, and setting the reaction free energy

Initial Estimates for Equilibrium Algorithms

changes to zero, we also set

$$\xi_j = \delta_{i\alpha} n_{t\alpha} \exp\left\{\frac{1}{RT}\left[-\mu_i^* - \sum_{k=1}^{C} \mu_k(\mathbf{n}^{LP})\nu_{kj}\right]\right\}; \quad i = C+1, C+2, \ldots, N',$$

(9.1-3)

where $j = i - C$ and $\delta_{i\alpha}$ is unity for a species in a multispecies phase and zero otherwise. Finally, we introduce a step-size parameter ω and set

$$n_i^{(0)} = n_i^{LP} + \omega \sum_{j=1}^{R} \nu_{ij}\xi_j; \quad i = 1, 2, \ldots, N',$$

(9.1-4)

where ω is chosen so that all mole numbers remain positive, and G for the system is approximately minimized [see Smith and Missen (1968) for details].

The second type is characterized by fewer than C positive values of n_i^{LP}, with the rest being zero. In this case we set $n_i^{(0)}$ equal to the greater of n_i^{LP} and an arbitrarily small positive number, such as 10^{-15}.

9.1.2 Estimate of Lagrange Multipliers

We describe two methods used for the nonstoichiometric programs presented in Appendix B, one for the HP-41C program and one for the BASIC program.

9.1.2.1 HP-41C Program

In the HP-41C program the estimate is based on the assumption that species with relatively low values of μ^* tend to be more abundant at equilibrium. Thus we usually choose (see *User's Guide* in Appendix B) the set of C species with the lowest μ_i^* values as component species. The mole fractions of these species are set to unity in equation 4.4-39, and we obtain the following set of linear equations in $\ln z_l^{(0)}$:

$$\sum_{l=1}^{C} \alpha_{li} \ln z_l^{(0)} = \frac{\mu_i^*}{RT}; \quad i = 1, 2, \ldots, C.$$

(9.1-5)

The resulting value of $\boldsymbol{\theta}^{(0)}$ obtained from equations 4.4-35 and 4.4-36 is used as an initial estimate in the Newton-Raphson procedure to solve equations 4.4-33.

9.1.2.2 BASIC Program

In the BASIC program equations 4.4-39 are used for the entire set of N' species, with each mole fraction set to $1/N'$. This yields the following (overdetermined) set of linear equations in $\ln z_l^{(0)}$:

$$\sum_{l=1}^{C} \alpha_{li} \ln z_l^{(0)} = \ln \frac{1}{N'} + \frac{\mu_i^*}{RT}; \qquad i = 1, 2, \ldots, N'. \qquad (9.1\text{-}6)$$

A weighted least-squares solution to equation 9.1-6 is obtained by minimizing

$$\phi(\ln \mathbf{z}) = \frac{1}{2} \sum_{i=1}^{N'} w_i \left(\sum_{l=1}^{C} \alpha_{li} \ln z_l - \frac{1}{N'} - \frac{\mu_i^*}{RT} \right)^2. \qquad (9.1\text{-}7)$$

Setting $\partial \phi / \partial \ln z_k = 0$; $k = 1, 2, \ldots, C$, we obtain the following set of C linear equations in $\ln \mathbf{z}^{(0)}$:

$$\sum_{l=1}^{C} \ln z_l^{(0)} \sum_{i=1}^{N'} w_i \alpha_{li} \alpha_{ki} = \sum_{i=1}^{N'} w_i \alpha_{ki} \left(\frac{1}{N'} + \frac{\mu_i^*}{RT} \right); \qquad k = 1, 2, \ldots, C.$$

$$(9.1\text{-}8)$$

The weights are determined by an heuristic procedure incorporated in the program.

9.2 NUMERICAL SINGULARITIES AND THE NONNEGATIVITY CONSTRAINT

In the use of equilibrium algorithms that involve the solution of sets of linear equations in the iterative process (e.g., the BNR algorithm) there may occasionally arise situations in which the coefficient matrix of the equations becomes singular or may be numerically singular due to rounding errors. In such cases the algorithm fails, unless appropriate steps are taken. Such difficulties have been noted in the literature and have been discussed by Smith (1980a). In this section we analyze the sources of these difficulties and suggest corrective action when they occur.

From Section 3.8, we recall that the number of moles of a species is zero if and only if $n_{l\alpha}$ is zero. We discuss the two possible cases for $n_i = 0$, corresponding to the species being in a single-species phase or in a multispecies phase, as an ideal solution.

9.2.1 Single-Species Phases

We consider the case of any nonzero number of single-species phases in the species list, the amounts of which may or may not be zero at equilibrium, together with any number (including zero) of multispecies phases known to be present at equilibrium. Without loss of generality, we discuss in detail problems involving at most only one multispecies phase. Such problems involving single-species phases are particularly important in metallurgical systems (Madeley and Toguri, 1973b).

The coefficient matrix of the linear equations in the BNR algorithm at a given iteration is in this case

$$\beta = \begin{pmatrix} \mathbf{A}_1 \mathbf{\Lambda} \mathbf{A}_1^T & \mathbf{b}_1 & \mathbf{A}_2 \\ \mathbf{b}_1^T & \mathbf{Q} & \\ \mathbf{A}_2^T & & \end{pmatrix}, \tag{9.2-1}$$

where $\mathbf{A}_1$ and $\mathbf{A}_2$ are those parts of $\mathbf{A}$ referring to the multispecies phase and the single-species phases, respectively; that is, $\mathbf{A} = (\mathbf{A}_1, \mathbf{A}_2)$, where $\mathbf{A}_2$ is of order $C \times \pi_s$; $\mathbf{\Lambda} = \mathrm{diag}(n_1, n_2, \ldots, n_h)$, where h is the number of species in the multispecies phase, $(N' - \pi_s)$; $\mathbf{b}_1$ is the element-abundance vector for the multispecies phase,

$$\mathbf{b}_1 = \mathbf{A}_1 \mathbf{n}_1, \tag{9.2-2}$$

where $\mathbf{n}_1 = (n_1, n_2, \ldots, n_h)^T$; and $\mathbf{Q}$ is the null matrix of order $\pi \times \pi$, except that $Q_{11} = -n_z$.

Because all the entries in β that refer to the multispecies phase (entries with subscript 1) are replaced by zeros, β is always singular when $h = 0$. The BNR algorithm is thus not applicable to such cases, which are linear programming problems. Also, β is singular whenever the block of zeros in the lower right corner given by $\mathbf{Q}$ is too large. To prevent this, we must have $\pi_s \leq C - \pi_m$.

Since $\mathbf{A}_1 \mathbf{\Lambda} \mathbf{A}_1^T$ can be expressed as the product of a nonsingular matrix and its transpose, it is never singular (Smith, 1980a). Thus, for problems consisting of only one multispecies phase, β is never singular in principle. However, finite computer word lengths can cause β to be *numerically* singular in some cases. When $n_z = 0$, β is singular only if the columns of $(\mathbf{b}, \mathbf{A}_2)$ are linearly dependent. When $n_z \neq 0$, β is singular only if the columns of $\mathbf{A}_2$ are linearly dependent.

In the following, we classify the different types of situation that can give rise to a singular coefficient matrix β. We are primarily concerned with situations involving single-species phases, although we also discuss the case in which β may be numerically singular. We classify difficulties concerning single-species phases into two groups, those in which the columns of $(\mathbf{b}, \mathbf{A}_2)$ are linearly independent and linearly dependent.

9.2.1.1 Numerical Singularity When $\pi_s = 0$

Numerical singularity of β when $\pi_s = 0$ occurs, for example, when $\mathbf{b}$ is nearly all bound up in fewer than C species. Gordon and McBride (1971, 1976) point out that for the system $\{(O_2, H_2, H_2O), (H, O)\}$, with $\mathbf{b} = (2, 1)^T$, T and P may be such that the only species present with $x_i > 10^{-8}$ is H_2O. In this case β becomes

$$\beta = \begin{pmatrix} 4+\varepsilon_1 & 2+\varepsilon_2 & 2 \\ 2+\varepsilon_2 & 1+\varepsilon_3 & 1 \\ 2 & 1 & 0 \end{pmatrix},$$

where ε_i are small numbers of the order of magnitude of 10^{-8}. In a numerical calculation β may be singular. Gordon and McBride suggest treating such difficulties by altering **b** slightly.

9.2.1.2 Columns of (b, A_2) Linearly Independent

When the columns of (b, A_2) are linearly independent, two cases arise: (2.1) the system can be made up of the species in the multispecies phase only; that is, a vector n_1 can be found with all $n_{1i} > 0$, such that $A_1 n_1 = b$; and (2.2) the system cannot be made up in such a way. A concise way to distinguish these cases is whether the set $W = \{n_1: A_1 n_1 = b, n_{1i} > 0\}$ is nonempty (2.1) or empty (2.2).

9.2.1.2.1 W Nonempty

Equilibrium is first calculated with $n_i = 0$ for all single-species phases. Then the quantities

$$\Delta G_i = \mu_i - \sum_{k=1}^{C} a_{ki} \lambda_k \tag{9.2-3}$$

are calculated for each species in a single-species phase. If any $\Delta G_i < 0$, the species with the smallest value of ΔG_i is used in a recalculation of the equilibrium. This procedure is repeated until no negative ΔG_i values are obtained. Then all single-species phases actually present ($n_i > 0$) have been included in the equilibrium calculation. Equation 9.2-3 utilizes equation 3.8-1b to test whether the inclusion of species i marginally lowers G for the system.

9.2.1.2.2 W Empty

If W is empty, n_i must be positive at equilibrium for at least one of the single-species phases, but which of these may not be known *a priori*. Difficulties arise because there is no explicit procedure in the BNR algorithm for ensuring that $n_i > 0$ for single-species phases. If a single-species phase is incorrectly assumed to be present, its equilibrium n_i is calculated to be negative. When π_s is small, it may be feasible to insert and remove single-species phases in the course of a calculation in accordance with equation 9.2-3 (Lahiri, 1979), but this is tedious. Each problem must be treated as a special case.

For both cases described in Sections 9.2.1.2.1 and 9.2.1.2.2, numerical singularities may occur as described in Section 9.2.1.1. As Oliver et al. (1962)

and Eriksson (1975) point out, if portions of **b** are linearly dependent on one, or more than one, column of $\mathbf{A}_2$, all but a very small amount of those portions may be combined in the single-species phases, resulting in $\boldsymbol{\beta}$ appearing numerically singular. As an example (Oliver et al., 1962), for the system consisting of $Al_2O_3(s)$ together with several gaseous species containing the elements oxygen and hydrogen, when $(b_{Al}, b_O) = (2, 3)$, we may have

$$\boldsymbol{\beta} = \begin{pmatrix} \tau & \varepsilon_2 & 0 & b_H & 0 \\ \varepsilon_2 & \varepsilon_3 & 0 & \varepsilon_1 & 3 \\ 0 & 0 & 0 & 0 & 2 \\ b_H & \varepsilon_1 & 0 & 0 & 0 \\ 0 & 3 & 2 & 0 & 0 \end{pmatrix},$$

where τ is a relatively large number. The second and third columns/rows of $\boldsymbol{\beta}$ can result in numerical singularities. In such cases Oliver et al. (1962) alter **b** slightly.

9.2.1.3 Columns of $(\mathbf{b}, \mathbf{A}_2)$ Linearly Dependent

When the columns of $(\mathbf{b}, \mathbf{A}_2)$ are linearly dependent, the equilibrium problem may not have a unique solution, in contrast to the situation described in Sections 9.2.1.1 and 9.2.1.2 (Hancock and Motzkin, 1960). Four cases arise: (1) columns of $\mathbf{A}_2$ are linearly independent and W is nonempty; (2) as for case 1, but W is empty; (3) columns of $\mathbf{A}_2$ are linearly dependent and $\pi_m > 0$; and (4) as for case 3, but $\pi_m = 0$.

9.2.1.3.1 Columns of $\mathbf{A}_2$ Linearly Independent; W Nonempty

Provided that $n_z \neq 0$, this case may be treated in the same way as in the case described in Section 9.2.1.2.1. When $n_z = 0$, if any subset of the single-species phases with formula vectors linearly dependent on **b** is included in the calculation, $\boldsymbol{\beta}$ may be singular. This would occur when, at some stage of the calculation, a fraction of the element-abundance vector is distributed among the species in the multispecies phase and the remainder is distributed among the single-species phases. In such a situation $\mathbf{b}_1$ is linearly dependent on the columns of $\mathbf{A}_2$, so that $\boldsymbol{\beta}$ is singular, and the algorithm fails.

9.2.1.3.2 Columns of $\mathbf{A}_2$ Linearly Independent; W Empty

In this case the situation is similar to that described in Section 9.2.1.2.2.

9.2.1.3.3 Columns of $\mathbf{A}_2$ Linearly Dependent; $\pi_m > 0$

In this case $\boldsymbol{\beta}$ is singular if more than $C - 1$ single-species phases are included in the calculation. If W is nonempty, it may be feasible to use the procedure discussed in Section 9.2.1.2.1. If W is empty, at least one single-species phase must be removed, and the difficulties of the type described in Section 9.2.1.2.2 apply in addition.

9.2.1.3.4 Columns of $\mathbf{A}_2$ Linearly Dependent; $\pi_m = 0$

In this case $\boldsymbol{\beta}$ is always singular, and the problems are linear programming problems, to which the BNR algorithm is not applicable.

We summarize in Table 9.1 the classification of possible numerical singularities in the BNR algorithm, together with examples of the various cases. We note that, although we have discussed in detail only the case $\pi_m = 1$, Table 9.1 essentially applies when $\pi_m > 1$, provided that all multispecies phases are actually present at equilibrium. For problems of the types described in Sections 9.2.1.1, 9.2.1.2.1, and 9.2.1.3.1, numerical singularities may be obviated in the BNR algorithm by slight alteration of $\mathbf{b}$ (Oliver et al., 1962; Gordon and McBride, 1971, 1976; Eriksson, 1975).

The stoichiometric algorithms are ideally suited to problems involving single-species phases since they do not give rise to numerical singularities. The general procedure has been described by Cruise (1964). If n_i for a single-species phase tends to become negative during the course of the calculation, it is set to zero. On each subsequent iteration, ΔG_j for the stoichiometric equation forming one mole of the species is examined, where ΔG_j is given by

$$\Delta G_j = \mu_i + \sum_{k=1}^{C} \nu_{kj}\mu_k, \qquad (9.2\text{-}4)$$

and $j = i - C$. If $\Delta G_j < 0$, the species is reintroduced into the calculation. By this means, single-species phases may readily enter and leave the main calculation during the course of the iterative procedure. Even case 3.4 in Table 9.1 may be treated in this way.

Other approaches to treating multispecies phases have been described by Bigelow (1970), Ma and Shipman (1972), and Madeley and Toguri (1973a, 1973b).

Example 9.1 To illustrate a problem of type 3.3 in Table 9.1, consider the system $\{(O_2(g), H_2O(g), CH_4(g), CO(g), CO_2(g), H_2(g), CHO(g), CH_2O(g), OH(g), Fe_3O_4(s), FeO(s), Fe(s), C(s), CaCO_3(s), CaO(s), N_2(g)), (C, O, H, Fe, Ca, N_2)\}$ (Madeley and Toguri, 1973a, 1973b). Calculate the equilibrium mole numbers of the species at 1050 K and 1 atm, using the free-energy data given by Smith (1976) and both the VCS and BNR algorithms.

Solution Using the VCS algorithm, we obtain the solution shown in Figure 9.1. Only the two solid species CaO(s) and Fe(s) have $n_i > 0$ at equilibrium.

Since the columns of $\mathbf{A}_2$ are linearly dependent, the BNR algorithm fails if all solids are included in the computation because $\boldsymbol{\beta}$ is then singular. Since W is empty, at least one iron-containing and one calcium-containing species must have $n_i > 0$ at equilibrium, but it is not apparent *a priori* for which species this is the case. One way of using the BNR algorithm is to solve the following six subproblems, in each of which the presence at equilibrium of the two solids

Table 9.1 Classification of Possible Numerical Singularities in BNR Algorithm (Ideal Solution; $\pi_s \geq 0$)

Type	Example	Comment
1. $\pi_s = 0$	$\{(O_2, H_2, H_2O), (H, C)\}$	Many experience numerical singularity
2. Columns of $(\mathbf{b}, \mathbf{A}_2)$ linearly independent		
2.1 W'^a nonempty	$\{(O_2, CO, CO_2, C(s)), (C, O)\}$	Procedure described in text; may experience numerical singularity
2.2 W'^a empty	$\{(H_2, O_2, CO, CO_2, H_2O, C(s), Fe(s), FeO(s)), (H, O, C, Fe)\}$	At least one single-species phase has $n_i > 0$ at equilibrium; each problem a special case
3. Columns of $(\mathbf{b}, \mathbf{A}_2)$ linearly dependent		
3.1 Columns of $\mathbf{A}_2$ linearly independent; W'^a nonempty	$\{(H_2, O_2, H_2O(g), H_2O(\ell)), (H, O)\}$	As for type 2.1[b]
3.2 As in type 3.1 but W'^a empty	See Example 9.1	As for type 2.2[b]
3.3 Columns of $\mathbf{A}_2$ linearly dependent $\pi_m > 0$	$\{(O_2, CO, CO_2, Fe(s), FeO(s), Fe_2O_3(s)), (O, C, Fe)\}$	At least one single-species phase has $n_i = 0$ at equilibrium; if W empty, type 2.2[b] applies in addition
3.4 $\pi_m = 0$	$\{(CaCO_3(s), CaO(s), CO_2(g)), (Ca, C, O)\}$	Linear programming problem; BNR algorithm not applicable

[a] $W = \{\mathbf{n}_1; \mathbf{A}_1\mathbf{n}_1 = \mathbf{b}, n_{1i} > 0\}$, where $\mathbf{n}_1$ are the species in the multispecies phases.
[b] Types listed in column 1 of this table.

VCS CALCULATION METHOD

BLAST FURNACE PROBLEM

```
    15 SPECIES         5 ELEMENTS              5 COMPONENTS
     9 PHASE1 SPECIES           0 PHASE2 SPECIES       6 SINGLE SPECIES PHASES

    PRESSURE                 1.000 ATM
    TEMPERATURE           1050.000 K
    PHASE1 INERTS          187.100

    ELEMENTAL ABUNDANCES          CORRECT              FROM ESTIMATE

                         O    9.6151000000000D 01    9.6151000000000D 01
                         H    1.3766000000000D 01    1.3766000000000D 01
                         C    8.8294000000000D 01    8.8294000000000D 01
                         FE   4.2827000000000D 01    4.2827000000000D 01
                         CA   7.5620000000000D-01    7.5620000000000D-01

    USER ESTIMATE OF EQUILIBRIUM
    STAN. CHEM. POT. IN KJ./MOLE

    SPECIES       FORMULA VECTOR           STAN. CHEM. POT.    EQUILIBRIUM EST.

                  O  H  C FE CA SI (I)
    C (S)         0  0  1  0  0  0              0.0              8.82880D 01
    O2            2  0  0  0  0  1              0.0              4.76904D 01
    FE (S)        0  0  0  1  0  0              0.0              4.28230D 01
    H2            0  2  0  0  0  1              0.0              6.87800D 00
    CAO (S)       1  0  0  0  1  0          -5.29190D 02         7.55200D-01
    CH2O          1  2  1  0  0  1          -8.61100D 01         1.00000D-03
    CHO           1  1  1  0  0  1          -6.25500D 01         1.00000D-03
    OH            1  1  0  0  0  1           2.25900D 01         1.00000D-03
    CO2           2  0  1  0  0  1          -3.95970D 02         1.00000D-03
    H2O           1  2  0  0  0  1          -1.89870D 02         1.00000D-03
    CH4           0  4  1  0  0  1           2.48500D 01         1.00000D-03
    CO            1  0  1  0  0  1          -2.04640D 02         1.00000D-03
    FE3O4 (S)     4  0  0  3  0  0          -7.62660D 02         1.00000D-03
    CACO3 (S)     3  0  1  0  1  0          -9.42450D 02         1.00000D-03
    FEO (S)       1  0  0  1  0  0          -1.93930D 02         1.00000D-03

    ITERATIONS =    17
    EVALUATIONS OF STOICHIOMETRY =   10

    SPECIES         EQUILIBRIUM MOLES      MOLE FRACTION       DG/RT REACTION

    CO               8.1623915D 01         2.8917589D-01
    FE (S)           4.2827000D 01         1.0000000D 00
    H2               6.4270838D 00         2.2766226D-02
    CO2              6.6635330D 00         2.3607457D-02
    CAO (S)          7.5620000D-01         1.0000000D 00
    H2O              4.4381579D-01         1.5723434D-03        5.8795D-07
    CH4              6.5487405D-03         2.3200772D-05        3.4745D-06
    CH2O             2.3581110D-06         8.3542776D-09        1.4143D-06
    CHO              1.0516532D-06         3.7257802D-09        7.5988D-07
    OH               7.9295075D-11         2.8092531D-13       -9.4406D-06
    O2               1.7303566D-19         6.1302795D-22       -1.4721D-06
    FEO (S)          0.0                   0.0                  2.2076D 00
    FE3O4 (S)        0.0                   0.0                  1.0327D 01
    CACO3 (S)        0.0                   0.0                  1.7657D 00
    C (S)            0.0                   0.0                  2.5987D-01

    G/RT =   -2.5014683D 03
    TOTAL PHASE1 MOLES =     2.8226D 02

    ELEMENTAL ABUNDANCES       O      9.61510000D 01
                               H      1.37660000D 01
                               C      8.82940000D 01
                               FE     4.28270000D 01
                               CA     7.56200000D-01
```

Figure 9.1 VCS-algorithm solution for Example 9.1.

Table 9.2 BNR Algorithm Solutions for the Six Subproblems Described in Example 9.1

Species	Equilibrium Mole Numbers for Subproblem					
	1	2	3	4	5	6
$O_2 \times 10^{20}$	17.30	13.79	4.14	3.95	2.41	2.24
$N_2 \times 10^{-2}$	1.871	1.871	1.871	1.871	1.871	1.871
H_2O	0.444	0.399	0.241	0.236	0.190	0.184
$CH_4 \times 10^3$	6.55	7.460	10.42	10.50	11.19	11.28
CO	81.62	81.58	48.19	46.88	35.70	34.36
CO_2	6.66	5.95	2.07	1.97	1.20	1.12
H_2	6.43	6.47	6.62	6.63	6.67	6.68
$CHO \times 10^7$	10.52	10.56	6.78	6.61	5.18	5.00
$CH_2O \times 10^6$	2.36	2.38	1.66	1.62	1.31	1.27
$OH \times 10^{11}$	7.93	7.10	3.94	3.85	3.01	2.91
C(s)	0	0	38.03	38.68	51.38	52.05
Fe_3O_4(s)	—	—	—	—	14.28	14.28
FeO(s)	—	—	42.83	42.83	—	—
Fe(s)	42.83	42.83	—	—	—	—
$CaCO_3$(s)	—	0.756	—	0.756	—	0.756
CaO(s)	0.756	—	0.756	—	0.756	—
$G/RT \times 10^{-3}$	−2.5015	−2.5001	−2.3895	−2.3873	2.3258	−2.3233

indicated is assumed, in addition to that of carbon(s): (1) [Fe(s), CaO(s)]; (2) [Fe(s), $CaCO_3$(s)]; (3) [FeO(s), CaO(s)]; (4) [FeO(s), $CaCO_3$(s)]; (5) [Fe_3O_4(s), CaO(s)]; and (6) [Fe_3O_4(s), $CaCO_3$(s)].

The solutions to these six subproblems, obtained from the BNR algorithm of Appendix C, are given in Table 9.2. The one with lowest value of G/RT is problem 1 (in the preceding paragraph), and hence this is the desired solution, which agrees with the results obtained with use of the VCS algorithm.

Although the BNR algorithm has been used successfully on an *ad hoc* basis for this problem, the solution procedure would become more tedious as π_s increases. In contrast, the VCS algorithm readily treats such problems, regardless of the value of π_s.

9.2.2 Multispecies Phases

9.2.2.1 Ideal Solution

In Section 9.2.1 we essentially assumed that any multispecies phases in the problem formulation are actually present at equilibrium (i.e., $n_{t\alpha} > 0$). In this section, for $\pi_m > 0$, we do not assume any prior knowledge as to whether $n_{t\alpha} > 0$, and we allow for the possibility of $n_{t\alpha} = 0$. Extensions of the use of equations 9.2-3 and 9.2-4 to a multispecies phase involve examining the sign of a quantity T_α given by (cf. equations 3.8-4 and 3.8-5)

$$T_\alpha = 1 - \sum_{(\alpha)} \exp\left(-\frac{\mu_i^*}{RT} + \sum_{k=1}^{C} a_{ki}\psi_k\right) \qquad (9.2\text{-}5)$$

and

$$T_\alpha = 1 - \sum_{(\alpha)} \exp\left[\frac{1}{RT}\left(-\mu_i^* - \sum_{k=1}^{C} \nu_{kj}\mu_k\right)\right], \qquad (9.2\text{-}6)$$

where $j = (i - C)$ and (α) denotes summation over all species in phase α. When $n_{t\alpha}$ becomes very small in the calculations, it is set to zero. On subsequent iterations, if $T_\alpha < 0$, phase α is reintroduced into the calculations by setting $n_{i\alpha}$ to some arbitrarily small amount; if $T_\alpha \geq 0$, all $n_{i\alpha}$ remain zero. This procedure ensures that for all phases with $n_{t\alpha} = 0$ at equilibrium, $T_\alpha \geq 0$. The motivation for this procedure was discussed in Section 3.8.

The BNR algorithm in Appendix C does not incorporate the use of equation 9.2-5, although this could be done. If it were done, however, we note that numerical singularities may arise in addition to those discussed in Section 9.2.2.1.

The VCS algorithm given in Appendix D incorporates the use of equation 9.2-6. This does not lead to numerical singularities.

Boll (1961) has suggested the use of equation 9.2-5 in the context of the Brinkley algorithm, and the equation has also been used by Eriksson (1975).

9.2.2.2 Nonideal Solution

When a multispecies phase is nonideal, an extension of equations 9.2-5 and 9.2-6 must be used to test for its presence. For example, for the BNR algorithm, one possibility is to use, for all species in a potential phase,

$$\gamma_i(\mathbf{x})x_i = \exp\left[\frac{1}{RT}\left(-\mu_i^* + \sum_{k=1}^{M} a_{ki}\lambda_k\right)\right] \qquad (3.8\text{-}6)$$

at the given value of λ and then examine whether the resulting $\sum x_i$ is greater or less than unity. The difficulty with the use of equation 3.8-6, as opposed to that of equation 9.2-5, is that the former may have more than one solution, and it is not generally obvious how to find all of them. Use of equation 3.8-6 would require a separate computer subroutine that would compute the compositions of potential phases having the same value of λ. This approach has not been implemented in the literature, although an approximate version has been used by Eriksson (1975). A careful evaluation of its effectiveness has not been undertaken.

Provided that phases actually present at equilibrium have nonzero initial estimates in an algorithm, the correct equilibrium composition may be obtained, depending on these initial estimates, if they are sufficiently close to the solution. Otherwise, a local solution may be obtained with one, or more than one, of the phases absent.

Another approach to test for the presence of nonideal, multispecies phases has been implemented by Gautam and Seider (1979), in conjunction with the nonideal version of the BNR algorithm, which they call "phase-splitting." At a given stage in the calculation, they introduce tentative pairs of new phases α and β whose compositions satisfy

$$\gamma_i(\mathbf{x}_\alpha)x_{i\alpha} = \gamma_i(\mathbf{x}_\beta)x_{i\beta}. \tag{9.2-7}$$

The resulting phases are included in the calculations for several iterations to examine whether G decreases. If G does not decrease, the original composition of the system is maintained. Although Gautam and Seider indicate that this approach may fail when the equilibrium amount of a nonideal, multispecies phase is very small, it apparently works well in other cases.

9.3 THE CASE OF $C \equiv$ Rank (A) $\neq M$

The quantity $C \equiv$ rank (A) is the number of linearly independent element-abundance constraints. Usually $C = M$, but when they differ, C must be used rather than M. In this latter case, for the BNR algorithm and for the algorithms of Appendix B, the formula matrix $\mathbf{A}$ must be modified to avoid having linearly dependent element-abundance constraints, which would lead to a singular coefficient matrix; for the VCS algorithm, on the other hand, C is automatically determined from $\mathbf{A}$ during the course of the calculation, and the user of this computer program need not be concerned about whether $C = M$.

When the formula matrix must be modified, the methods for finding linearly dependent rows of $\mathbf{A}$ may or may not be obvious; in general, $M -$ rank (A) rows must be removed. For relatively simple systems, this may be done by inspection. For example, in the system $\{(NH_3, HCl, NH_4Cl), (N, H, Cl)\}$, $M = 3$ and rank (A) $= 2$. Since the third row of $\mathbf{A}$ is equal to the second minus three times the first, any one of the three rows may be deleted. A general procedure is to obtain a modified $\mathbf{A}$ by means of the procedure described in Section 2.3.3, as implemented in the algorithms given in Appendix A. The nonzero portion of the unit matrix form of $\mathbf{A}, \mathbf{A}'$, given by equation 2.3-11, is a suitable modified formula matrix. This modified matrix is then used as input to the BNR algorithm, along with (a modified) $\mathbf{b}'$ obtained from $\mathbf{b}' = \mathbf{A}'\mathbf{n}°$.

Example 9.2 For the system $\{(H_3PO_4, H_2PO_4^-, HPO_4^{2-}, PO_4^{3-}, H^+, OH^-, H_2O), (H, O, P, p)\}$ in Problem 2.1, determine C and a suitable modified $\mathbf{A}$ and modified $\mathbf{b}$.

Solution From the procedure discussed in Section 2.3.3, or the stoichiometry algorithm, we obtain the unit matrix form of $\mathbf{A}$:

$$\mathbf{A}^* = \begin{pmatrix} \overset{(1)}{1} & \overset{(2)}{0} & \overset{(6)}{0} & \overset{(4)}{-2} & \overset{(5)}{1} & \overset{(3)}{-1} & \overset{(7)}{1} \\ 0 & 1 & 0 & 3 & -1 & 2 & -1 \\ 0 & 0 & 1 & 0 & 0 & 0 & 1 \\ 0 & 0 & 0 & 0 & 0 & 0 & 0 \end{pmatrix},$$

where the species have been reordered as indicated by the numbers over the columns of the matrix. Then C is 3, and the modified formula matrix is

$$\mathbf{A}' = \begin{pmatrix} 1 & 0 & 0 & -2 & 1 & -1 & 1 \\ 0 & 1 & 0 & 3 & -1 & 2 & -1 \\ 0 & 0 & 1 & 0 & 0 & 0 & 1 \end{pmatrix}.$$

The modified element-abundance vector is given by

$$\mathbf{b}' = \mathbf{A}'(2,0,0,0,0,0,1)^T = (3, -1, 1)^T.$$

9.4 STANDARD FREE-ENERGY DATA FROM EQUILIBRIUM CONSTANTS

The computer programs in the appendixes utilize free-energy data in the form of standard chemical potentials, μ_i° (or μ_i^*) of the individual species, and we have described in Section 3.12 various ways in which μ_i° is obtained from tabulated data. It is sometimes the case, however, that free-energy data are available only as equilibrium constants for certain stoichiometric equations involving species of the system. This is frequently the case for aqueous systems of inorganic species. This, in turn, has led to the widespread use of "equilibrium-constant" stoichiometric algorithms for computing equilibrium compositions for such systems (Nordstrom et al., 1979), since it is not obvious how such information should be utilized in algorithms of the types given in the appendixes. In this section we describe how an arbitrary set of equilibrium constants may be used to generate a consistent set of individual standard chemical potentials of the species involved. We begin with the case of a single stoichiometric equation ($R = 1$) and then consider the general case.

9.4.1 The Case $R = 1$

Consider a system represented by the stoichiometric equation (equation 2.3-8)

$$\sum_{i=1}^{N} \nu_i A_i = 0. \tag{9.4-1}$$

Standard Free-Energy Data from Equilibrium Constants

The equilibrium constant K_a is related to the standard free energy of reaction by (cf. equation 3.10-1)

$$\Delta G° = -RT \ln K_a, \qquad (9.4\text{-}2)$$

and $\Delta G°$, in turn, satisfies

$$\Delta G° = \sum_{i=1}^{N} \nu_i \mu_i°. \qquad (9.4\text{-}3)$$

We wish to determine a suitable $\mu°$ from K_a or $\Delta G°$.

From the discussion in Section 3.12, we note that different sets of $\mu°$ give rise to the same $\Delta G°$ for any stoichiometric equation involving subsets of the species and hence give rise to identical solutions to equilibrium problems involving such species. This means that *any* $\mu°$ that satisfies equation 9.4-3 may be used to compute the equilibrium composition. Any vector $\mu°$ that satisfies equation 9.4-3 is said to be *consistent* with the given $\Delta G°$.

Since equation 9.4-3 is a single equation in N unknowns $\mu°$, a simple solution of it is $(0, 0, \ldots, \Delta G°/\nu_N)^T$. Thus a consistent $\mu°$ is obtained by setting all $\mu_i°$ except one to zero. The remaining $\mu_i°$ is then chosen to satisfy equation 9.4-3.

Example 9.3 Obtain a consistent $\mu°$ for the species in the system $\{(O_2, SO_2, SO_3), (O, S)\}$, if the equilibrium constant K_a for the stoichiometric equation

$$2SO_2 + O_2 = 2SO_3$$

is 3.276 atm^{-1} at 1000 K.

Solution From equation 9.4-2, $\Delta G° = -9866$ J. We then arbitrarily choose, for SO_3, $\mu_3° = \Delta G°/\nu_3 = -4933$ J and $\mu_1° = \mu_2° = 0$.

9.4.2 The General Case

In general, we wish to find a $\mu°$ that satisfies (cf. equation 3.10-2)

$$\sum_{i=1}^{N} \nu_{ij} \mu_i° = \Delta G_j°; \qquad j = 1, 2, \ldots, R, \qquad (9.4\text{-}4)$$

where we assume here that the number of stoichiometric equations R satisfies $R = \text{rank}(\mathbf{N}) = N - \text{rank}(\mathbf{A})$. Equations 9.4-4 are a set of R equations in N unknowns with $N > R$. These equations have an infinite number of solutions. Any particular one of these is a consistent $\mu°$.

A particular solution of equations 9.4-4 may be obtained by using any suitable computer program that solves underdetermined sets of linear equations. Alternatively, since equations 9.4-4 may be written as

$$(\mathbf{N}^T, -\boldsymbol{\Delta G}^\circ)\begin{pmatrix}\boldsymbol{\mu}^\circ \\ 1\end{pmatrix} = \mathbf{0}, \tag{9.4-5}$$

the stoichiometry programs in Appendix A may be used according to the following procedure:

1. Use the $R \times (N + 1)$ matrix $(\mathbf{N}^T, -\boldsymbol{\Delta G}^\circ)$ as input to the stoichiometry algorithm in Appendix A. Use R and $(N + 1)$ as the respective values of M and N in the program.
2. The values of $\{\mu_i^\circ; i = 1, 2, \ldots, R\}$ are given by the first R coefficients of the final stoichiometric equation produced by the program, and the remainder are set to zero.

Example 9.4 For the system described in Example 9.2, assume that data are given in the form

$$H_3PO_4 = H_2PO_4^- + H^+ \;;\; K_1$$

$$H_2PO_4^- = HPO_4^{2-} + H^+ \;;\; K_2$$

$$HPO_4^{2-} = PO_4^{3-} + H^+ \;;\; K_3$$

$$H_2O = OH^- + H^+ \;;\; K_4 = \frac{K_w}{a_{H_2O}},$$

where K_1 to K_4 are the equilibrium constants for the reactions as written, and K_w is the ion product for water. We assume values of these are known as indicated below. Find a consistent $\boldsymbol{\mu}^\circ$ for the seven species of the system.

Solution We first calculate $\Delta G_i^\circ = -RT \ln K_i$ and form the matrix

$$(\mathbf{N}^T, -\boldsymbol{\Delta G}^\circ) = \begin{pmatrix} -1 & 1 & 0 & 0 & 1 & 0 & 0 & -\Delta G_1^\circ \\ 0 & -1 & 1 & 0 & 1 & 0 & 0 & -\Delta G_2^\circ \\ 0 & 0 & -1 & 1 & 1 & 0 & 0 & -\Delta G_3^\circ \\ 0 & 0 & 0 & 0 & 1 & 1 & -1 & -\Delta G_4^\circ \end{pmatrix}.$$

The unit matrix form of this is

$$\begin{pmatrix} (1) & (2) & (3) & (6) & (5) & (4) & (7) & \\ 1 & 0 & 0 & 0 & -3 & -1 & 0 & \Delta G_1^\circ + \Delta G_2^\circ + \Delta G_3^\circ \\ 0 & 1 & 0 & 0 & -2 & -1 & 0 & \Delta G_2^\circ + \Delta G_3^\circ \\ 0 & 0 & 1 & 0 & -1 & -1 & 0 & \Delta G_3^\circ \\ 0 & 0 & 0 & 1 & 1 & 0 & -1 & -\Delta G_4^\circ \end{pmatrix}.$$

Species Present in Small Amounts

A consistent $\boldsymbol{\mu}^\circ$, with the species ordered as in the problem statement of Example 9.2, is $(\Delta G_1^\circ + \Delta G_2^\circ + \Delta G_3^\circ, \Delta G_2^\circ + \Delta G_3^\circ, \Delta G_3^\circ, 0, 0, -\Delta G_4^\circ, 0)^T$.

If we write the R ($= 4$) stoichiometric equations represented by the first N ($= 7$) entries in each of the rows of the unit matrix form, we see that the method in effect sets standard chemical potentials or free energies of formation for R noncomponent species (here H_3PO_4, $H_2PO_4^-$, HPO_4^{2-}, and OH^-) relative to a set of C ($= 3$) component species (here H^+, PO_4^{3-}, and H_2O), which are assigned zero values. This is analogous to the usual assignment of free energies of formation of molecular species relative to the elements, the ultimate components.

9.4.3 Treatment of Possibly Incompatible Data

We consider the case

$$\sum_{i=1}^{N} \nu_{ij}\mu_i^\circ = \Delta G_j^\circ; \quad j = 1, 2, \ldots, S, \qquad (9.4\text{-}6)$$

where $S \geqslant R$. In this case there may exist no solutions $\boldsymbol{\mu}^\circ$ to equation 9.4-6. For example, if two different values of ΔG° are given for a single stoichiometric equation, this is equivalent to $S = 2$ and $R = 1$. In such a case we may compromise by seeking a least-squares solution of equation 9.4-6 by minimizing the function

$$f(\boldsymbol{\mu}^\circ) = \tfrac{1}{2} \sum_{j=1}^{S} w_j \left(\sum_{i=1}^{N} \nu_{ij}\mu_i^\circ - \Delta G_j^\circ \right)^2, \qquad (9.4\text{-}7)$$

where the w_j are statistical weights normally taken as inversely proportional to the respective variances of the uncertainties of ΔG_j°, assuming that these uncertainties are uncorrelated; in the absence of such information, w_j may be taken as unity. The value of $\boldsymbol{\mu}^\circ$ is determined by a particular solution of the N linear equations

$$\sum_{i=1}^{N} \mu_i^\circ \sum_{j=1}^{S} w_j \nu_{ij} \nu_{kj} = \sum_{j=1}^{S} w_j \nu_{kj} \Delta G_j^\circ; \quad k = 1, 2, \ldots, N. \qquad (9.4\text{-}8)$$

9.5 SPECIES PRESENT IN SMALL AMOUNTS

When a species in a multispecies phase is present in a very small amount, the step-size parameter ω in equation 6.3-2 may be forced to be very small, to satisfy the nonnegativity constraint. This would cause the VCS algorithm to be slow. The programs in Appendix D avoid this difficulty by removing such a minor species from the main calculation and computing its amount separately. The equations used for this purpose are

$$n_i^{(m+1)} = n_i^{(m)} \exp \frac{\Delta G_j}{RT}. \tag{9.5-1}$$

These equations can also be used to compute the amounts of species not included in an equilibrium calculation, provided that these are present in only very small amounts. For the BNR algorithm, a minor species not included in the main calculation may be computed by means of

$$n_i = n_{t\alpha} \exp\left(-\frac{\mu_i^*}{RT} + \sum_{k=1}^{C} a_{ki}\psi_k\right) \tag{9.5-2}$$

9.6 RESTRICTED EQUILIBRIUM PROBLEMS

Thus far, we have concentrated on the solution of the standard form of equilibrium problem of minimizing G subject to the element-abundance constraints. Constraints in addition to these may be placed on the equilibrium composition, and here we show how such a problem can be converted into one of standard form, allowing its solution by the algorithms in the appendixes.

We consider additional compositional constraints of three types: (1) linear combinations of equilibrium mole numbers specified explicitly; (2) as for (1), but constraints specified implicitly; and (3) products of equilibrium mole fractions specified. The first two types have been discussed by Smith (1980a), and an example of the third has been considered by White and Seider (1981):

1. When additional constraints are specified for the equilibrium mole numbers of the form

$$\mathbf{Dn} = \mathbf{d}, \tag{2.4-14}$$

   a modified formula matrix $\mathbf{A}'$ and an element-abundance vector $\mathbf{b}'$ may be found by the procedure discussed in Section 2.4.4, so that the entire set of constraints is expressed in the form $\mathbf{A}'\mathbf{n} = \mathbf{b}'$ (equation 2.4-13).

2. When additional constraints are specified implicitly by a stoichiometric matrix $\mathbf{N}$ with rank $(\mathbf{N}) < R$, a compatible formula matrix and element-abundance vector may be found by the procedure described in Section 2.4.5. Problems involving phase equilibrium only are of this type.

3. When additional constraints are specified of the form

$$\prod_{i=1}^{N} x_i^{\eta_{ij}} = \kappa_j \tag{9.6-1}$$

where η and κ are specified, such a problem may be converted to standard form by rewriting equation 9.6-1 as

$$\ln \kappa_j = \sum_{i=1}^{N} \eta_{ij} \ln x_i. \qquad (9.6\text{-}2)$$

Equation 9.6-2 is satisfied if the standard chemical potentials of the species satisfy

$$\sum_{i=1}^{N} \eta_{ij} \mu_i^* = -RT \ln \kappa_j. \qquad (9.6\text{-}3)$$

Such a μ^* may be found by using the methods described in Section 9.4.

Example 9.5 For the system discussed in Example 2.5 (Björnbom, 1975; Smith, 1976), calculate the equilibrium composition at 980 K and 43 atm for an initial feed of 4 moles of H_2 and 1 mole of $C_6H_5CH_3$.

Solution From Example 2.5, we obtain $\mathbf{A}'$ and $\mathbf{b}'$, and using the data given in Example 8.5, together with any of the equilibrium algorithms in the appendixes, we obtain $\mathbf{n} = (0.002, 3.002, 0.998, 0.998)^T$. We note, for comparison, that the unrestricted solution is $\mathbf{n} = (0.118, 0.079, 0.544, 2.908)^T$.

9.7 TREATMENT OF ISOMERS

Isomers are chemical species with the same formula vector that are distinguishable by their molecular structures and that (normally) have different standard chemical potentials. Distinguishability leads to an important effect of isomerization on chemical equilibrium (Missen, 1963; Smith and Missen, 1974). Furthermore, the calculation of equilibrium in a system involving isomers can be simplified considerably by considering all such isomers as a single "equivalent" species with an appropriate chemical potential, thus reducing the total number of species that must be considered in an equilibrium calculation. The composition with respect to the individual isomers is then calculated subsequently and directly from the composition of the equivalent species. This approach has been used, in effect, by Smith (1959), Dantzig and DeHaven (1962), Duff and Bauer (1962), and Missen (1963).

We first derive the forms of the standard chemical potential of the equivalent species and of the equilibrium distribution of isomers. We then use the former to show a general effect of isomerization on chemical equilibrium. We restrict the treatment and the proof to a system that is an ideal solution, but the effect derived is not restricted by this.

Consider substance I to occur in m_I accessible isomeric forms. We may refer to m_I as the isomeric degeneracy of the substance. For example, there are three isomers of pentane (C_5H_{12}): n-pentane, isopentane, and neopentane; pentane then has a maximum possible degeneracy (m_I) of 3. The system may contain species other than the isomers of I, but if it is ideal, the chemical potential of the ith isomer of I, μ_{I_i}, is, from equation 3.7-15a,

$$\mu_{I_i} = \mu_{I_i}^* + RT \ln \frac{n_{I_i}}{n_t}, \qquad (9.7\text{-}1)$$

where $\mu_{I_i}^*$ and n_{I_i} are the standard chemical potential and number of moles of the ith isomer, respectively. Since the chemical potentials of all the isomers are equal at equilibrium,

$$\mu_I = \mu_{I_i} = \mu_{I_i}^* + RT \ln \frac{n_{I_i}}{n_t}, \qquad (9.7\text{-}2)$$

where μ_I is the common chemical potential of all the isomers, and, from this,

$$n_{I_i} = n_t \exp\left(\frac{\mu_I}{RT}\right) \exp\left(-\frac{\mu_{I_i}^*}{RT}\right). \qquad (9.7\text{-}3)$$

If we sum equation 9.7-3 over all isomers, we obtain an expression for the total number of moles of I, n_I,

$$n_I = \sum_{i=1}^{m_I} n_{I_i} = n_t \exp\left(\frac{\mu_I}{RT}\right) \sum_{i=1}^{m_I} \exp\left(-\frac{\mu_{I_i}^*}{RT}\right). \qquad (9.7\text{-}4)$$

Equation 9.7-4 may be written as the chemical potential of a single "equivalent species":

$$\mu_I = \mu_I^* + RT \ln \frac{n_I}{n_t}, \qquad (9.7\text{-}5)$$

where

$$\mu_I^* = -RT \sum_{i=1}^{m_I} \exp\left(-\frac{\mu_{I_i}^*}{RT}\right). \qquad (9.7\text{-}6)$$

Equation 9.7-6 gives the standard chemical potential of the equivalent species I, μ_I^*, in terms of the standard chemical potentials of the individual isomers. This quantity may be used to represent substance I in any reaction in which it is involved in an ideal solution, and the computation of equilibrium in such a system leads to the value of n_I. The mole numbers n_{I_i} of the individual isomers can then be obtained from n_I by dividing equation 9.7-3 by equation 9.7-4.

This leads to

$$n_{I_i} = \frac{n_I \exp(-\mu_{I_i}^*/RT)}{\Sigma \exp(-\mu_{I_i}^*/RT)}. \tag{9.7-7}$$

We note that the equilibrium distribution of isomers n_{I_i}/n_I may be calculated directly from equation 9.7-7. This result relates only to equilibrium among the isomers, is independent of any other equilibrium considerations, and can be carried out separately if desired.

Two important consequences may be derived from equation 9.7-6:

$$\mu_I^* < \mu_{I_i}^*; \quad i = 1, 2, \ldots, m_I \tag{9.7-8}$$

and

$$\delta\mu_I^* \, \delta m_I < 0. \tag{9.7-9}$$

The first of these means that the standard chemical potential of the equivalent species I is less than that of any of the isomers, and the second means that this standard chemical potential decreases (increases) as the number of accessible isomers increases (decreases).

If we couple relation 9.7-9 with relation 8.4-41a, we obtain an interesting result about the isomeric degeneracy of a species (Smith and Missen, 1974). Relation 8.4-41a is not restricted to a system that is an ideal solution, provided that the solution to the chemical equilibrium problem is unique. We may exploit this relation here by considering the change in n_I, δn_I, resulting from a change in μ_I^*, $\delta\mu_I^*$, as a result of a change in m_I, the degeneracy of species I. Since we are focussing on I only, we rewrite relation 8.4-41a as

$$\delta n_I \, \delta\mu_I^* < 0. \tag{9.7-10}$$

Combining relations 9.7-9 and 9.7-10, we obtain

$$\delta n_I \, \delta m_I > 0. \tag{9.7-11}$$

This result leads to the following statement (Smith and Missen, 1974): "If a set of isomeric forms is accessible to a chemical substance at equilibrium in a closed system (at given T and P, say), then the (relative) amount of the substance present at equilibrium is greater than it would be if only a strict subset of these forms were accessible."

Examples of interpretations of this statement are: (1) the equilibrium conversion of n-butane (C_4H_{10}) to n-butene (C_4H_8) by dehydrogenation increases as the number of isomers (three are possible) of n-butene present increases, other conditions being the same; and (2) conversely, the equilibrium conversion of n-butene to 1,3-butadiene (C_4H_6) by dehydrogenation decreases

Table 9.3 Effect of Isomeric Degeneracy (m_I) on Equilibrium Conversion (f) for Dehydrogenation of *n*-Butane and *n*-Butene

m_I [Isomer(s) present]	Equilibrium Conversion (f)	
	n-C_4H_{10} Dehydrogenation[a]	n-C_4H_8 Dehydrogenation[b]
1 (1-C_4H_8)	0.505	0.859
1 (*cis*-2-C_4H_8)	0.512	0.855
1 (*trans*-2-C_4H_8)	0.575	0.811
3 (all three isomers)	0.740	0.684

[a] $T = 900$ K; $P = 1$ atm; no inert species.
[b] $T = 900$ K; $P = 1.25$ atm; $n_z/n°(n\text{-}C_4H_8) = 20$.

as the number of isomers of *n*-butene present increases (Missen, 1963). These apparently conflicting results are due to the fact that the degenerate substance, *n*-butene, is a product in the first case and a reactant in the second case. Results for these examples are illustrated in Table 9.3.

9.8 TREATMENT OF ISOTOPES

When an element in a species exists in two or more isotopic forms, the question arises as to whether isotopic distinguishability must be taken into account in determining the composition of a system containing the species at equilibrium. Ordinarily this need not be done because the natural distribution of isotopes is not disturbed by chemical reaction. In isotope-exchange reactions, however, used for the concentration and separation of isotopes, the distribution of isotopes over the molecular species *is* disturbed, and the change must be taken into account in determining the equilibrium distribution of the distinguishable molecular species.

There are three methods for the calculation of this distribution at equilibrium: (1) by probability theory based on the assumption of random distribution of isotopes over the distinguishable molecular forms; (2) by means of the algorithms in the appendixes, using properties of isotopic molecular species calculated, for example, by the methods of statistical thermodynamics; and (3) from equilibrium constants of isotope-exchange reactions, determined experimentally or calculated as described in method 2. (The distribution can also, of course, be measured directly by mass spectrometry.) These three methods of calculation give rise to two types of results: a distribution that either does or does not take into account the effect of temperature. In the language of statistical mechanics, a distribution that does not allow for this results from consideration only of configurational randomness, and not of the dependence of the energy on either the configuration or on the isotopic composition. The difference between these two types of result is called the *equilibrium isotope effect* (EIE) or the thermodynamic isotope effect (TIE) (Skorobogatov, 1961). Approach 1 does not take EIE into account, but approaches 2 and 3 do.

Isotope effects (whether equilibrium or kinetic) are due primarily to relative mass differences (McKay, 1971, pp. 294–295). Difference in mass affects vibrational energy levels. This is most pronounced for the lightest elements at low temperatures. The EIE is generally small and may be negligible for the heavier elements or at a sufficiently high temperature for the lighter elements as the classical energy distribution is approached.

We consider next the three ways in which the equilibrium distribution of isotopic molecular forms may be calculated, although much of the detail is left to the literature sources.

9.8.1 Calculation by Probability Theory

Consider a species i in which M is the number of elements; L_k is the number of isotopes of the kth element E_k, a_{ki} is the subscript to E_k in the molecular formula A_i of the species, and a_{li} is the subscript to isotope l, with mass number m_{lk}, of the kth element $E_k^{m_{lk}}$, such that

$$\sum_{l=1}^{L_k} a_{li} = a_{ki}; \qquad k = 1, 2, \ldots, M. \tag{9.8-1}$$

The molecular formula of the species may be represented by

$$A_i \equiv (E_1)_{a_{1i}} \cdots (E_k)_{a_{ki}} \cdots (E_M)_{a_{Mi}}$$

$$\equiv (E_1)_{a_{1i}}^{m_{11}} \cdots (E_1)_{a_{li}}^{m_{l1}} \cdots (E_1)_{a_{L_1 i}}^{m_{L_1 1}} \cdots$$

$$(E_k)_{a_{1i}}^{m_{1k}} \cdots (E_k)_{a_{li}}^{m_{lk}} \cdots (E_k)_{a_{L_k i}}^{m_{L_k k}} \cdots$$

$$(E_M)_{a_{1i}}^{m_{1M}} \cdots (E_M)_{a_{li}}^{m_{lM}} \cdots (E_M)_{a_{L_M i}}^{m_{L_M M}}. \tag{9.8-2}$$

The number N_I of distinguishable isotopic molecular forms of A_i is given by (Skorobogatov, 1961)

$$N_I = \prod_{k=1}^{M} \frac{(a_{ki} + L_k - 1)!}{a_{ki}!(L_k - 1)!} \tag{9.8-3}$$

As an example, consider the species C_2H_6O, in which the isotopes C^{12}, C^{13}, H, H^2 (or D), O^{16}, O^{17}, and O^{18} appear. The number of distinguishable forms of

C_2H_6O due to isotopic distinguishability is

$$N_I = \frac{(2+2-1)!}{2!(2-1)!} \frac{(6+2-1)!}{6!(2-1)!} \frac{(1+3-1)!}{1!(3-1)!} = 63,$$

one of which is $C^{12}C^{13}H_4H_2^2O^{16}$.

The relative molar abundance or mole fraction of the molecular form represented by the expanded form of equation 9.8-2, among the N_I forms, is (Skorobogatov, 1961; Margrave and Polansky, 1962)

$$x(\text{any form}) = \prod_{k=1}^{M} \frac{a_{ki}! \left[\prod_{l=1}^{L_k} p_{kl}^{a_{lj}}\right]}{\prod_{l=1}^{L_k} a_{li}!} \tag{9.8-4}$$

where p_{kl} is the relative molar (atomic) abundance of the lth isotope of element k, such that

$$\sum_{l=1}^{L_k} p_{kl} = 1; \quad k = 1, 2, \ldots, M. \tag{9.8-5}$$

According to equation 9.8-4, the mole fraction of $C^{12}C^{13}H_4H_2^2O^{16}$ in C_2H_6O [based on relative abundances of 0.9889, 0.0111, 0.99985, 0.00015, and 0.99759 for the isotopes in the order indicated (Weast, 1979–1980)] is 7.387×10^{-5}.

9.8.2 Calculation by Equilibrium Algorithms

The equilibrium distribution can be calculated by any of the algorithms described in Chapters 4 and 6, provided that the appropriate free-energy information is available. Often this is in the form of equilibrium constants for exchange reactions (see Section 9.8.3), and the method described in Section 9.4 may be used to obtain a consistent $\mu°$ or $\mu*$. Jones and McDowell (1959) have described the calculation of thermodynamic properties, including the free-energy function, for isotopic methanes from spectroscopic data and have used these to calculate equilibrium constants for some isotope-exchange reactions. Apse and Missen (1967) have used their results to calculate the natural equilibrium distribution of the deuteromethanes. In this case, since the relative abundances of isotopes H and D are greatly different, simplifications can be introduced that have the effect of reducing the calculations to ones that can be done "by hand."

9.8.3 Calculation from Equilibrium Constants

The equilibrium distribution of isotopic species can be calculated from a knowledge of the equilibrium constants of appropriate isotope-exchange reactions. The calculation of such equilibrium constants has been described by Urey (1947) and Jones and McDowell (1959).

An example of an isotope-exchange reaction is

$$CH_4 + CHD_3 = CH_3D + CH_2D_2. \tag{A}$$

Values of the equilibrium constant K_a calculated (Apse and Missen, 1967) from the data due to Jones and McDowell (1959), are shown in Table 9.4. The values tend to a high-temperature value of about 6, which is consistent with the value obtained by the distribution calculated by the method described in Section 9.8.1, from which we thus obtain

$$K_a = \frac{a_{CH_3D} a_{CH_2D_2}}{a_{CH_4} a_{CHD_3}} = \frac{4p_H^3 p_D 6 p_H^2 p_D^2}{p_H^4 4 p_H p_D^3} = 6,$$

where the activities of the species have been replaced by their relative abundances (equivalent to the assumption of an ideal solution), as determined from equation 9.8-4 in terms of the relative abundances p_H and p_D of isotopes H and D. The equilibrium distributions calculated by the method described in Section 9.8.1 are thus the "high-temperature" results of the methods described in Sections 9.8.2 and 9.8.3.

An alternative way of regarding the difference between the methods described in Sections 9.8.1, 9.8.2, and 9.8.3 is by means of the equation

$$\Delta G° = \Delta H° - T \Delta S°. \tag{3.12-6}$$

The method based on probability theory ignores the enthalpy term, and, in effect, determines the entropy change entirely as a configurational quantity. The enthalpy of isotopic exchange is generally quite small. From the data given in Table 9.4, by means of the van't Hoff equation (equation 3.10-4), we obtain an average value of about 250 J for $\Delta H°$ for equation A.

Skorobogatov (1961) has provided a general method of calculating the equilibrium distribution of isotopic species based on (known) equilibrium constants of appropriate isotope-exchange reactions. Since we do not emphasize methods based on equilibrium constants in this book, we do not describe this approach, but note some of the results of his calculations. As a measure of the EIE, he used the percentage difference from the result of equation 9.8-4. For hydrogen isotopes in exchanges involving H, D, and T (tritium or H^3) in H_2, D_2, and T_2 at 298 K, he showed effects as large as 40%.

Table 9.4 Equilibrium Constant K_a for the Reaction $CH_4 + CHD_3 = CH_3D + CH_2D_2$[a]

T, K	93.2	173.2	298.2	373.2	573.2	773.2	1273.2
K_a	4.57	5.34	5.76	5.91	5.96	5.99	6.06

[a] From data due to Jones and McDowell (1959).

For exchanges involving H and D in the deuteromethanes at 739 K (a relatively high temperature), however, his results showing significant effects should probably by replaced by results based on the data due to Jones and McDowell (1959), which indicate almost negligible effects. His calculations also show that there can be significant effects for exchanges involving nitrogen isotopes in N_2 and in deuterized ammonia. A general conclusion is that the EIE has the effect of enhancing the relative amount of homoisotopic molecules at equilibrium at the expense of heteroisotopic molecules.

9.9 THERMODYNAMIC CONSTRAINTS OTHER THAN (T, P)

9.9.1 Equilibrium Computation

As noted in Chapter 1, there are many types of chemical equilibrium problem, characterized by their thermodynamic constraint functions (Table 1.1). The most common type, as considered to this point, is that specified by T and P. However, any two thermodynamic constraint functions may be chosen from the set $\{P, V, T, S, U, H, A, G\}$ (or, indeed, any two functionally independent combinations of members of this set). Such problems may be solved by determining the values of T and P that obtain under the given constraint conditions and then calculating the equilibrium composition at this (T, P) point. We show that knowledge of the derivatives of the equilibrium composition of a standard (T, P) problem with respect to T and P enables this to be done.

To illustrate the approach, we use the thermodynamic constraints (H, P). In this case we must find T and the corresponding equilibrium mole numbers $\mathbf{n}(T)$ that satisfy

$$f(T) = \sum_{i=1}^{N} n_i(T)\bar{h}_i[T, \mathbf{n}(T)] - H^1 = 0, \qquad (9.9\text{-}1)$$

where H^1 is the specified enthalpy. Equation 9.9-1 is a nonlinear equation in T. We may solve this equation by any of several methods (Ralston and Rabinowitz, 1978); using the Newton-Raphson method, for example, we obtain the iteration equations

$$T^{(m+1)} = T^{(m)} - \frac{\omega^{(m)} f(T^{(m)})}{f'(T^{(m)})}. \qquad (9.9\text{-}2)$$

The derivative f' is given by

$$f'(T) = \sum_{i=1}^{N} \left\{ \bar{h}_i \left(\frac{\partial n_i}{\partial T} \right) + n_i \left[\frac{\partial \bar{h}_i}{\partial T} + \sum_{j=1}^{N} \left(\frac{\partial \bar{h}_i}{\partial n_j} \right) \left(\frac{\partial n_j}{\partial T} \right) \right] \right\}. \qquad (9.9\text{-}3)$$

We assume that we have available information to determine $\partial \bar{h}_i/\partial T$, which is the partial molar heat capacity $\bar{c}_{pi}$, and to determine $\partial \bar{h}_i/\partial n_j$; **n** is the equilibrium composition at T, and $\partial \mathbf{n}/\partial T$ is the temperature sensitivity coefficient described in Section 8.4. Our approach in solving equation 9.9-1 thus consists of an *inner iteration loop* in which the equilibrium composition is determined at $T^{(m)}$ by means of a chemical equilibrium algorithm and an *outer iteration loop* in which $T^{(m)}$ is refined by means of equation 9.9-2.

In general, we consider any pair of constraints η and χ from the set (excluding T and P) and write the equations corresponding, respectively, to equations 9.9-1 and 9.9-3 as

$$f_1(T, P) = \sum_{i=1}^{N} n_i(T, P)\bar{\eta}_i[T, P, \mathbf{n}(T, P)] - \eta^1 = 0,$$

$$f_2(T, P) = \sum_{i=1}^{N} n_i(T, P)\bar{\chi}_i[T, P, \mathbf{n}(T, P)] - \chi^1 = 0; \quad (9.9\text{-}4)$$

$$\begin{pmatrix} T^{(m+1)} \\ P^{(m+1)} \end{pmatrix} = \begin{pmatrix} T^{(m)} \\ P^{(m)} \end{pmatrix} - \omega^{(m)} \begin{pmatrix} \frac{\partial f_1}{\partial T} & \frac{\partial f_1}{\partial P} \\ \frac{\partial f_2}{\partial T} & \frac{\partial f_2}{\partial P} \end{pmatrix}_{(m)}^{-1} \begin{pmatrix} f_1 \\ f_2 \end{pmatrix}_{(m)} ; \quad (9.9\text{-}5)$$

$$\frac{\partial f_1}{\partial T} = \sum_{i=1}^{N} \left\{ \bar{\eta}_i \left(\frac{\partial n_i}{\partial T} \right) + n_i \left[\frac{\partial \bar{\eta}_i}{\partial T} + \sum_{j=1}^{N} \left(\frac{\partial \bar{\eta}_i}{\partial n_j} \right) \left(\frac{\partial n_j}{\partial T} \right) \right] \right\}, \quad (9.9\text{-}6a)$$

$$\frac{\partial f_1}{\partial P} = \sum_{i=1}^{N} \left\{ \bar{\eta}_i \left(\frac{\partial n_i}{\partial P} \right) + n_i \left[\frac{\partial \bar{\eta}_i}{\partial P} + \sum_{j=1}^{N} \left(\frac{\partial \bar{\eta}_i}{\partial n_j} \right) \left(\frac{\partial n_j}{\partial P} \right) \right] \right\}, \quad (9.9\text{-}6b)$$

$$\frac{\partial f_2}{\partial T} = \sum_{i=1}^{N} \left\{ \bar{\chi}_i \left(\frac{\partial n_i}{\partial T} \right) + n_i \left[\frac{\partial \bar{\chi}_i}{\partial T} + \sum_{j=1}^{N} \left(\frac{\partial \bar{\chi}_i}{\partial n_j} \right) \left(\frac{\partial n_j}{\partial T} \right) \right] \right\}, \quad (9.9\text{-}6c)$$

$$\frac{\partial f_2}{\partial P} = \sum_{i=1}^{N} \left\{ \bar{\chi}_i \left(\frac{\partial n_i}{\partial P} \right) + n_i \left[\frac{\partial \bar{\chi}_i}{\partial P} + \sum_{j=1}^{N} \left(\frac{\partial \bar{\chi}_i}{\partial n_j} \right) \left(\frac{\partial n_j}{\partial P} \right) \right] \right\}. \quad (9.9\text{-}6d)$$

We assume that we have available information to determine the first partial derivatives of $\bar{\eta}_i$ and $\bar{\chi}_i$ with respect to T, P, and n_j; **n** is the equilibrium composition at (T, P), and $\partial \mathbf{n}/\partial T$ and $\partial \mathbf{n}/\partial P$ are the temperature and pressure sensitivity coefficients. Our general approach in solving equations 9.9-4 again involves inner and outer iterations.

In contrast to this "inner-outer" (IO) approach, an iteration scheme may be used in which the thermodynamic constraint functions are incorporated directly into a single iteration. Such an approach considers the equilibrium conditions and thermodynamic constraint given by equations 9.9-4 as a single set of nonlinear equations in the unknowns $(\mathbf{n}, T, P)$. The Newton-Raphson iteration equations resulting from this set of equations are then used (e.g., Huff, et al., 1951; Brinkley, 1960).

Convergence difficulties have been reportedly experienced with such methods (Barnhard and Hawkins, 1963). The IO approach decouples the thermodynamic constraint equations from the equilibrium conditions. One advantage of this is that the overall problem is split into two manageable parts. Also, the approach is quite general since the most complex part of the algorithm is the single computer subprogram that calculates equilibrium compositions at given T and P (the inner loop). For ideal systems the programs in Appendixes C and D may be used for this purpose. The outer loop is a very short computer program that varies T and P to effect a solution to equations 9.9-4.

9.9.2 Sensitivity Analysis

We can derive sensitivity matrices for $\mathbf{n}$ with respect to any of the problem parameters, which are given in general by $\{\eta, \chi, \boldsymbol{\mu}^*, \mathbf{b}, \mathbf{n}_z\}$, by straightforward extensions of the methods described in Section 8.4.

To illustrate the general approach, we consider the case of an (H, P) problem, as given by equation 9.9-1. To determine the effect of H^1 on T and $\mathbf{n}$, we differentiate equation 9.9-1 to obtain

$$\frac{\partial T}{\partial H^1} = \frac{1}{f'(T)}, \qquad (9.9\text{-}7)$$

where $f'(T)$ is given by equation 9.9-3. The sensitivity coefficients $\partial n_i / \partial H^1$ are given by

$$\frac{\partial n_i}{\partial H^1} = \frac{\partial n_i / \partial T}{f'(T)}. \qquad (9.9\text{-}8)$$

In the special case of a single stoichiometric equation, we have (Pings, 1963)

$$\frac{\partial \xi}{\partial H^1} = \left(\frac{\partial \xi}{\partial T}\right)\left(\frac{\partial T}{\partial H^1}\right)$$

$$= \frac{\partial \xi / \partial T}{(\partial \xi / \partial T)\left[\sum_i \sum_j \nu_j \left(\partial \bar{h}_i / \partial n_j\right) + \Delta H\right] + C_P}, \qquad (9.9\text{-}9)$$

where

$$C_P = \sum n_i \bar{c}_{P_i}. \qquad (9.9\text{-}10)$$

PROBLEMS

9.1 Calculate the equilibrium mole numbers and mole fractions resulting from the chlorination at 323 K and 1 atm of ferrophosphorus (FeP), containing 4.1 mole % Cr, 4.4% V, and 5% Si, using the species list and data tabulated in the following list (Holman, 1980). Assume that $\mathbf{b} = (b_{Fe}, b_P, b_{Cr}, b_V, b_{Si}, b_{Cl})^T = (55.7, 30.8, 4.1, 4.4, 5.0, 82.51)^T$ and that the solid species are immiscible and the liquid species are miscible and form an ideal solution. Data for the system ($M = 6$, $N = 25$) are as follows:

Species	$\mu°/RT$	Species	$\mu°/RT$
Fe(s)	0	$PCl_3(\ell)$	−144.897
P(s)	0	$SiCl_4(\ell)$	−295.789
Cr(s)	0	$VCl_4(\ell)$	−185.571
V(s)	0	PCl(g)	25.85
Si(s)	0	$PCl_3(g)$	−132.591
$Cl_2(g)$	0	$PCl_5(g)$	−171.448
$FeCl_2(s)$	−215.644	FeCl(g)	62.416
$CrCl_2(s)$	−174.204	$Fe_2Cl(g)$	−308.216
$CrCl_3(s)$	−426.957	$CrCl_3(g)$	69.190
$VCl_2(s)$	−146.578	$CrCl_4(g)$	32.615
$VCl_3(s)$	−190.192	SiCl(g)	46.336
$PCl_5(s)$	−182.664	$SiCl_2(g)$	−92.481
		$SiCl_4(g)$	−284.573

9.2 Continue Example 9.4 by calculating the concentration (molarity) of each species at equilibrium at 25°C (a) assuming ideal-solution behavior and (b) using the Davies equation (equation 7.2-38) to estimate activity coefficients. The values of the equilibrium constants at 25°C are as follows (Feenstra, 1979): $K_1 = 7.112 \times 10^{-3}$; $K_2 = 6.339 \times 10^{-8}$; $K_3 = 4.3 \times 10^{-13}$. Use $K_w = 1.002 \times 10^{-14}$ (Lewis and Randall, 1961, p. 367).

9.3 In Example 4.3 the species H, O, and OH were neglected as species present in relatively small amounts, and this assumption was essentially confirmed in Example 4.7. Compare this assumption and the numerical results of the latter example with the treatment given in Section 9.5. Use the data from Figure 4.7.

9.4 Metallic zinc can be made by reducing ZnO with carbon in a vessel from which the products of reaction may be continuously removed but from which air is excluded. Under the conditions of reaction, ZnO and C are solid, and the products are Zn (vapor, but may be partly liquefied), CO, and CO_2. Assuming that the reaction is carried out at 1 atm and that equilibrium is attained at the outlet (a) calculate the temperature and composition (mole fractions) of the product gas, and (b) investigate whether any zinc is liquefied. Standard free energies of formation for

CO and CO_2 (Zwolinski, et al., 1974), and for ZnO, together with the vapor pressure of zinc (ℓ) (Mathewson, 1959) at several temperatures are as follows:

T, K	ΔG_f°, kJ mole^{-1}			$p_{Zn}^*(\ell)$, atm
	CO(g)	CO_2(g)	ZnO(s)	
1100	−209.42	−395.96	−243.51	0.43
1200	−218.20	−396.06	−223.22	1.20
1300	−226.92	−396.13	−202.92	3.06
1400	−235.59	−396.18	−182.84	6.33

The standard free energy of formation of ZnO refers to $Zn(g) + \frac{1}{2}O_2(g) = ZnO(s)$ (cf. Hougen et al., 1959, pp. 1054–1058; Denbigh, 1981, pp. 191–193).

9.5 Calculate the mole fractions of the five isomers of hexane (C_6H_{14}) at equilibrium with respect to isomerization at 600 K. The five isomers are *n*-hexane, 2-methylpentane, 3-methylpentane, 2,2-dimethylbutane, and 2,3-dimethylbutane. The standard free energies of formation (Stull et al., 1969) are $(180.00, 177.36, 180.71, 179.66, 183.13)^T$, in kJ mole^{-1}.

9.6 Verify the results from Table 9.3. Standard free energies of formation at 900 K (Stull et al., 1969) for *n*-butane, 1-butene, *cis*-2-butene, *trans*-2-butene, and 1,3-butadiene are 227.32, 235.35, 235.06, 232.30, and 244.35 kJ mole^{-1}, respectively.

9.7 Consider liquid-vapor equilibrium in the system $\{(C_6H_6(\ell), C_6H_6(g), C_7H_8(\ell), C_7H_8(g), o\text{-}C_8H_{10}(\ell), o\text{-}C_8H_{10}(g), m\text{-}C_8H_{10}(\ell), m\text{-}C_8H_{10}(g), p\text{-}C_8H_{10}(\ell), p\text{-}C_8H_{10}(g)), (C, H)\}$. From the data given in the following list, calculate the composition of each phase at equilibrium at 100°C and 1 atm, if the liquid phase contains 20% xylenes. Assume that each phase is an ideal solution and that the three xylenes were formed originally at 800 K, such that isomerization equilibrium was attained and not altered subsequently. Data (some superfluous) are as follows [p^* and ΔG_f° from Rossini, et al. (1953) and $v^*(\ell)$ and B from Timmermans (1950, 1965)]:

Species	p^*, atm (100°C)	$v^*(\ell)$ cm^3 mole^{-1} (100°C)	B (100°C)	ΔG_f°, kJ mole^{-1} (800 K)
C_6H_6	1.776	100	−850	221.07
C_7H_8	0.732	120	−1360	260.40
$o\text{-}C_8H_{10}$	0.262	130	−2220	314.26
$m\text{-}C_8H_{10}$	0.308	135	−2260	309.64
$p\text{-}C_8H_{10}$	0.316	135	−2210	314.95

9.8 Mihail et al., (1978) have considered equilibrium in the formation of isoprene (C_5H_8) from isopentane (C_5H_{12}) by the two-stage dehydrogenation process involving the three isomeric isopentenes (C_5H_{10}), 2-

methyl-1-butene, 3-methyl-1-butene, and 2-methyl-2-butene, at various temperatures and pressures. Calculate the composition of this six-species system at equilibrium at 800 K, assuming that the standard free energies of formation (Mihail et al., 1978) of the five hydrocarbons, in the order cited and in kJ mole^{-1}, are $\Delta G_f^{\circ} = (275.68, 242.17, 252.59, 266.27, 251.25)^T$.

9.9 (a) Determine the magnitude of the equilibrium isotope effect (EIE) for the system $\{(CH_4, CH_3D, CH_2D_2, CHD_3, CD_4), (C, H, D)\}$, assuming that the relative abundance of H : D is 3 : 1 at (i) $-100°C$ and (ii) $500°C$. Equilibrium constants K_a for the following three exchange reactions at these temperatures, based on the data due to Jones and McDowell (1959), are as indicated (Apse and Missen, 1967):

Reaction	K_a	
	$-100°C$	$500°C$
$CH_4 + CH_2D_2 = 2CH_3D$	2.51	2.67
$CH_4 + CHD_3 = CH_3D + CH_2D_2$	5.34	5.99
$CH_4 + CD_4 = CH_3D + CHD_3$	13.67	15.96

(b) Estimate $\Delta H°$ for each of the three exchange reactions.

9.10 The manufacture of sulfuric acid from elemental sulfur involves, in part, burning molten sulfur completely with excess air (assume 79 mole % N_2 and 21% O_2) to SO_2 and then oxidizing SO_2 catalytically in several stages to SO_3. Assuming that the gas from the sulfur burner contains 9.5 mole % SO_2 and enters the catalytic converter at 700 K and 1 atm, calculate the fraction of SO_2 oxidized to SO_3 and the temperature at the outlet of the first stage. Assume also that equilibrium is attained, reaction occurs adiabatically, and pressure drop is negligible. The following data are taken from JANAF (1971):

	T, K			
	700	800	900	1000
ΔG_f°, kJ mole^{-1}				
SO_2	-299.42	-303.65	-296.32	-288.99
SO_3	-332.38	-327.24	-310.58	-293.97
ΔH_f°, kJ mole^{-1}				
SO_2	-306.29	-362.31	-362.24	-362.14
SO_3	-405.02	-460.70	-460.28	-459.78
C_p°, J mole^{-1} K^{-1}				
SO_2	50.96	52.43	53.58	54.48
SO_3	70.39	72.76	74.57	75.97
O_2	32.98	33.74	34.36	34.88
N_2	30.75	31.43	32.09	32.70

9.11 Calculate the adiabatic reaction or flame temperature for the combustion of propane at 1 atm with 20% excess air (79 mole % N_2, 21% O_2), assuming that the air and the fuel are initially at 27°C. Include the species listed in Example 4.3. Constants in the heat-capacity equation $C_P = (a + bT + cT^{-2})$, where C_P is in J mole^{-1} K^{-1}, are as follows (Smith and Van Ness, 1975, p. 107):

Species	a	$b \times 10^3$	$c \times 10^{-5}$
CO	28.41	4.10	-0.46
CO_2	44.22	8.79	-8.62
H_2	27.28	3.26	0.50
N_2	28.58	3.77	-0.50
NO	29.41	3.85	-0.59
O_2	29.96	4.18	-1.67
H_2O(g)	30.54	10.29	0.00

Standard enthalpies and free energies of formation at 300 K, in kJ mole^{-1}, in the order given in the preceding list (JANAF, 1971), are $\Delta H_f^\circ = (-110.52, -393.53, 0, 0, 90.29, 0, -241.85)^T$, and $\Delta G_f^\circ = (-137.33, -394.41, 0, 0, 86.58, 0, -228.52)^T$. The enthalpy of combustion of propane (to H_2O(g)) is -2044 kJ mole^{-1} (calculated from Smith and Van Ness, 1975, p. 120) at 25°C (assume it to be 27°C).

9.12 Vonka and Holub (1975) (see Problems 6.8 and 7.1) have studied the effect of using different models (ideal-gas solution, ideal solution, and nonideal solution) on the calculation of equilibrium in the reaction of ethylene and ammonia to produce ethylamines at various temperatures and pressures. For this purpose, they used the Redlich-Kwong equation of state (equation 7.2-25) to calculate the fugacity coefficient ϕ as described in Problems 6.8 and 7.1.

Using this same procedure, the same three models, and the data tabulated in the following list, calculate the equilibrium composition at 600 K and 100 atm for the system $\{(C_2H_4, NH_3, C_2H_5NH_2, (C_2H_5)_2NH, (C_2H_5)_3N), (C, H, N)\}$, assuming that $\mathbf{b} = (2, 7, 1)^T$, corresponding to a feed equimolar in the two reactants. Free-energy data (Stull et al., 1969) and critical constants (Kobe and Lynn, 1953) are as follows:

Species	ΔG_f°, kJ mole^{-1}	T_C, K	P_C, atm
C_2H_4	87.53	282.4	50.5
NH_3	16.07	405.5	111.3
$C_2H_5NH_2$	127.61	456	55.5
$(C_2H_5)_2NH$	227.57	496	36.6
$(C_2H_5)_3N$	334.93	532	30

9.13 Repeat Problem 6.12 at 1 atm without the assumption that the liquid phase is an ideal solution. Also use the data from Problems 4.10 and 7.4

9.14 Mehrotra et al., (1979) have carried out a thermodynamic study of metal recovery from coal ash by chlorination accompanied by reduction by carbon or carbon monoxide. On the basis of typical analyses of coal ash, they considered eight metals (aluminum, calcium, iron, potassium, manganese, sodium, silicon and titanium), together with oxygen, chlorine, and carbon ($M = 11$). They also considered up to 135 species, many of which may be solid or liquid at the conditions imposed—they assumed that all liquid and solid phases were single-species phases. They also assumed that any substance is present in only one phase. The element abundance was based on a fixed average coal-ash composition in terms of the metal oxides and variable amounts of Cl_2 and C (or CO): $\mathbf{b} \equiv (Al_2O_3, CaO, Fe_2O_3, K_2O, MgO, Na_2O, SiO_2, TiO_2, Cl, C) = 0.2409, 0.0749, 0.0627, 0.0192, 0.0617, 0.0087, 0.5340, 0.0167, b_{Cl}, b_C)^T$. Assume here that b_{Cl} and b_C are the stoichiometric amounts required for complete chlorination and removal of oxygen as CO_2: $b_{Cl} = 4.3532$ and $b_C = 1.0883$. Calculate the equilibrium mole numbers at 1200 K and 1 atm based on the following list of species, each accompanied by the value of ΔG_f° in kJ mole^{-1} (JANAF, 1971):

Species	ΔG_f°	Species	ΔG_f°	Species	ΔG_f°
Al(ℓ):	0	K(g):	0	$SiCl_2$(g):	−212.26
AlCl(g):	−147.16	$K_2CO_3(\ell)$:	−784.98	$SiCl_3$(g):	−360.31
$AlCl_2$(g):	−316.62	KCl(ℓ):	−315.72	$SiCl_4$(g):	−505.41
$AlCl_3$(g):	−520.56	$KClO_4$(g):	63.81	SiO_2(s):	−695.55
Al_2Cl_6(g):	−995.99	K_2O(s):	−176.08	Ti(s):	0
K_3AlCl_6(s):	−1452.27	K_2SiO_3(s):	−1152.5	TiC(s):	−170.55
Na_3AlCl_6(s):	−1379.35	Mg(ℓ):	0	$TiCl_2$(s):	−325.58
$NaAlO_2$(s):	−865.93	$MgAl_2O_4$(s):	−1813.3	$TiCl_3$(g):	−480.36
$Al_2O_3(\alpha)$:	−1295.1	$MgCl_2(\ell)$:	−460.02	$TiCl_4$(g):	−618.93
Al_2SiO_5(s):	−1995.1	MgO(s):	−469.59	$TiO(\alpha)$:	−428.14
$Al_6Si_2O_{13}$(s):	−5289.2	$MgSiO_3$(s):	−1199.0	TiO_2(s):	−727.45
Ca(ℓ):	0	Mg_2SiO_4(s):	−1696.6	Ti_2O_3(s):	−1190.4
CaCl(g):	−202.01	$MgTiO_3$(s):	−1221.0	$Ti_3O_5(\alpha)$:	−1910.4
$CaCl_2(\ell)$:	−616.99	$MgTi_2O_5$(s):	−1951.6	Ti_4O_7(s):	−2660.1
CaO(s):	−510.0	Mg_2TiO_4(s):	−1696.6	C(gr):	0
Fe(s):	0	Mg_2Si(s):	−68.32	CO(g):	−217.77
$FeCl_2(\ell)$:	−206.00	Na(g):	0	CO_2(g):	−396.15
$FeCl_3$(g):	−228.75	NaCl(ℓ):	−302.86	Cl(g):	53.48
Fe_2Cl_4(g):	−384.90	Na_2O(s):	−250.34	Cl_2(g):	0
Fe_2Cl_6(g):	−437.58	Na_2SiO_3(s):	−1173.34	O(g):	174.69
FeO(s):	−194.45	$Na_2Si_2O_5(\ell)$:	−1882.44	O_2(g):	0
Fe_2O_3(s):	−512.71	Si(s):	0	O_3(g):	225.31
Fe_3O_4(s):	−732.60	$SiC(\alpha)$:	−62.00		

9.15 Dunsmore and Midgley (1974) have considered equilibrium in metal (M)–liquid (A) equilibria in terms of equilibrium stability constants defined by

$$K_{mha} = \frac{(M_m H_h A_a)}{(M)^m (H)^h (A)^a},$$

where $M_m H_h A_a$ is a complex, H denotes hydrogen, and the parentheses indicate (molar) concentration. For the system (with charges omitted) $\{(M, H, A, HA, H_2A, MA, MA_2, MHA, OH, H_2O), (M, H, A, O, p)\}$, calculate the composition at equilibrium, assuming that the element abundance corresponds to 1 kg of H_2O, 5×10^{-3} moles of M, and 5×10^{-3} moles of A. The logarithms (to base 10) of equilibrium constants K_{011}, K_{021}, K_{101}, K_{102}, K_{111}, and K_w are 4.0, 7.0, 3.0, 6.0, 6.0, and -14.0, respectively.

9.16 Cumme (1973) has considered the calculation of the equilibrium concentrations of complexing ligands and metals that cannot be measured directly. He gives an example of a biological application involving human erythrocyte, in which the most important constituents are ATP, 2,3-DPG, magnesium, potassium and sodium ions, and the 16 possible complexes formed from them. The system then is $\{$(DPG, ATP, Mg, K, Na, H, HDPG, H_2DPG, H_3DPG, KDPG, HKDPG, NaDPG, HNaDPG, MgDPG, HMgDPG, Mg_2DPG, MgATP, HMgATP, KATP, NaATP, HATP, H_2ATP), (H, K, Mg, Na, ATP, DPG)$\}$. Logarithms of cumulative formation constants (K_k) are $(8.21, 14.84, 18.36, 1.93, 9.16, 1.93, 9.19, 3.87, 10.92, 6.12, 4.96, 9.80, 1.15, 1.18, 6.95, 10.88)^T$. If the kth complex species is represented by $M_{NM(m,k)} H_{NH(k)} L_{NL(l,k)}$, where the subscripts represent the number of ions of the mth metal, the number of hydrogen ions, and the number of particles of the lth ligand, respectively, the cumulative formation constant is defined by

$$K_k = \frac{(k)}{(H^+)^{NH(k)} \prod_{m=1}^{M} (M_m)^{NM(m,k)} \prod_{l=1}^{L} (L_l)^{NL(l,k)}},$$

where the parentheses denote "concentration of." Calculate the concentrations of all the species, assuming that the total concentrations (molarities) of DPG, ATP, Mg, K, and Na are 0.0072, 0.002, 0.0035, 0.0130, and 0.020, respectively, and the pH is 7.2. Assume that the system is an ideal solution.

APPENDIX A

Computer Programs for Generating Stoichiometric Equations

A.1 HP-41C PROGRAM

A.1.1 Program Listing

```
01♦LBL "STOCOF"      58 STO IND 00      115 STO 83        174 FRC           233 RCL 89
02♦LBL C             59 ISG 00          116 STO 85        175 1 E3          234 STO 88
03 STO 82            60 GTO 14          117♦LBL "CL"      176 *             235 RCL 92
04 X<>Y              61 RCL 82          118 1             177 STO 89        236 STO 89
05 STO 81            62 1               119 ST+ 85        178 RCL 90        237 XEQ "CID"
06 1                 63 RCL 90          120 RCL 85        179 RCL 88        238 STO IND 94
07 -                 64 +               121 RCL 82        180 -             239 RCL 89
08 RCL 82            65 X<=Y?           122 -             181 RCL 82        240 RCL 93
09 *                 66 GTO 13          123 RCL 85        182 /             241 /
10 1 E3              67 "END DATA"      124 RCL 82        183 STO 88        242 ST+ IND 94
11 /                 68 AVIEW           125 *             184 X=0?          243 RCL 92
12 RCL 82            69 RTN             126 +             185 GTO "SC"      244 RCL 90
13 1 E5                                 127 LASTX         186 RCL 87        245 /
14 /                 70♦LBL D           128 1 E3          187 STO 00        246 ST+ IND 91
15 +                 71 1               129 /             188♦LBL "SR"      247♦LBL "ND"
16 STO 99            72♦LBL I           130 +             189 RCL 88        248 RCL 87
17 FIX 0             73♦LBL 11          131 STO 00        190 RCL 82        249 STO 00
18 "M = "            74 STO 90          132 STO 87        191 *             250 RCL IND 00
19 ARCL 81           75 1 E3            133 RCL IND 00    192 RCL 00        251 STO 90
20 "⊢, N = "         76 /               134 X≠0?          193 +             252♦LBL 01
21 ARCL 82           77 RCL 90          135 GTO "ND"      194 STO 90        253 RCL 90
22 AVIEW             78 +               136 RCL 81        195 RCL IND 90    254 ST/ IND 00
23 RTN               79 RCL 99          137 1             196 RCL IND 00    255 ISG 00
                     80 +               138 -             197 STO IND 90    256 GTO 01
24♦LBL B             81 STO 00          139 RCL 82        198 X<>Y          257 1
25 CLX               82 FIX 0           140 *             199 STO IND 00    258 ST+ 83
26 STO 91            83 RCL 00          141 RCL 85        200 ISG 00        259 CLX
27♦LBL 10            84 INT             142 +             201 GTO "SR"      260 STO 86
28 1                 85 STO 92          143 1 E3          202 RCL 89        261♦LBL "RC"
29 ST+ 91            86 "<"             144 /             203 X=0?          262 RCL 87
30 FIX 0             87 ARCL 92         145 RCL 82        204 GTO "ND"      263 RCL 89
31 "<"               88 "⊢)"            146 1 E5          205♦LBL "SC"      264 1
32 ARCL 91           89♦LBL 12          147 /             206 RCL 91        265 ST+ 86
33 ">: "             90 ARCL IND 00     148 +             207 FRC           266 RCL 81
34 AVIEW             91 "⊢ "            149 RCL 00        208 RCL 85        267 RCL 86
35 STOP              92 ISG 00          150 INT           209 +             268 X>Y?
36 GTO 10            93 GTO 12          151 +             210 STO 00        269 GTO "ET"
37 RTN               94 AVIEW           152 STO 88        211♦LBL 00        270 RCL 85
                     95 STOP            153 STO 91        212 RCL 00        271 X=Y?
38♦LBL E             96 RCL 82          154♦LBL "SNZ"     213 RCL 89        272 GTO "RO"
39 1                 97 1               155 RCL 88        214 +             273 RCL 82
40♦LBL J             98 RCL 90          156 STO 90        215 STO 90        274 -
41♦LBL 13            99 +               157♦LBL "RZ"      216 RCL IND 90    275 RCL 86
42 STO 90            100 X<=Y?          158 RCL IND 90    217 RCL IND 00    276 RCL 82
43 1 E3              101 GTO 11         159 X≠0?          218 STO IND 90    277 *
44 /                 102 "END VIEW"     160 GTO "INZ"     219 X<>Y          278 +
45 RCL 90            103 AVIEW          161 ISG 90        220 STO IND 00    279 LASTX
46 +                 104 RTN            162 GTO "RZ"      221 ISG 00        280 1 E3
47 RCL 99                               163 1 E-3         222 GTO 00        281 /
48 + ·               105♦LBL A          164 ST+ 88        223 RCL 85        282 +
49 STO 00            106 .0102030405    165 1             224 STO 88        283 STO 00
50 FIX 0             107 STO 95         166 ST+ 88        225 ST+ 89        284 RCL IND 00
51 "COL: "           108 .0607080910    167 ISG 00        226 XEQ "CID"     285 STO 90
52 ARCL 00           109 STO 96         168 GTO "SNZ"     227 RCL 94        286♦LBL "RD"
53 AVIEW             110 .1112131415    169 GTO "ET"      228 STO 91        287 RCL 90
54 STOP              111 STO 97         170♦LBL "INZ"     229 X<>Y          288 RCL IND 89
55♦LBL 14            112 .1617181920    171 RCL 90        230 STO IND 91    289 *
56 CLX               113 STO 98         172 RCL 91        231 RCL 93        290 ST- IND 00
57 STOP              114 CLX            173 -             232 STO 90        291 CF 06
```

A.1 HP-41C Program

```
292 RCL IND 00
293 ABS
294 1 E-6
295 X>Y?
296 SF 06
297 CLX
298 FS? 06
299 STO IND 00
300 1
301 ST+ 89
302 ISG 00
303 GTO "RD"
304 RCL 86
305 RCL 81
306 X>Y?
307 GTO "RO"
308 RCL 85
309 RCL 81
310 X>Y?
311 GTO "CL"
312 GTO "ET"
313◆LBL "UID"
314 94
315 STO 94
316 RCL 88
317 5
318 /
319 INT
320 ST+ 94
321 LASTX
322 FRC
323 STO 93
324 X≠0?
325 1
326 ST+ 94
327 RCL 93
328 10
329 *
330 X=0?
331 10
332 10↑X
333 STO 93
334 RCL IND 94
335 *
336 FRC
337 LASTX
338 INT
339 1 E2
340 /
341 ENTER↑
342 FRC
343 STO 92
344 -
345 1 E2
346 ST* 92
347 *
348 +
349 RCL 93
350 /

351 RTN
352◆LBL "ET"
353 FS? 01
354 GTO F
355 OFF

356◆LBL F
357 CLA
358 FIX 0
359 "SPECS. : "
360 ARCL 82
361 XEQ "VW"
362 "COMP. "
363 ARCL 83
364 XEQ "VW"
365 RCL 82
366 RCL 83
367 -
368 STO 86
369 "EQUA. "
370 ARCL 86
371 XEQ "VW"

372◆LBL G
373 94
374 STO 94
375 1
376 STO 84
377◆LBL 02
378 5
379 STO 85
380 1
381 ST+ 94
382 RCL IND 94
383 STO 93
384◆LBL 03
385 RCL 93
386 RCL 84
387 X>Y?
388 GTO 04
389 FIX 0
390 CLA
391 "("
392 RCL 93
393 1 E2
394 *
395 FRC
396 STO 93
397 LASTX
398 INT
399 STO IND 84
400 ARCL IND 84
401 ")"
402 ASTO IND 84
403 1
404 ST- 85
405 ST+ 84
406 RCL 85
407 X=0?

408 GTO 02
409 GTO 03
410◆LBL 04
411 RCL 81
412 1
413 STO 90
414 -
415 RCL 82
416 *
417 RCL 84
418 +
419 1 E3
420 /
421 ST+ 84
422 RCL 83
423 LASTX
424 /
425 ST+ 90
426 RCL 82
427 1 E5
428 /
429 ST+ 84
430◆LBL H
431 RCL 93
432 STO 87
433 RCL 85
434 STO 89
435 RCL 90
436 STO 00
437 RCL 84
438 STO 92
439 RCL 94
440 STO 88
441◆LBL 05
442 RCL 87
443 STO 93
444 RCL 89
445 X≠0?
446 GTO 06
447 5
448 STO 89
449 1
450 ST+ 88
451 RCL IND 88
452 STO 87
453◆LBL 06
454 CF 06
455 RCL 83
456 RCL 91
457 CLA
458 "("
459 FIX 0
460 RCL 87
461 1 E2
462 *
463 FRC
464 STO 87
465 LASTX
466 INT

467 STO 86
468 ARCL 86
469 ")="
470◆LBL "DEC"
471 FIX 2
472 ARCL IND 92
473 ARCL IND 00
474◆LBL 07
475 AVIEW
476 CLA
477 CF 05
478◆LBL 08
479 1
480 ST- 91
481 RCL 91
482 X=Y?
483 SF 06
484 X=0?
485 GTO 09
486 ISG 00
487 ISG 92
488 RCL IND 92
489 0
490 X<=Y?
491 "+"
492 ARCL IND 92
493 ARCL IND 00
494 FS? 06
495 GTO 07
496 FS? 05
497 GTO 07
498 SF 05
499 GTO 08
500◆LBL 09
501 BEEP
502 ADV
503 RCL 90
504 STO 00
505 1
506 ST- 85
507 RCL 85
508 STO 89
509 1.001
510 ST+ 84
511 RCL 82
512 RCL 84
513 STO 92
514 INT
515 X<=Y?
516 GTO 05
517 OFF

518◆LBL "VW"
519 AVIEW
520 FS? 02
521 STOP
522 ADV
523 RTN
524 END
```

A.1.2 User's Guide

1. Enter the program.
2. Key XEQ Alpha S I Z E Alpha 1 0 0.
3. Set calculator in USER mode.
4. Press GTO Alpha S T O C O F Alpha.
5. Specify number of elements (M) and number of species (N): (M and N must satify MN ≤ 80.)

Keystroke	Display
(value of M)	(value)
ENTER ↑	(value)
(value of N)	(value)
$\sqrt{x}$	M = ___, N = ___.

6. Specify formula vector for each species:

Keystroke	Display
LN	COL: 1.

For i = 1,2,...,N, enter the following:

Keystroke	Display
R/S	0
(value of a_{1i})	(value)
R/S	0
(value of a_{2i})	(value)
R/S	0
.	.
.	.
.	.
(value of a_{mi})	(value)
R/S	COL: (value of i+1)
	or END DATA when i = N

A.1 HP-41C Program

After all data have been entered:
   To display all data that have been entered, press LOG.
   Press R/S to display next column of $\underline{A}$.
   To display data in a particular column, key in column number n and press COS.
   To revise data in a specific column, key in column number n and press TAN; enter the correct data for that column only.

7. To start the calculation, press SF $\underline{0}$ $\underline{1}$ $\Sigma+$. If CF $\underline{0}$ $\underline{1}$ is keyed in, rather than SF $\underline{0}$ $\underline{1}$, the calculator shows no display and shuts off automatically when the calculations are complete. To view results, turn on calculator, press SF $\underline{0}$ $\underline{1}$, and see step 8.

8. When the calculation is completed, the following display is seen:

   SPECS: (value of N)

   COMP: (value of C)

   EQUA: (value of R)

   (N - M chemical equations)

When the chemical equations are displayed, each species is indicated by a coded number inside brackets < >.
The end of each equation is signalled by a beep.
When all equations have been displayed, the calculator shuts off automatically.
To allow for ample viewing time of the display, press SF $\underline{0}$ $\underline{2}$; to advance to the next display each time, press R/S.
To repeat the display of N, C, and R, press x $\rightleftarrows$ y.
To repeat the display of the preceding chemical equation press SIN.
To repeat display of all equations once the calculator has shut off, turn on the calculator and press R $\downarrow$.

A.2 BASIC PROGRAM

A.2.1 Program Listing

```
100 REM STOICHIOMETRY ALGORITHM
110 INIT
120 PAGE
130 PRINT "ENTER DEVICE NUMBER FOR PRINTING"
140 INPUT D
150 D1=31
160 PRINT "ENTER NUMBER OF SPECIES AND ELEMENTS:"
170 INPUT @D1:N,M
180 DIM W(M),B2(M,N),S$(12*N)
190 S$=""
200 FOR J=1 TO N
210 PRINT "ENTER SPECIES NAME"
220 INPUT @D1:E$
230 PRINT "ENTER FORMULA VECTOR" ";M;" NUMBERS"
240 INPUT @D1:W
250 F$="            "
260 F$=REP(E$,1,LEN(E$))
270 S$=S$&"            "
280 S$=REP(F$,(J-1)*12+1,12)
290 FOR I=1 TO M
300 B2(I,J)=W(I)
310 NEXT I
320 NEXT J
330 DELETE W
340 E$=""
350 LET M1=M+1
360 FOR I=1 TO M
370 LET I1=I+1
380 FOR K=I TO N
390 FOR L=I TO M
400 IF B2(L,K)<>0 THEN 460
410 NEXT L
420 NEXT K
430 LET M=I-1
440 LET M1=I
450 GO TO 620
460 IF K=I THEN 480
470 GOSUB 900
480 IF L=I THEN 520
490 FOR J=I TO N
500 LET B2(I,J)=B2(I,J)+B2(L,J)
510 NEXT J
520 IF I1>N THEN 600
530 FOR L=1 TO M
540 IF L=I OR B2(L,I)=0 THEN 590
550 LET R=B2(L,I)/B2(I,I)
560 FOR J=I1 TO N
570 LET B2(L,J)=B2(L,J)-B2(I,J)*R
580 NEXT J
590 NEXT L
```

A.2 Basic Program

```
600 NEXT I
610 PRINT @D:RESULTS FOLLOW
620 IF M<N THEN 650
630 PRINT NO REACTIONS
640 GO TO 1020
650 FOR I=1 TO M
660 FOR J=M1 TO N
670 LET B2(I,J)=B2(I,J)/B2(I,I)
680 NEXT J
690 NEXT I
700 R1=N-M
710 PRINT @D:"JSPECIES ";N,"  COMPONENTS "," REACTIONS ";R1
720 PRINT @D:
730 DELETE F$
740 DIM F$(200)
750 FOR J=M1 TO N
760 F$=""
770 LET C$=SEG(S$,(J-1)*12+1,12)
780 FOR I=1 TO M
790 IF B2(I,J)=0 THEN 840
800 V$=STR(B2(I,J))
810 LET F$=F$&V$
820 LET B$=SEG(S$,(I-1)*12+1,12)
830 LET F$=F$&B$
840 NEXT I
850 PRINT@D:C$;"=";F$
860 PRINT @D:
870 NEXT J
880 GO TO 1020
890 REM SUBROUTINE SWITCH
900 LET J1=12*K-11
910 LET J2=12*I-11
920 LET B$=SEG(S$,J1,12)
930 LET C$=SEG(S$,J2,12)
940 LET S$=REP(C$,J1,12)
950 LET S$=REP(B$,J2,12)
960 FOR J3=1 TO M
970 LET T=B2(J3,K)
980 LET B2(J3,K)=B2(J3,I)
990 LET B2(J3,I)=T
1000 NEXT J3
1010 RETURN
1020 END
```

A.2.2 Sample Input and Output

```
ENTER DEVICE NUMBER FOR PRINTING
61

ENTER NUMBER OF SPECIES AND ELEMENTS
10,6

ENTER SPECIES NAME (UPTO CHARACTERS)
FE(C204)+

ENTER FORMULA VECTOR 6 NUMBERS
1,2,4,0,0,1

ENTER SPECIES NAME
FE(C204)2-

ENTER FORMULA VECTOR 6 NUMBERS
1,4,8,0,0,-1

ENTER SPECIES NAME
FE(C204)3---

ENTER FORMULA VECTOR 6 NUMBERS
1,6,12,0,0,-3

ENTER SPECIES NAME
FE+++

ENTER FORMULA VECTOR 6 NUMBERS
1,0,0,0,0,3

ENTER SPECIES NAME
S04--

ENTER FORMULA VECTOR 6 NUMBERS
0,0,4,1,0,-2

ENTER SPECIES NAME
HS04-

ENTER FORMULA VECTOR 6 NUMBERS
0,0,4,1,1,-1

ENTER SPECIES NAME
H+

ENTER FORMULA VECTOR 6 NUMBERS
0,0,0,0,1,1

ENTER SPECIES NAME
HC204-
```

A.2 Basic Program

```
ENTER FORMULA VECTOR 6 NUMBERS
0,2,4,0,1,-1

ENTER SPECIES NAME
H2C2O4

ENTER FORMULA VECTOR 6 NUMBERS
0,2,4,0,2,0

ENTER SPECIES NAME
C2O4--

ENTER FORMULA VECTOR 6 NUMBERS
0,2,4,0,0,-2
```


```
RESULTS FOLLOW

SPECIES 10   COMPONENTS 4   REACTIONS 6

   FE(C2O4)3---= -1FE(C2O4)+    2FE(C2O4)2-

   FE+++       = 2FE(C2O4)+     1FE(C2O4)2-

   H+          = -1SO4--        1HSO4-

   HC2O4-      = -1FE(C2O4)+    1FE(C2O4)2-    -1SO4--    1HSO4-

   H2C2O4      = -1FE(C2O4)+    1FE(C2O4)2-    -2SO4--    2HSO4-

   C2O4--      = -1FE(C2O4)+    1FE(C2O4)2-
```

A.3 FORTRAN PROGRAM WITH SAMPLE INPUT AND OUTPUT

```
 1              DIMENSION X(25),A(10,25)
 2              DOUBLE PRECISION X
 3        100   FORMAT(A7, 7F2.0)
 4              READ 102,NS,M
 5        102   FORMAT(5I2)
 6              DO 1  I=1,NS
 7        01    READ 100,X(I),(A(J,I), J=1,M)
 8              PRINT 103
 9        103   FORMAT(/17H INPUT DATA CARDS/)
10              DO 11 I=1,NS
11        11    PRINT 101,X(I),(A(J,I), J=1,M)
12        101   FORMAT(1X,A7,7F3.0)
13              CALL STOICH(X,A,NS,M,10,25)
14              STOP
15              END

16              SUBROUTINE STOICH(X,C,NS,M,I1,I2)
17              DIMENSION X(1),C(I1,I2)
18              DOUBLE PRECISION X
19              M1=M+1
20              DO 58 I=1,M
21              IP1=I+1
        C
        C       FIND NONZERO PIVOT
        C
22              DO 53 K=I,NS
23              DO 53 L=I,M
24              IF(C(L,K).NE.0.)GO to 54
25        53    CONTINUE
26              M=I-1
27              M1=I
28              GO TO 165
29        54    IF(K.EQ.I)GO TO 13
        C
        C       SWITCH COLUMNS
        C
30              TEMP=X(K)
31              X(K)=X(I)
32              X(I)=TEMP
33              DO 166 J=1,M
34              TEMP=C(J,K)
35              C(J,K)=C(J,I)
36       166    C(J,I)=TEMP
37        13    IF(L.EQ.I)GO to 51
        C
        C       ADJUST ROWS
        C
38              DO 57 J=I,NS
39        57    C(I,J)=C(I,J)+C(L,J)
```

A.3 Fortran Program with Sample Input and Output

```
                 C
                 C            ZERO OUT COLUMN
                 C
40        51     IF(IP1.GT.NS)GO TO 58
41               DO 59 L=1,M
42               IF(L.EQ.I.OR.C(L,I).EQ.0.)GO TO 59
43               R=C(L,I)/C(I,I)
44               DO 62 J=IP1,NS
45               IF(ABS(C(L,J).LT.1.E-6)C(L,J) = 0.
46        62     CONTINUE
47        59     CONTINUE
48        58     CONTINUE
                 C
                 C            SOLVE FOR UNKNOWNS
                 C
49               PRINT 105
50        105    FORMAT(//1 5H RESULTS FOLLOW)
51        165    IF(M.LT.NS)GO TO 167
52               PRINT 103,NS,M
53        103    FORMAT(/I5,8H SPECIES, I5,11H COMPONENTS/5X,13H
                                              NO REACTIONS/)
54        104    FORMAT(10A7)
55               PRINT 104,(X(J),J=1,NS)
56               RETURN
57        167    DO 63 I=1,M
58               DO 63 J=M1,NS
59        63     C(I,J)=C(I,J)/C(I,I)
                 C
                 C            PRINT REACTIONS
                 C
60               NR=NS-M
61               PRINT 102,NS,M,NR
62        102    FORMAT(/I5,8H SPECIES,I5,11H COMPONENTS, I5,10H
                                              REACTIONS/)
63               DO 64 J=1M1,NS
64               PRINT 100,X(J), (C(I,J),X(I), I=1,M)
65        100    FORMAT(2X,A7,3H = ,5(F5.2,1X,A8)
66        64     CONTINUE
67               RETURN
68               END

         DATA

INPUT DATA CARDS

H2OL        0.   2.   1.
C2H6OL      2.   6.   1.
C2H4O2L     2.   4.   2.
C4H8O2L     4.   8.   2.
H2OG        0.   2.   1.
C2H6OG      2.   6.   1.
C2H4O2G     2.   4.   2.
C4H8O2G     4.   8.   2.
C2H4O2D     4.   8.   4.
```

```
RESULTS FOLLOW

       9  SPECIES         3  COMPONENTS        6  REACTIONS

           C4H8O2L   =   -1.00  H2OL      1.00  C2H6OL      1.00  C2H4O2L
           H2OG      =    1.00  H2OL      0.00  C2H6OL      0.00  C2H4O2L
           C2H6OG    =    0.00  H2OL      1.00  C2H6OL      0.00  C2H4O2L
           C2H4O2G   =    0.00  H2OL      0.00  C2H6OL      1.00  C2H4O2L
           C4H8O2G   =   -1.00  H2OL      1.00  C2H6OL      1.00  C2H4O2L
           C2H4O2D   =    0.00  H2OL      0.00  C2H6OL      2.00  C2H4O2L
```

Input Data:

1. The first card specifies NS (number of species) and M (number of elements); see lines 4 and 5 of the program listing.
2. The remaining input data, one card for each species, are as indicated in the output on p. 245; see lines 3, 6, and 7 of the listing.

APPENDIX B

Computer Programs for Calculating Equilibrium for Relatively Simple Systems

B.1 STOICHIOMETRIC ALGORITHMS

B.1.1 HP-41C Program

B.1.1.1 Program Listing

```
01♦LBL "STOIEQ"
02♦LBL C
03 STO 90
04 STO 92
05 STO 93
06 STO 94
07 STO 98
08 X<>Y
09 ST* 90
10 STO 91
11 ST- 93
12 ST- 94
13 STO 97
14 1 E3
15 ST/ 90
16 ST/ 94
17 ST/ 97
18 ST/ 98
19 RCL 92
20 1 E5
21 /
22 ST+ 90
23 1
24 ST+ 94
25 ST+ 97
26 ST+ 98
27 FIX 0
28 "M ="
29 ARCL 91
30 "⊦, N ="
31 ARCL 92
32 AVIEW
33 RTN

34♦LBL E
35 1
36♦LBL J
37♦LBL 10
38 STO 99
39 1 E3
40 /
41 RCL 99
42 +
43 RCL 90
44 +
45 STO 00
46 FIX 0
47 "COL= "
48 ARCL 00
49 AVIEW
50 STOP
51♦LBL 11
52 CLX
53 STOP
54 STO IND 00
55 ISG 00
56 GTO 11
57 RCL 92
58 RCL 99
59 1
60 +
61 X<=Y?
```

```
62 GTO 10
63 XEQ "**"
64 RTN

65♦LBL D
66 1
67♦LBL I
68♦LBL 12
69 STO 99
70 1 E3
71 /
72 RCL 99
73 +
74 RCL 90
75 +
76 STO 00
77 FIX 0
78 RCL 00
79 INT
80 STO 89
81 "("
82 ARCL 89
83 ")="
84♦LBL 13
85 RCL 91
86 RCL 92
87 *
88 RCL 00
89 INT
90 X>Y?
91 FIX 3
92 ARCL IND 00
93 "⊦ "
94 FIX 0
95 ISG 00
96 GTO 13
97 AVIEW
98 STOP
99 RCL 92
100 1
101 RCL 99
102 +
103 X<=Y?
104 GTO 12
105 XEQ "**"
106 RTN
107♦LBL B
108 "T ? K"
109 AVIEW
110 STOP
111 STO 81
112 "P ? ATM"
113 AVIEW
114 STOP
115 STO 82
116 RCL 91
117 RCL 92
118 *
119 ENTER↑
120 ENTER↑
121 RCL 92
122 +
123 1 E3
```

```
124 /
125 +
126 1
127 +
128 STO 00
129 RCL 82
130 LN
131 RCL 81
132 1 E3
133 /
134 1.9872
135 *
136♦LBL 14
137 ST/ IND 00
138 X<>Y
139 ST+ IND 00
140 X<>Y
141 ISG 00
142 GTO 14
143 XEQ "**"
144 RTN

145♦LBL F
146 RCL 91
147 1
148 +
149 RCL 92
150 *
151 STO 89
152 RCL 98
153 STO 00
154♦LBL 15
155 1
156 ST+ 89
157 FIX 0
158 "N<"
159 ARCL 00
160 "⊦> ?"
161 AVIEW
162 STOP
163 STO IND 89
164 ISG 00
165 GTO 15
166 CLX
167 STO 99
168 "INE. ?"
169 AVIEW
170 STOP
171 STO 81
172 "END DATA"
173 AVIEW
174 RTN

175♦LBL "**"
176 " **"
177 AVIEW
178 RTN

179♦LBL A
180 CF 05
181 XEQ "LEQ"
182 XEQ "VIJ"
```

```
183♦LBL "AA"
184 FIX 0
185 "ITER. "
186 ARCL 99
187 FS? 01
188 AVIEW
189 XEQ "MCF"
190 XEQ "GJF"
191 XEQ "JCM"
192 BEEP
193 FS? 02
194 STOP
195♦LBL 00
196 ADV
197 ADV
198 1
199 ST+ 99
200 SF 05
201 XEQ "LEQ"
202 XEQ "AJM"
203 GTO "AA"

204♦LBL "VIJ"
205 RCL 94
206 STO 00
207♦LBL "CV"
208 72
209 RCL 00
210 +
211 STO 89
212 CLX
213 STO IND 89
214 ISG 00
215 GTO "CV"
216 RCL 94
217 STO 00
218♦LBL "VJ"
219 RCL 97
220 STO 88
221 RCL 00
222 RCL 91
223 +
224 STO 89
225 CLX
226 STO 87
227♦LBL "VI"
228 RCL IND 89
229 ST+ 87
230 RCL 92
231 ST+ 89
232 ISG 88
233 GTO "VI"
234 72
235 RCL 00
236 +
237 STO 89
238 RCL 87
239 STO IND 89
240 ISG 00
241 GTO "VJ"
242 RTN
```

B.1 Stoichiometric Algorithms

```
243◆LBL "MCF"
244 RCL 81
245 STO 95
246 RCL 98
247 STO 00
248 RCL 91
249 1
250 +
251 RCL 92
252 *
253 STO 89
254 *
255◆LBL 01
256 1
257 ST+ 89
258 FIX 0
259 "N<"
260 ARCL 00
261 "⊢>: "
262 SCI 7
263 ARCL IND 89
264 FS? 01
265 AVIEW
266 RCL IND 89
267 ST+ 95
268 ISG 00
269 GTO 01
270 RTN

271◆LBL "GJF"
272 CLX
273 STO 83
274 1
275 STO 82
276 RCL 91
277 +
278 STO 85
279◆LBL "FJ"
280 RCL 91
281 RCL 92
282 *
283 STO 84
284 RCL 92
285 +
286 STO 86
287 RCL 91
288 +
289 STO 87
290 RCL 82
291 1
292 -
293 RCL 92
294 *
295 RCL 91
296 +
297 STO 88
298 CLX
299 STO IND 88
300 RCL 97
301 STO 00
302◆LBL "FI"
303 1
```

```
304 ST+ 84
305 ST+ 86
306 RCL IND 86
307 RCL 95
308 /
309 LN
310 PCL IND 84
311 +
312 RCL IND 85
313 *
314 ST- IND 88
315 RCL 92
316 ST+ 85
317 ISG 00
318 GTO "FI"
319 RCL 87
320 RCL 82
321 +
322 STO 89
323 RCL IND 89
324 RCL 95
325 /
326 LN
327 RCL 92
328 ST- 89
329 RDN
330 RCL IND 89
331 +
332 ST+ IND 88
333 FIX 0
334 "GJ<"
335 ARCL 82
336 "⊢>: "
337 SCI 7
338 ARCL IND 88
339 FS? 01
340 AVIEW
341 RCL IND 88
342 ABS
343 ST+ 83
344 1
345 ST+ 82
346 RCL 82
347 RCL 91
348 +
349 STO 85
350 RCL 93
351 RCL 82
352 X<=Y?
353 GTO "FJ"
354 RCL 83
355 1 E-5
356 X>Y?
357 OFF
358 RTN

359◆LBL "JCM"
360 1
361 STO 82
362 RCL 91
363 +
364 RCL 92
```

```
365 *
366 STO 88
367 RCL 91
368 +
369 STO 87
370◆LBL "JL"
371 1
372 STO 83
373 ST+ 87
374◆LBL "KL"
375 RCL 88
376 STO 86
377 RCL 92
378 1
379 -
380 RCL 92
381 *
382 RCL 83
383 +
384 STO 80
385 RCL 83
386 RCL 82
387 X>Y?
388 GTO "JS"
389 RCL 91
390 RCL 82
391 +
392 STO 84
393 RCL 91
394 RCL 83
395 +
396 STO 85
397 CLX
398 STO IND 80
399 RCL 82
400 RCL 83
401 X=Y?
402 SF 07
403 RCL IND 87
404 1/X
405 FS? 07
406 ST- IND 80
407 CF 07
408 72
409 RCL 83
410 +
411 STO 89
412 RCL IND 89
413 72
414 RCL 82
415 +
416 STO 89
417 RDN
418 RCL IND 89
419 1
420 -
421 *
422 RCL 95
423 /
424 ST+ IND 80
425 RCL 97
426 STO 00
```

```
427◆LBL "IL"
428 1
429 ST+ 86
430 RCL IND 84
431 RCL IND 85
432 *
433 RCL IND 86
434 /
435 ST- IND 80
436 RCL 92
437 ST+ 84
438 ST+ 85
439 ISG 00
440 GTO "IL"
441◆LBL "SB"
442 1
443 ST+ 83
444 RCL 93
445 RCL 83
446 X<=Y?
447 GTO "KL"
448 1
449 ST+ 82
450 RCL 93
451 RCL 82
452 X<=Y?
453 GTO "JL"
454 RTN
455◆LBL "JS"
456 RCL 83
457 1
458 -
459 RCL 92
460 *
461 RCL 82
462 +
463 STO 89
464 RCL IND 89
465 72
466 RCL 82
467 +
468 STO 89
469 RDN
470 RCL IND 89
471 72
472 RCL 83
473 +
474 STO 89
475 RDN
476 RCL IND 89
477 -
478 RCL 95
479 /
480 +
481 STO IND 80
482 GTO "SB"

483◆LBL "LEQ"
484 CLX
485 STO 85
486◆LBL "LI"
487 1
```

Appendix B

```
488 ST+ 85           550 +               611 STO 85          673 RCL IND 89
489 RCL 85           551 STO 00          612◆LBL "NJ"        674 RCL 83
490 RCL 92           552 RCL 91          613 RCL 91          675 RCL 96
491 -                553 1               614 STO 87          676 *
492 RCL 85           554 -               615 RCL 94          677 X>Y?
493 RCL 92           555 1 E3            616 STO 88          678 GTO "NE"
494 *                556 /               617◆LBL 04          679 FS? 06
495 +                557 FS? 05          618 RCL 85          680 ST- IND 89
496 LASTX            558 ST- 00          619 RCL 91          681 ISG 00
497 1 E3             559 RCL IND 00      620 +               682 GTO "MI"
498 /                560 STO 83          621 RCL 88          683 FS? 06
499 +                561◆LBL "RI"        622 +               684 RTN
500 STO 87           562 RCL 83          623 STO 89          685 SF 06
501 RCL 91           563 RCL IND 89      624 RCL IND 89      686 GTO "NJ"
502 1                564 *               625 RCL IND 87      687◆LBL "NE"
503 -                565 ST- IND 00      626 RCL 96          688 .1
504 1 E3             566 1               627 *               689 ST- 96
505 /                567 ST+ 89          628 CHS             690 RCL 96
506 FS? 05           568 ISG 00          629 X>Y?            691 X=0?
507 ST- 87           569 GTO "RI"        630 GTO "NE"        692 OFF
508◆LBL "ND"         570 RCL 93          631 FS? 06          693 GTO "NJ"
509 RCL 87           571 RCL 91          632 ST- IND 89
510 STO 00           572 FS? 05          633 RCL 92          694◆LBL "EQU"
511 INT              573 X<>Y            634 ST+ 87          695 RCL 94
512 STO 88           574 RCL 86          635 ISG 88          696 STO 00
513 RCL 85           575 X<>Y            636 GTO 04          697◆LBL 08
514 STO 89           576 X>Y?            637 RCL 91          698 " EQU "
515 RCL IND 00       577 GTO "LJ"        638 STO 87          699 FIX 0
516 X=0?             578 RCL 85          639 RCL 97          700 ARCL 00
517 GTO "ZS"         579 X<>Y            640 STO 00          701 "⊦ :"
518 STO 83           580 X>Y?            641◆LBL "MI"        702 AVIEW
519◆LBL 06           581 GTO "LI"        642 CLX             703 RCL 00
520 RCL 83           582◆LBL "ZS"        643 STO 83          704 RCL 91
521 ST/ IND 00       583 RCL 89          644 RCL 94          705 +
522 ISG 00           584 RCL 91          645 STO 88          706 STO 89
523 GTO 06           585 X=Y?            646 RCL 91          707
524 CLX              586 OFF             647 STO 87          708 ARCL 89
525 STO 86           587 RCL 92          648 1               709 "⊦> ="
526◆LBL "LJ"         588 ST+ 88          649 +               710 RCL 97
527 RCL 87           589 RCL IND 88      650 RCL 00          711 STO 80
528 STO 89           590 X=0?            651 INT             712◆LBL 09
529 1                591 GTO "ZS"        652 1               713 RCL IND 89
530 ST+ 86           592 RCL 87          653 -               714 0
531 RCL 91           593 STO 00          654 RCL 92          715 X<=Y?
532 FS? 05           594◆LBL "SR"        655 *               716 "⊦ +
533 RCL 93           595 RCL IND 88      656 +               717 FIX 3
534 RCL 86           596 ST+ IND 00      657 STO 82          718 ARCL IND 89
535 X>Y?             597 1               658◆LBL "MJ"        719 FIX 0
536 RTN              598 ST+ 88          659 RCL IND 87      720 "⊦<"
537 RCL 85           599 ISG 00          660 RCL IND 82      721 ARCL 80
538 X=Y?             600 GTO "SR"        661 *               722 "⊦>"
539 GTO "LJ"         601 GTO "ND"        662 ST+ 83          723 AVIEW
540 RCL 85                               663 RCL 92          724 CLA
541 RCL 92           602◆LBL "AJM"       664 ST+ 87          725 RCL 92
542 -                603 1               665 1               726 ST+ 89
543 RCL 86           604 STO 96          666 ST+ 82          727 ISG 80
544 RCL 92           605 CF 06           667 ISG 88          728 GTO 09
545 *                606 RCL 92          668 GTO "MJ"        729 BEEP
546 +                607 PCL 91          669 RCL 85          730 ISG 00
547 LASTX            608 *               670 RCL 00          731 GTO 08
548 1 E3             609 RCL 92          671 +               732 XEQ "**"
549 /                610 +               672 STO 89          733 END
```

B.1 Stoichiometric Algorithms

B.1.1.2 User's Guide

1. Enter the program.
2. Key XEQ Alpha S I Z E Alpha 1 0 0.
3. Set calculator in USER mode.
4. Press GTO Alpha S T O I E Q Alpha.
5. Specify the number of elements (M) and the number of species (N) (M and N must satisfy M ≥ 2N, and N(M+2) ≤ 75.):

Keystroke	Display
(value of M)	(value)
ENTER ↑	(value)
(value of N)	(value)
$\sqrt{x}$	M = ___ , N = ___ .

6. Specify the formula vectors and μ^o or μ^o/RT values for each species* (The first M species must have linearly independent formula vectors.):

Keystroke	Display
LN	COL: 1.

For i = 1,2,...,N, enter the following:

Keystroke	Display
R/S	0
(value of a_{1i})	(value)
R/S	0
(value of a_{2i})	(value)
R/S	0
.	.
.	.
.	.

---

*If μ^o entered, it may be in kJ mole^{-1} or kcal mole^{-1}; consistent units must be used for all species.

(value of a_{mi})	(value)
R/S	0
(value of μ_i^o or μ_i^o/RT)	(value)
R/S	COL: (value of i+1)
	or **[when i = N]

After all species data have been entered:
To display all data that have been entered, press LOG. Press R/S to display next column of data.
To display data in a particular column, key in the column number, then press COS.
To revise data in a specific column, key in the column number, then press TAN and enter the correct data for that column only.

7. To enter T and P, complete the following steps:

   (1) If the chemical potential values previously entered are in kJ mole^{-1}, the following instructions must be used:

Keystroke	Display
1/x	T ? K
(value of T in K)	(value)
ENTER	(value)
4 . 1 8 4 0	4.1840
X	(product RT)
R/S	P ? ATM
(value of P in atm)	(value)
R/S	**[after a short pause]

   (2) If the chemical potential values previously entered are in kcal mole^{-1}, the following instructions must be used:

Keystroke	Display
1/x	T ? K
(value of T in K)	(value)
R/S	P ? ATM

B.1 Stoichiometric Algorithms

(value of P in atm)	(value)
R/S	**[after a short pause]

(3) If the chemical protential values previously entered are in the form μ^o/RT, the following instructions must be used:

Keystroke	Display
1/x	T ? K
EEX 3	1000.00
ENTER	1000.00
1 . 9 8 7 2	1.99
÷	503.22
R/S	P ? ATM
(value of P in atm)	(value)
R/S	** after a short pause

8. To enter moles of inert species and initial estimate of equilibrium mole numbers, use the following steps:

Keystroke	Display
x ⇄ y	N<1.>
(moles of species 1)	(value)
R/S	N<2.>
(moles of species 2)	(value)
.	.
.	.
.	.
(moles of species N)	(value)
R/S	INE. ?
(moles of inert species)	(value)
R/S	END DATA

The estimate must satisfy the element abundances. In order to ensure convergence, the species should be ordered so that those that are likely to be the most abundant at equilibrium are listed first. A guide to this is the relative values of μ_i^o, with species having more negative values likely to be more abundant.

9. To start the calculation, press SF $\underline{0}$ $\underline{1}$ $\Sigma+$. If CF $\underline{0}$ $\underline{1}$ is keyed in, rather than SF $\underline{0}$ $\underline{1}$, the calculator shows no display and shuts off automatically when the calculations are complete. To view results see step 12. To change the program running mode during the calculations, press R/S, key in the alternate mode (SF $\underline{0}$ $\underline{1}$ or CF $\underline{0}$ $\underline{1}$), and then press R/S to continue.
To stop the calculations at the end of each iteration, key in SF $\underline{0}$ $\underline{2}$. In this case, press R/S to advance to the next iteration.

10. The display shows the following for each iteration:

    ITER: (iteration number)
    N<1.>:   (value of n_1)
    N<2.>:   (value of n_2)
    .
    .
    .
    GJ<1.>:   (value of ΔG_1)
    GJ<2.>:   (value of ΔG_2)
    .
    .
    .

    The end of each iteration is signalled by a beep, and the calculation automatically proceeds to the next iteration (but see final remark in step 9).

11. The calculation terminates when either of the following criteria is satisfied:

    (a) $\Sigma |\Delta G_j| < 10^{-5}$

    In this case, the display indicates this value.

    (b) A suitable step-size parameter cannot be found in equation (4.3-1).

B.1 Stoichiometric Algorithms

In this case the display indicates zero. Repeat step 8 (enter a new set of estimates for the equilibrium mole numbers). To restart the calculations, press XEQ Alpha A A Alpha.

12. To display the results, first press SF 0 1.
    To display the equilibrium mole numbers, press XEQ Alpha M F C Alpha.
    To display the equilibrium values of ΔG_j, press XEQ Alpha G J F Alpha.
    To display the number of iterations, press RCL 9 9.
    To display the stoichiometric equations, press XEQ Alpha E Q U Alpha.
    The equations are displayed in the same manner as described in step 8 of the "STOCOF" User's Guide (section A.1.2)

B.1.2 BASIC Program

B.1.2.1 *Program Listing*

```
100 REM STOICHIOMETRIC ALGORITHM FOR CALCULATING THE EQUILIBRIUM
110 REM COMPOSITION OF SINGLE PHASE SYSTEMS.
120 INIT
130 PAGE
140 PRINT "JENTER DEVICE NUMBER FOR PRINTINGJ"
150 INPUT D
160 D1=31
180 IMAGE 7A,2X,14A,43T,16A,S
190 IMAGE 8A,2X,33A,7D.3D,2X,6D.4D
200 L1=0
210 I6=0
220 S1=0
230 F1=0
240 PRINT "ENTER NUMBER OF SPECIES AND ELEMENTS:J"
250 INPUT @D1:N,M
260 PRINT "JTYPE OF CHEM. POT. :   KJ, KCAL OR MU/RT J"
270 PRINT "ENTER 1, 2 OR 3 RESPECTIVELYJ"
280 INPUT @D1:I0
290 PRINT "JHOW MANY SIGNIFICANT FIGURESJ"
300 INPUT I2
310 DIM S$(8*N),W(M),B2(M,N),F0(N),B(M),E$(3*M)
320 S$=""
330 FOR J=1 TO N
340 PRINT "ENTER SPECIES NAMEJ"
350 INPUT @D1:E$
360 PRINT "ENTER FORMULA VECTOR,CHEM POTENTIAL   ";M+1;"   NUMBERSJ"
370 INPUT @D1:W,F0(J)
380 F$="        "
390 F$=REP(E$,1,LEN(E$))
400 S$=S$&"        "
410 S$=REP(F$,(J-1)*8+1,8)
420 FOR I=1 TO M
430 B2(I,J)=W(I)
440 NEXT I
450 NEXT J
460 DELETE W
470 DIM W(N)
480 PRINT "JENTER INITIAL ESTIMATES FOR THE MOLE NUMBERS ";N;" NUMBERSJ"
490 INPUT W
500 E$=""
510 FOR J=1 TO M
520 PRINT "JENTER ELEMENT ";J;"J"
530 INPUT @D1:G$
540 F$="    "
550 F$=REP(G$,4-LEN(G$),LEN(G$))
560 E$=E$&F$
570 NEXT J
580 N8=0
590 PRINT "JARE THERE ANY INERT SPECIES PRESENT (Y/N)"
600 INPUT L$
```

B.1 Stoichiometric Algorithms

```
610 IF L$="N" THEN 640
620 PRINT "ENTER NO OF MOLES OF INERT SPECIES"
630 INPUT N8
640 PRINT "ENTER TEMPERATURE IN K,PRESSURE IN ATM:J"
650 INPUT @D1:T,P
660 PRINT "JENTER TITLEJ"
670 INPUT @D1:T$
680 PRINT "JDO YOU WANT TO PRINT INTERMEDIATE RESULTS (Y/N)J"
690 INPUT I$
694 PRINT "JDO YOU WANT TO PRINT THE REACTIONS (Y/N)J"
696 INPUT R$
700 REM COMPUTE ELEMENTAL ABUNDANCES
710 FOR J=1 TO M
720 B(J)=0
730 FOR I=1 TO N
740 B(J)=B(J)+B2(J,I)*W(I)
750 NEXT I
760 NEXT J
770 REM          * PRINT DATA     *
780 PRINT @D:T$
790 PRINT @D:"J";N;" SPECIES",M;" ELEMENTS"
800 PRINT @D:"JPRESSURE    ";P;" ATM "
810 PRINT @D:"JTEMPERATURE  ";T;" K"
820 PRINT @D:"JELEMENTAL ABUNDANCES    ";
830 FOR J=1 TO M
840 G$="    "
850 G$=SEG(E$,(J-1)*3+1,3)
860 G$=G$&"  "
870 PRINT @D:G$,B(J)
880 PRINT @D:"                        ";
890 NEXT J
900 PRINT @D:""
910 PRINT @D: USING 180:"SPECIES","FORMULA VECTOR","STAN. CHEM. POT.";
920 PRINT @D:"   INIT. EST."
930 PRINT @D:"              ";E$
940 FOR J=1 TO N
950 O$=""
960 FOR I=1 TO M
970 E$="    "
980 G$=STR(B2(I,J))
990 E$=REP(G$,4-LEN(G$),LEN(G$))
1000 O$=O$&E$
1010 NEXT I
1020 G$=SEG(S$,(J-1)*8+1,8)
1030 PRINT @D: USING 190:G$,O$,F0(J),W(J)
1040 NEXT J
1050 REM        * ADJUST FOR SPECIFIED TYPE OF STAN. CHEM. POT.*
1060 R0=1.9872*T
1070 GO TO I0 OF 1110,1080,1150
1080 T1=1000/R0
1090 PRINT @D:"JSTAN. CHEM. POT. IS IN KCAL/MOLE"
1100 GO TO 1170
1110 T1=0.0083143*T
1120 T1=1/T1
```

```
1130 PRINT @D:"JSTAN. CHEM. POT. IS IN KJ/MOLEJ"
1140 GO TO 1170
1150 T1=0
1160 PRINT @D:"JSTAN. CHEM. POT. IS MU/RT"
1170 FOR I=1 TO N
1180 IF T1=0 THEN 1200
1190 F0(I)=F0(I)*T1
1200 F0(I)=F0(I)+LOG(P)
1210 NEXT I
1220 REM CALCULATE STOICHIOMETRY
1250 GOSUB 2510
1260 DIM B3(R1),A3(R1,R1)
1270 REM EQUILIBRIUM FUNCTION
1280 T9=SUM(W)+N8
1290 FOR J=1 TO R1
1300 X=W(M+J)/T9
1310 IF X=0 THEN 2090
1320 B3(J)=LOG(X)+F0(M+J)
1330 FOR I=1 TO M
1340 X=W(I)/T9
1350 IF X=0 THEN 2090
1360 B3(J)=B3(J)-A2(I,J)*(LOG(X)+F0(I))
1370 NEXT I
1380 NEXT J
1390 REM JACOBIAN MATRIX
1400 FOR K=1 TO R1
1410 FOR I=1 TO R1
1420 A3(K,I)=0
1430 FOR J=1 TO M
1440 A3(K,I)=A3(K,I)+A2(J,K)*A2(J,I)/W(J)
1450 NEXT J
1460 V1=0
1470 V2=0
1480 FOR J=1 TO M
1490 V1=V1+A2(J,K)
1500 V2=V2+A2(J,I)
1510 NEXT J
1520 V1=(V1-1)/T9
1530 V2=V2-1
1540 A3(K,I)=A3(K,I)-V1*V2
1550 IF K<>I THEN 1570
1560 A3(K,I)=A3(K,I)+1/W(M+I)
1570 NEXT I
1580 NEXT K
1590 DIM B4(N)
1600 REM SOLVE THE LINEAR SYSTEM SOL IN B3
1610 I6=I6+1
1612 IF I6 < 500 THEN 1620
1614 PRINT@D:"NO CONVERGENCE AFTER 500 ITERATIONS"
1616 PRINT@D:"CURRENT RESULTS FOLLOW"
1618 GO TO 2120
1620 GOSUB 2330
1630 P1=1
1640 F9=0
```

B.1 Stoichiometric Algorithms

```
1650 FOR I=1 TO M
1660 B4(I)=0
1670 FOR J=1 TO R1
1680 B4(I)=B4(I)+A2(I,J)*B3(J)*P1
1690 NEXT J
1700 NEXT I
1710 FOR J=1 TO R1
1720 B4(M+J)=-B3(J)*P1
1730 NEXT J
1740 IF I$="N" THEN 1820
1750 IMAGE 18A,9E,37T,9E
1760 PRINT @D:"JITERATION       ";I6;"   CONV FORCER    ";P1
1770 PRINT @D:"JSPECIES","OLD MOLES","DELTA MOLESJ"
1780 FOR I=1 TO N
1790 G$=SEG(S$,(I-1)*8+1,8)
1800 PRINT @D: USING 1750:G$,W(I),B4(I)
1810 NEXT I
1820 IF F9>0 THEN 1960
1830 P1=0.5
1840 FOR J=1 TO N
1850 X1=-B4(J)/W(J)
1860 IF P1<X1 THEN 1880
1870 GO TO 1890
1880 P1=X1
1890 NEXT J
1900 P1=1/P1
1910 IF P1<=1.01 AND P1>0 THEN 1930
1920 GO TO 1970
1930 P1=0.99*P1
1940 F9=1
1950 GO TO 1650
1960 F9=0
1970 FOR I=1 TO M
1980 W(I)=B4(I)+W(I)
1990 NEXT I
2000 FOR J=1 TO R1
2010 W(M+J)=W(M+J)+B4(M+J)
2020 NEXT J
2030 S9=0
2040 FOR I=1 TO R1
2050 S9=S9+ABS(B3(I))
2060 NEXT I
2070 IF S9>10^-12 THEN 1280
2080 GO TO 2120
2090 PRINT @D:"JZERO MOLE NUMBER. REARRANGE DATA IN DECREASING ORDER";
2100 PRINT @D:" OF MOLE NUMBER"
2110 PRINT @D:"ACCORDING TO CURRENT RESULTS WHICH FOLLOW"
2120 PRINT @D:"JNUMBER OF ITERATIONS:",I6
2130 PRINT @D:"JSPECIES","  MOLE FRACTION","   MOLESJ"
2140 IMAGE 18A,9E,37T,9E
2150 IMAGE"JTOTAL NUMBER OF MOLES:",4X,9E
2160 IMAGE "JINERTS",19T,9E,37T,9E
2170 FOR J=1 TO N
2180 G$=SEG(S$,(J-1)*8+1,8)
```

```
2190 PRINT @D: USING 2140:G$,W(J)/T9,W(J)
2200 NEXT J
2210 IF X=0 THEN 2300
2212 G=0
2215 IF N8=0 THEN 2230
2220 PRINT @D: USING 2160:N8/T9,N8
2230 PRINT @D: USING 2150:T9
2235 IF N8=0 THEN 2250
2240 G=N8*LOG(N8/T9)
2250 FOR J=1 TO N
2260 G=G+W(J)*(F0(J)+LOG(W(J)/T9))
2270 NEXT J
2280 IMAGE "JG/RT = ",9E
2290 PRINT @D: USING 2280:G
2300 END
2310 REM SUBROUTNE TO SOLVE LINEAR SYSTEM
2320 REM AUGMENTED MATRIX
2330 IF R1=1 THEN 2480
2340 DIM A4(R1,R1+1),A5(R1,R1+1)
2350 FOR I=1 TO R1
2360 FOR J=1 TO R1
2370 A4(I,J)=A3(I,J)
2380 NEXT J
2390 NEXT I
2400 FOR J=1 TO R1
2410 A4(J,R1+1)=B3(J)
2420 NEXT J
2430 A5=INV(A4)
2440 FOR J=1 TO R1
2450 B3(J)=A5(J,R1+1)
2460 NEXT J
2470 GO TO 2490
2480 B3(1)=B3(1)/A3(1,1)
2490 RETURN
2500 REM SUBROUTINE TO CALCULATE STOICHIOMETRY
2510 LET M1=M+1
2520 FOR I=1 TO M
2530 LET I1=I+1
2540 FOR K=I TO N
2550 FOR L=I TO M
2560 IF B2(L,K)<>0 THEN 2620
2570 NEXT L
2580 NEXT K
2590 LET M=I-1
2600 LET M1=I
2610 GO TO 2780
2620 IF K=I THEN 2640
2630 GOSUB 3150
2640 IF L=I THEN 2680
2650 FOR J=I TO N
2660 LET B2(I,J)=B2(I,J)+B2(L,J)
2670 NEXT J
2680 IF I1>N THEN 2760
2690 FOR L=1 TO M
```

B.1 Stoichiometric Algorithms

```
2700 IF L=I OR B2(L,I)=0 THEN 2750
2710 LET R=B2(L,I)/B2(I,I)
2720 FOR J=I1 TO N
2730 LET B2(L,J)=B2(L,J)-B2(I,J)*R
2740 NEXT J
2750 NEXT L
2760 NEXT I
2770 PRINT @D:"RESULTS FOLLOW"
2780 IF M<N THEN 2810
2790 PRINT "NO REACTIONS"
2800 GO TO 2300
2810 FOR I=1 TO M
2820 FOR J=M1 TO N
2830 LET B2(I,J)=B2(I,J)/B2(I,I)
2840 NEXT J
2850 NEXT I
2860 R1=N-M
2870 IF R$="N" THEN 3060
2880 PRINT @D:"JSPECIES ";N,"  COMPONENTS ";M,"  REACTIONS ";R1
2890 PRINT @D:
2900 DELETE F$
2910 DIM F$(8*N)
2920 FOR J=M1 TO N
2930 F$=""
2940 LET C$=SEG(S$,(J-1)*8+1,8)
2950 FOR I=1 TO M
2960 IF B2(I,J)=0 THEN 3010
2970 V$=STR(B2(I,J))
2980 LET F$=F$&V$
2990 LET B$=SEG(S$,(I-1)*8+1,8)
3000 LET F$=F$&B$
3010 NEXT I
3020 PRINT @D:C$;"=";F$
3030 PRINT @D:
3040 NEXT J
3050 REM SAVE STOICH. MATRIX
3060 DELETE A2
3070 DIM A2(M,R1)
3080 FOR J=1 TO M
3090 FOR I=1 TO R1
3100 A2(J,I)=B2(J,M+I)
3110 NEXT I
3120 NEXT J
3130 RETURN
3140 REM SUBROUTINE SWITCH
3150 LET J1=8*K-7
3160 LET J2=8*I-7
3170 LET B$=SEG(S$,J1,8)
3180 LET C$=SEG(S$,J2,8)
3190 LET S$=REP(C$,J1,8)
3200 LET S$=REP(B$,J2,8)
3210 FOR J3=1 TO M
3220 LET T=B2(J3,K)
3230 LET B2(J3,K)=B2(J3,I)
3240 LET B2(J3,I)=T
3250 NEXT J3
3260 T1=F0(K)
3270 T2=W(K)
3280 F0(K)=F0(I)
3290 W(K)=W(I)
3300 F0(I)=T1
3310 W(I)=T2
3320 RETURN
```

B.1.2.2 Sample Input and Output

```
(N.B. The species must be entered in increasing
order of their chemical potentials.)

ENTER DEVICE NUMBER FOR PRINTING
61

ENTER NUMBER OF SPECIES AND ELEMENTS:
11,4

TYPE OF CHEM. POT. :KJ, KCAL OR MU/RT
ENTER 1, 2 OR 3 RESPECTIVELY
1

HOW MANY SIGNIFICANT FIGURES?
8

ENTER SPECIES NAME
H2O

ENTER FORMULA VECTOR,CHEM POTENTIAL   5 NUMBERS
0,2,1,0,-214.05

ENTER SPECIES NAME
CH3CHO

ENTER FORMULA VECTOR,CHEM POTENTIAL   5 NUMBERS
2,4,1,0,-96.11

ENTER SPECIES NAME
C2H5OH

ENTER FORMULA VECTOR,CHEM POTENTIAL   5 NUMBERS
2,6,1,0,-95.52

ENTER SPECIES NAME
H2S

ENTER FORMULA VECTOR,CHEM POTENTIAL   5 NUMBERS
0,2,0,1,-42.22

ENTER SPECIES NAME
H2

ENTER FORMULA VECTOR,CHEM POTENTIAL   5 NUMBERS
0,2,0,0,0

ENTER SPECIES NAME
S2

ENTER FORMULA VECTOR,CHEM POTENTIAL   5 NUMBERS
0,0,0,2,35.26
```

B.1 Stoichiometric Algorithms

```
ENTER SPECIES NAME
(C2H5)2O

ENTER FORMULA VECTOR,CHEM POTENTIAL   5 NUMBERS
4,10,1,0,17.53

ENTER SPECIES NAME
C2H6

ENTER FORMULA VECTOR,CHEM POTENTIAL   5 NUMBERS
2,6,0,0,24.94

ENTER SPECIES NAME
C2H5SH

ENTER FORMULA VECTOR,CHEM POTENTIAL   5 NUMBERS
2,6,0,1,44.77

ENTER SPECIES NAME
C2H4

ENTER FORMULA VECTOR,CHEM POTENTIAL   5 NUMBERS
2,4,0,0,87.53

ENTER SPECIES NAME
(C2H5)2S

ENTER FORMULA VECTOR,CHEM POTENTIAL   5 NUMBERS
4,10,0,1,130.83

ENTER INITIAL ESTIMATES FOR THE MOLE NUMBERS 11 NUMBERS
0.5,0.4,0.001,0.891,0.2755,0.001,0.099,0.1265,0.058,0.1185,0.049

ENTER ELEMENT 1
C

ENTER ELEMENT 2
H

ENTER ELEMENT 3
O

ENTER ELEMENT 4
S

ARE THERE ANY INERT SPECIES PRESENT (Y/N)
N

ENTER TEMPERATURE IN K, PRESSURE IN ATM:
600,80

ENTER TITLE
FORMATION OF SULFUR COMPOUNDS (VONKA AND HOLUB 1975)
```

DO YOU WANT TO PRINT INTERMEDIATE RESULTS (Y/N)
N

DO YOU WANT TO PRINT THE REACTIONS (Y/N)
N

FORMATION OF SULFUR COMPOUNDS (VONKA AND HOLUB 1975)

11 SPECIES   4 ELEMENTS

PRESSURE   80 ATM

TEMPERATURE   600 K

ELEMENTAL ABUNDANCES   C   2
                       H   8
                       O   1
                       S   1

SPECIES	FORMULA VECTOR				STAN. CHEM. POT.	INIT. EST.
	C	H	O	S		
H2O	0	2	1	0	-214.050	0.5000
CH3CHO	2	4	1	0	-96.110	0.4000
C2H5OH	2	6	1	0	-95.520	0.0010
H2S	0	2	0	1	-42.220	0.8910
H2	0	2	0	0	0.000	0.2755
S2	0	0	0	2	35.260	0.0010
(C2H5)2O	4	10	1	0	17.530	0.0990
C2H6	2	6	0	0	24.940	0.1265
C2H5SH	2	6	0	1	44.770	0.0580
C2H4	2	4	0	0	87.530	0.1185
(C2H5)2S	4	10	0	1	130.830	0.0490

STAN. CHEM. POT. IS IN KJ/MOLE

RESULTS FOLLOW

NUMBER OF ITERATIONS: 12

SPECIES	MOLE FRACTION	MOLES
H2O	2.420950355E-001	5.865427155E-001
CH3CHO	1.705936057E-001	4.133105684E-001
C2H5OH	6.051159406E-005	1.466062062E-004
H2S	3.509999115E-001	8.503951384E-001
H2	4.990567367E-006	1.209104073E-005
S2	2.346459988E-003	5.684953473E-003
(C2H5)2O	4.536988708E-008	1.099211998E-007
C2H6	1.752815351E-001	4.246683843E-001
C2H5SH	4.886081630E-002	1.183789490E-001
C2H4	1.561537962E-003	3.783261044E-003
(C2H5)2S	8.195550380E-003	1.985600557E-002

TOTAL NUMBER OF MOLES: 2.422778783E+000

G/RT = -2.960665528E+001

B.2 NONSTOICHIOMETRIC ALGORITHMS

B.2.1 HP-41C Program

B.2.1.1 Program Listing

```
01♦LBL "NONSTO"        54 +                   106 " P ? ATM"         159 RCL 71
02♦LBL C               55 X<=Y?               107 AVIEW              160 ST/ 81
03 STO 90              56 GTO 10              108 STOP               161 RCL 92
04 STO 92              57 XEQ "**"            109 STO 82             162 1
05 STO 98              58 RTN                 110 RCL 91             163 +
06 X<>Y                                       111 RCL 92             164 STO 85
07 ST* 90              59♦LBL D               112 *                  165 RCL 98
08 STO 91              60 1                   113 ENTER↑             166 STO 00
09 STO 97              61♦LBL I               114 ENTER↑             167 2
10 1 E3                62♦LBL 12              115 RCL 92             168 STO 80
11 ST/ 90              63 STO 99              116 +                  169♦LBL 15
12 ST/ 97              64 1 E3                117 1 E3               170 70
13 ST/ 98              65 /                   118 /                  171 RCL 08
14 RCL 92              66 RCL 99              119 +                  172 +
15 1 E5                67 +                   120 1                  173 STO 70
16 /                   68 RCL 90              121 +                  174♦LBL 16
17 ST+ 90              69 +                   122 STO 00             175 RCL 71
18 1                   70 STO 00              123 RCL 82             176 CHS
19 ST+ 97              71 FIX 0               124 LN                 177 ST* IND 85
20 ST+ 98              72 RCL 00              125 RCL 81             178 RCL IND 00
21 FIX 0               73 INT                 126 1 E3               179 RCL IND 70
22 " M = "             74 STO 89              127 /                  180 *
23 ARCL 91             75 "<"                 128 1.9872             181 ST+ IND 85
24 "⊢, N ="            76 ARCL 89             129 *                  182 1
25 ARCL 92             77 "⊢>"                130♦LBL 14             183 ST+ 85
26 AVIEW               78♦LBL 13              131 ST/ IND 00         184 ISG 00
27 RTN                 79 RCL 91              132 X<>Y               185 GTO 16
                       80 RCL 92              133 ST+ IND 00         186 RCL 98
28♦LBL E               81 *                   134 X<>Y               187 STO 00
29 1                   82 RCL 00              135 ISG 00             188 1
30♦LBL J               83 INT                 136 GTO 14             189 ST+ 80
31♦LBL 10              84 X>Y?                137 XEQ "**"           190 RCL 91
32 STO 99              85 FIX 3               138 RTN                191 RCL 80
33 1 E3                86 ARCL IND 00                                192 X<=Y?
34 /                   87 "⊢ "                139♦LBL F              193 GTO 15
35 RCL 99              88 FIX 0               140 RCL 97             194 XEQ "**"
36 +                   89 ISG 00              141 STO 00             195 RTN
37 RCL 90              90 GTO 13              142♦LBL 17
38 +                   91 AVIEW               143 70                 196♦LBL H
39 STO 00              92 STOP                144 RCL 00             197 RCL 91
40 FIX 0               93 RCL 92              145 +                  198 2
41 "COL: "             94 1                   146 STO 80             199 +
42 ARCL 00             95 RCL 99              147 FIX 0              200 RCL 92
43 AVIEW               96 +                   148 "ELEM"             201 *
44 STOP                97 X<=Y?               149 ARCL 00            202 STO 00
45♦LBL 11              98 GTO 12              150 AVIEW              203 RCL 97
46 CLX                 99 XEQ "**"            151 STOP               204 STO 87
47 STOP                100 RTN                152 STO IND 80         205♦LBL 18
48 STO IND 00                                 153 ISG 00             206 "COL ?"
49 ISG 00              101♦LBL B              154 GTO 17             207 AVIEW
50 GTO 11              102 " T ? K"           155 "INERT ?"          208 STOP
51 RCL 92              103 AVIEW              156 AVIEW              209 STO 89
52 RCL 99              104 STOP               157 STOP               210 RCL 91
53 1                   105 STO 81             158 STO 81             211 RCL 92
                                                                     212 *
```

```
213 +                 266♦LBL "MFC"        321 "⊢):              375 RCL 00
214 1 E3              267 RCL 98           322 SCI 7             376 +
215 /                 268 STO 00           323 ARCL IND 88       377 STO 88
216 RCL 92            269 CLX              324 FS? 01            378 RCL 83
217 1 E5              270 STO 85           325 AVIEW             379 RCL 92
218 /                 271♦LBL 01           326 ISG 00            380 *
219 +                 272 RCL 97           327 GTO 01            381 RCL 00
220 ST+ 89            273 STO 84           328 RTN               382 +
221♦LBL 19            274 RCL 00                                 383 STO 87
222 1                 275 INT              329♦LBL "JCM"         384 RCL IND 88
223 ST+ 00            276 ENTER↑           330 RCL 91            385 RCL IND 87
224 RCL IND 89        277 ENTER↑           331 2                 386 *
225 STO IND 00        278 1 E3             332 +                 387 FS? 07
226 ISG 89            279 /                333 RCL 92            388 ST+ 82
227 GTO 19            280 +                334 *                 389 FS? 05
228 ISG 87            281 RCL 90           335 STO 89            390 GTO 04
229 GTO 18            282 +                336 STO 88            391 RCL 86
230 SF 04             283 STO 87           337 RCL 91            392 RCL 00
231 XEQ "LEQ"         284 CLX              338 1                 393 +
232 CF 04             285 STO 86           339 +                 394 STO 88
233 "END DATA"        286♦LBL 02           340 RCL 91            395 RDN
234 AVIEW             287 92               341 *                 396 RCL IND 88
235 RTN               288 RCL 84           342 +                 397 *
                      289 +                343 1 E3              398 GTO 05
236♦LBL "**"          290 STO 89           344 /                 399♦LBL 04
237 " **"             291 RCL IND 87       345 1                 400 RCL 83
238 AVIEW             292 RCL IND 89       346 +                 401 RCL 92
239 RTN               293 *                347 ST+ 88            402 *
                      294 ST+ 86           348♦LBL 03            403 ST- 87
240♦LBL A             295 ISG 87           349 CLX               404 RDN
241 CLX               296 ISG 84           350 STO IND 88        405 RCL IND 87
242 STO 99            297 GTO 02           351 ISG 88            406 RCL 81
243♦LBL "AA"          298 RCL IND 87       352 GTO 03            407 *
244 FIX 0             299 ST- 86           353 RCL 98            408 1
245 "ITER."           300 RCL 87           354 STO 00            409 +
246 "⊢ "              301 RCL 92           355 RCL 97            410 X<>Y
247 ARCL 99           302 +                356 STO 84            411 *
248 FS? 01            303 STO 88           357 .1                412 FS? 06
249 AVIEW             304 RCL 86           358 ST+ 89            413 GTO 05
250 FS? 01            305 E↑X              359 CLX               414 LASTX
251 XEQ "LAM"         306 1 E2             360 STO 82            415 RCL 83
252 XEQ "MFC"         307 X<=Y?            361 STO 86            416 1
253 XEQ "JCM"         308 OFF              362 SF 05             417 +
254 XEQ "FUN"         309 X<>Y             363 SF 06             418 RCL 91
255 BEEP              310 STO IND 88       364 SF 07             419 *
256 FS? 02            311 RCL IND 00       365♦LBL "JR"          420 RCL 89
257 STOP              312 RCL 81           366 CLX               421 +
258♦LBL 00            313 *                367 STO 83            422 STO 88
259 ADV               314 1                368 ISG 89            423 RDN
260 ADV               315 +                369♦LBL "JC"          424 ST- IND 88
261 !                 316 *                370 RCL 91            425 RDN
262 ST+ 99            317 ST+ 85           371 1                 426♦LBL 05
263 CF 04             318 FIX 0            372 +                 427 ST+ IND 89
264 XEQ "LEQ"         319 "X<"              373 RCL 92            428 ISG 00
265 GTO "AA"          320 ARCL 00          374 *                 429 GTO "JC"
```

B.2 Nonstoichiometric Algorithms

```
430 CF 06
431 CF 07
432 RCL 98
433 STO 00
434 ISG 89
435 1
436 ST+ 83
437 RCL 83
438 RCL 91
439 X>Y?
440 GTO "JC"
441 CF 05
442 RCL 92
443 ST+ 86
444 ISG 84
445 GTO "JR"
446 RTN

447♦LBL "FUN"
448 RCL 91
449 2
450 +
451 RCL 92
452 *
453 1
454 STO 00
455 +
456 RCL 91
457 +
458 STO 89
459 1
460 RCL 85
461 -
462 STO IND 89
463 CLX
464 STO 83
465♦LBL 06
466 FIX 0
467 "FX"
468 "⊦<"
469 ARCL 00
470 "⊦>: "
471 SCI 7
472 ARCL IND 89
473 FS? 01
474 AVIEW
475 RCL IND 89
476 ABS
477 ST+ 83
478 1
479 ST+ 00
480 RCL 91
481 1
482 +
483 ST+ 89
484 RCL 91
485 RCL 00
486 X<=Y?

487 GTO 06
488 RCL 83
489 1 E-5
490 X>Y?
491 OFF
492 RTN

493♦LBL "LEQ"
494 CLX
495 STO 85
496♦LBL "LI"
497 1
498 ST+ 85
499 RCL 85
500 RCL 91
501 1
502 +
503 -
504 RCL 85
505 LASTX
506 *
507 LASTX
508 1
509 +
510 RCL 92
511 *
512 +
513 +
514 LASTX
515 1 E3
516 /
517 +
518 STO 87
519♦LBL "ND"
520 RCL 87
521 STO 00
522 INT
523 STO 88
524 RCL 85
525 STO 89
526 RCL IND 00
527 X=0?
528 GTO "ZS"
529 STO 83
530♦LBL 07
531 RCL 83
532 ST/ IND 00
533 ISG 00
534 GTO 07
535 CLX
536 STO 86
537♦LBL "LJ"
538 RCL 87
539 STO 89
540 1
541 ST+ 86
542 RCL 91
543 RCL 86

544 X>Y?
545 GTO "ET"
546 RCL 85
547 X=Y?
548 GTO "LJ"
549 RCL 91
550 1
551 +
552 -
553 RCL 86
554 LASTX
555 *
556 LASTX
557 1
558 +
559 RCL 92
560 *
561 +
562 +
563 LASTX
564 1 E3
565 /
566 +
567 STO 00
568 RCL IND 00
569 STO 03
570♦LBL "RI"
571 RCL 83
572 RCL IND 89
573 *
574 ST- IND 00
575 1
576 ST+ 89
577 ISG 00
578 GTO "RI"
579 RCL 86
580 RCL 91
581 X>Y?
582 GTO "LJ"
583 RCL 85
584 RCL 91
585 X>Y?
586 GTO "LI"
587♦LBL "ZS"
588 RCL 89
589 RCL 91
590 X=Y?
591 OFF
592 RCL 91
593 1
594 +
595 ST+ 88
596 RCL IND 88
597 X=0?
598 GTO "ZS"
599 RCL 87
600 STO 00
601♦LBL "SR"

602 RCL IND 88
603 ST+ IND 00
604 1
605 ST+ 88
606 ISG 00
607 GTO "SR"
608 GTO "ND"
609♦LBL "ET"
610 RCL 97
611 STO 00
612 RCL 91
613 2
614 +
615 RCL 92
616 *
617 STO 89
618♦LBL 08
619 92
620 RCL 00
621 +
622 STO 87
623 RCL 00
624 RCL 91
625 +
626 RCL 00
627 +
628 RCL 89
629 +
630 STO 88
631 RCL IND 87
632 FS? 04
633 ST- IND 87
634 RCL IND 88
635 ST+ IND 87
636 ISG 00
637 GTO 08
638 RTN

639♦LBL "LAM"
640 RCL 97
641 STO 00
642♦LBL 09
643 92
644 RCL 00
645 +
646 STO 89
647 FIX 0
648 "LAM "
649 ARCL 00
650 "⊦: "
651 SCI 7
652 ARCL IND 89
653 AVIEW
654 ISG 00
655 GTO 09
656 RTN
657 .END.
```

B.2.1.2 User's Guide

1. Follow steps 1-7, inclusive, from "STOIEQ" User's Guide of section B.1.1.2 except that in step 5, M and N must satisfy $(N+M)(M+2) \leq 84$ and $M \geq 2$, and step 4 is replaced by:

4. Press GTO Alpha N O N S T O Alpha.

8. To enter element abundances and moles of inert species, complete the following steps:

Keystroke	Display
x ⇄ y	ELEM 1.
(value of b)	(value)
R/S	ELEM 2.
(value of b)	(value)
.	.
.	.
.	.
(value of b)	(value)
R/S	INE. ?
(moles of inert species)	(value)
R/S	** after a short pause

9. To enter initial estimates of Lagrange multipliers, λ, choose a set of M species which have linearly independent formula vectors, and which have the lowest values of μ_i^o :

Keystroke	Display
SIN	COL ?
(column number)	(value)
R/S	COL ?
(column number)	(value)
.	.
.	.
.	.
R/S	END DATA after a long pause

B.2 Nonstoichiometric Algorithms

10. Use step 9 from "STOIEQ" User's Guide of section B.1.1.2.

11. The display shows the following for each iteration:

    ITER.:   (iteration number)
    LAM⟨1.⟩:   (value of λ_1)
    LAM⟨2.⟩:   (value of λ_2)
    .
    .
    .
    X⟨1.⟩:   (value of x_1)
    X⟨2.⟩:   (value of x_2)
    .
    .
    .
    FX⟨1.⟩:   (values of residuals,
                         $f_i(\lambda)$ for
    FX⟨2.⟩:   equations (4.4-33))

    The end of each iteration is signalled by a beep, and the calculation automatically proceeds to the next iteration (but see final remarks in step 9 from "STOIEQ" User's Guide of section B.1.1.2).

12. The calculation terminates when either of the following criteria is satisfied:

    (a) $\Sigma |f_i(\lambda)| < 10$

    In this case, the display indicates this value.

    (b) $x > 100$

    In this case, the display indicates a large number. Repeat step 9 (choose a new set of M species). To restart the calculations, press XEQ Alpha A̲ Alpha.

13. To display the results, first press SF 0̲ 1̲.
    To display the equilibrium mole fractions, press XEQ Alpha M̲ F̲ C̲ Alpha.
    To display λ, press XEQ Alpha L̲ A̲ M̲ Alpha.
    To display the values of $f_i(\lambda)$, press XEQ Alpha F̲ U̲ N̲ Alpha.
    To display the number of iterations, press RCL 9̲ 9̲.

B.2.2 BASIC Program

The listing of the program follows. A set of sample input and output is given in Example 4-7 of the text. The number of elements must be two or greater.

```
100 REM NON-STOICHIOMETRC ALGORITHM FOR CALCULATING THE EQUILIBRIUM
110 REM OF SINGLE PHASE SYSTEMS
120 INIT
130 PAGE
140 PRINT "JENTER DEVICE NUMBER FOR PRINTINGJ"
150 INPUT D
160 D1=31
180 IMAGE 7A,2X,14A,55T,16A
190 IMAGE 8A,2X,46A,7D.3D
200 L1=0
210 I1=0
220 S1=0
230 PRI "ENTER NUMBER OF SPECIES AND ELEMENTS (MUST BE GREATER THAN 1)J"
240 INPUT @D1:N,M
250 PRINT "JTYPE OF CHEM. POT. : KJ, KCAL OR MU/RT J"
260 PRINT "ENTER 1, 2 OR 3 RESPECTIVELYJ"
270 INPUT @D1:I0
280 PRINT "JHOW MANY SIGNIFICANT FIGURESJ"
290 INPUT I2
300 DIM S$(8*N),W(M),A2(M,N),F0(N),B(M),E$(3*M),B2(M-1)
310 S$=""
320 FOR J=1 TO N
330 PRINT "JENTER SPECIES NAME"
340 INPUT @D1:E$
350 PRINT "JENTER FORMULA VECTOR,CHEM POTENTIAL   ";M+1;" NUMBERS"
360 INPUT @D1:W,F0(J)
370 F$="        "
380 F$=REP(E$,1,LEN(E$))
390 S$=S$&"        "
400 S$=REP(F$,(J-1)*8+1,8)
410 FOR I=1 TO M
420 A2(I,J)=W(I)
430 NEXT I
440 NEXT J
450 PRINT "ENTER ELEMENTAL ABUNDANCES:J"
460 INPUT @D1:B
470 PRINT "JARE THERE ANY INERT SPECIES PRESENT   (Y/N)J"
480 INPUT L$
490 IF L$="N" THEN 530
500 PRINT "JENTER NO OF MOLES OF INERT SPECIESJ"
510 INPUT N8
520 GO TO 540
530 N8=0
540 E$=""
550 FOR J=1 TO M
560 PRINT "JENTER ELEMENT ";J;"J"
570 INPUT @D1:G$
```

B.2 Nonstoichiometric Algorithms

```
580 F$="      "
590 F$=REP(G$,4-LEN(G$),LEN(G$))
600 E$=E$&F$
610 NEXT J
620 PRINT "ENTER TEMPERATURE IN K,PRESSURE IN ATM :J"
630 INPUT @D1:T,P
640 PRINT "JENTER TITLEJ"
650 INPUT @D1:T$
660 PRINT "JDO YOU WANT TO MAKE YOUR OWN INITIAL ESTIMATE FOR THE";
670 PRINT " LAGR. MULT. (Y/N)J"
680 INPUT L$
690 IF L$="N" THEN 720
700 PRINT "ENTER INITIAL ESTIMATES ";M;" NUMBERSJ"
710 INPUT W
720 PRINT "JDO YOU WANT TO PRINT INTERMEDIATE RESULTS (Y/N)J"
730 INPUT I$
740 REM       * PRINT DATA    *
750 PRINT @D:T$
760 PRINT @D."J",N," SPECIES",M," ELEMENTS"
770 PRINT @D:"JPRESSURE   ";P;" ATM "
780 PRINT @D:"JTEMPERATURE   ";T;" K"
790 PRINT @D:"JELEMENTAL ABUNDANCES     ";
800 FOR J=1 TO M
810 G$="   "
820 G$=SEG(E$,(J-1)*3+1,3)
830 PRINT @D:G$,B(J)
840 PRINT @D:"                       ";
850 NEXT J
850 PRINT @D:""
870 PRINT @D: USING 180:"SPECIES","FORMULA VECTOR","STAN. CHEM. POT."
880 PRINT @D:"            ";E$
890 FOR J=1 TO N
900 O$=""
910 FOR I=1 TO M
920 E$="    "
930 G$=STR(A2(I,J))
940 E$=REP(G$,4-LEN(G$),LEN(G$))
950 O$=O$&E$
960 NEXT I
970 G$=SEG(S$,(J-1)*8+1,8)
980 PRINT @D: USING 190:G$,O$,F0(J)
990 NEXT J
1000 REM      * ADJUST FOR SPECIFIED TYPE OF STAN. CHEM. POT.*
1010 R0=1.9872*T
1020 GO TO I0 OF 1060,1030,1100
1030 T1=1000/R0
1040 PRINT @D:"JSTAN. CHEM. POT. IS IN KCAL/MOLE"
1050 GO TO 1120
1060 T1=0.0083143*T
1070 T1=1/T1
1080 PRINT @D:"JSTAN. CHEM. POT. IS IN KJ/MOLE"
1090 GO TO 1120
1100 T1=0
1110 PRINT @D:"JSTAN. CHEM. POT. IS MU/RT "
```

```
1120 FOR I=1 TO N
1130 IF T1=0 THEN 1150
1140 F0(I)=F0(I)*T1
1150 F0(I)=F0(I)+LOG(P)
1160 NEXT I
1170 IF L$="N" THEN 1200
1180 PRINT @D:"JUSER ESTIMATE"
1190 GO TO 1220
1200 PRINT @D:"JMACHINE ESTIMATE"
1210 REM          COMPUTE ALPHAS
1220 FOR J=2 TO M
1230 FOR I=1 TO N
1240 A2(J,I)=A2(J,I)-B(J)/B(1)*A2(1,I)
1250 NEXT I
1260 B2(J-1)=B(J)
1270 NEXT J
1280 REM SAVE B(1)
1290 B1=B(1)
1300 R1=N8/B1
1310 DIM W0(N),W1(M,N),C(M,M),T2(N,M),X(N)
1320 IF L$="Y" THEN 1560
1330 REM       * SET UP LINEAR SYSTEM FOR LEAST SQUARES *
1340 REM       * ESTIMATES OF LAGRANGIAN MULTIPLIERS        *
1350 K=M+1
1360 REM CALL WEIGHTS
1370 GOSUB 3350
1380 FOR I=1 TO N
1390 FOR J=1 TO M
1400 W1(J,I)=W0(I)*A2(J,I)
1410 NEXT J
1420 NEXT I
1430 T2=TRN(W1)
1440 C=W1 MPY T2
1450 FOR I=1 TO M
1460 B(I)=0
1470 FOR J=1 TO N
1480 B(I)=B(I)+W1(I,J)*W0(J)*(LOG(1/N)+F0(J))
1490 NEXT J
1500 NEXT I
1510 REM CALL MLEQU(C,15,M,W,1)
1520 GOSUB 3060
1530 W=B
1540 I1=I1+1
1550 REM COMPUTE FIRST LAGR. MULT
1560 A1=0
1570 FOR J=1 TO M-1
1580 A1=A1+B2(J)/B1*W(J+1)
1590 NEXT J
1600 V1=W(1)-A1
1610 REM        * PRINT INITIAL EST. FOR LAGRANGE MULTIPL. *
1620 PRI @D:"JINITIAL ESTIMATES FOR THE LAGRANGE MULTIPLIERS (LAMBDA/RT)"
1630 IMAGE 9E
1640 PRINT @D: USING 1630:V1
1650 FOR I=2 TO M
```

B.2 Nonstoichiometric Algorithms

```
1660 PRINT @D: USING 1630:W(I)
1670 NEXT I
1680 S0=0
1690 ON SIZE THEN 1900
1700 FOR I=1 TO N
1710 X(I)=-F0(I)
1720 FOR J=1 TO M
1730 X(I)=X(I)+A2(J,I)*W(J)
1740 NEXT J
1750 IF X(I)<=170 THEN 1770
1760 S0=1
1770 X(I)=EXP(X(I))
1780 NEXT I
1790 IF I$="N" THEN 1850
1800 PRINT @D:"JITERATION   ",I1
1810 FOR I=1 TO N
1820 G$=SEG(S$,(I-1)*8+1,8)
1830 PRINT @D:G$,X(I)
1840 NEXT I
1850 IF S0=0 THEN 2130
1860 S1=S1+1
1870 REM IF MOLE FRACTION >EXP(170),TRY TO RECOVER
1880 REM IF SAME CONDITION OCCURS AGAIN,STOP EXECUTION.
1890 IF S1=1 THEN 2140
1900 IF L$="N" THEN 2030
1910 PRINT "JYOUR INITIAL ESTIMATE DID NOT CONVERGE,YOU HAVE THE ";
1920 PRINT "FOLLOWING OPTIONS"
1930 PRINT "JMAKE ANOTHER INITIAL ESTIMATE, ENTER 1J"
1940 PRINT "JHAVE THE MACHINE MAKE AN ESTIMATE, ENTER 2J"
1950 PRINT "JPRINT THE RESULTS AND STOP, ENTER 3J"
1960 INPUT I3
1970 S1=0
1980 I1=0
1990 GO TO I3 OF 2000,1350,2700
2000 PRINT "JENTER INITIAL ESTIMATES   ";M;" NUMBERSJ"
2010 INPUT W
2020 GO TO 1620
2030 PRINT "JMACHINE ESTIMATE DID NOT CONVERGE, YOU HAVE THE";
2040 PRINT " FOLLOWING OPTIONS:J"
2050 PRINT "JMAKE YOUR OWN ESTIMATE, ENTER 1J"
2060 PRINT "JPRINT THE RESULTS AND STOP, ENTER 2J"
2070 INPUT I3
2090 S1=0
2100 IF I3=2 THEN 2700
2110 GO TO 2000
2120 REM (*) CONTINUE
2130 IF I1>1 THEN 2310
2140 FOR J=1 TO N
2150 IF X(J)>5 THEN 2180
2160 NEXT J
2170 GO TO 2310
2180 FOR J=1 TO N
2190 IF X(J)<0.01 OR X(J)>10 THEN 2230
2200 W0(J)=X(J)
```

```
2210 NEXT J
2220 GO TO 1380
2230 FOR J=1 TO N
2240 W0(J)=X(J)
2250 IF NOT (X(J)<0.01) THEN 2270
2260 W0(J)=0.01
2270 IF NOT (X(J)>10) THEN 2290
2280 W0(J)=10+LOG(X(J))
2290 NEXT J
2300 GO TO 1380
2310 FOR J=1 TO M
2320 FOR I=1 TO N
2330 W1(J,I)=A2(J,I)*X(I)
2340 NEXT I
2350 NEXT J
2360 FOR J=1 TO M
2370 C(1,J)=0
2380 FOR I=1 TO N
2390 C(1,J)=C(1,J)+W1(J,I)*(1+R1*A2(1,I))
2400 NEXT I
2410 NEXT J
2420 FOR J=2 TO M
2430 FOR K=1 TO M
2440 C(J,K)=0
2450 FOR I=1 TO N
2460 C(J,K)=C(J,K)+A2(J,I)*W1(K,I)
2470 NEXT I
2480 NEXT K
2490 NEXT J
2500 B(1)=1
2510 FOR I=1 TO N
2520 B(1)=B(1)-X(I)*(1+R1*A2(1,I))
2530 NEXT I
2540 FOR J=2 TO M
2550 B(J)=0
2560 FOR I=1 TO N
2570 B(J)=B(J)-W1(J,I)
2580 NEXT I
2590 NEXT J
2600 REM      COUNT NUMBER OF ITRATIONS
2610 I1=I1+1
2612 IF I1<500 THEN 2630
2614 PRINT @D:"NO CONVERGENCE AFTER 500 ITERATIONS."
2616 PRINT @D:"CURRENT RESULTS FOLLOW"
2618 GO TO 2700
2620 REM CALL MLEQU(C,15,M,B,1)
2630 GOSUB 3060
2640 T3=0
2650 W=W+B
2660 FOR J=1 TO M
2670 T3=T3+ABS(B(J))
2680 NEXT J
2690 IF T3>10^-I2 THEN 1700
2700 REM CALCULATE NUMBER OF MOLES
```

B.2 Nonstoichiometric Algorithms

```
2710 N9=0
2720 FOR I=1 TO N
2730 N9=N9+A2(1,I)*X(I)
2740 NEXT I
2750 N9=B1/N9
2760 PRINT @D:"JNUMBER OF ITERATIONS",I1
2770 PRINT @D:"JSPECIES"," MOLE FRACTION"," MOLESJ"
2780 IMAGE 17A,9E,36T,9E
2790 IMAGE "JINERTS",18T,9E,36T,9E
2800 IMAGE 9E
2810 IMAGE "JTOTAL NUMBER OF MOLES:",4X,9E
2820 FOR J=1 TO N
2830 G$=SEG(S$,(J-1)*8+1,8)
2840 PRINT @D: USING 2780:G$,X(J),N9*X(J)
2850 NEXT J
2860 REM COMPUTE FIRST LAGR. MULT.
2870 A1=0
2880 FOR J=1 TO M-1
2890 A1=A1+B2(J)/B1*W(J+1)
2900 NEXT J
2910 V1=W(1)-A1
2915 G1=0
2920 IF N8=0 THEN 2950
2930 G1=N8*LOG(N8/N9)+B1*V1
2940 PRINT @D: USING 2790:N8/N9,N8
2950 PRINT @D: USING 2810:N9
2960 PRINT @D:"JFINAL LAGRANGE MULTIPLIERS (LAMBDA/RT)J"
2970 PRINT @D: USING 2800:V1
2980 FOR I=2 TO M
2990 G1=G1+B2(I-1)*W(I)
3000 PRINT @D: USING 2800:W(I)
3010 NEXT I
3020 IMAGE "JG/RT = ",9E
3030 PRINT @D: USING 3020:G1
3040 GO TO 3700
3050 REM SUBROUTINE MLEQU
3060 OFF SIZE
3070 ON SIZE THEN 3300
3080 DIM D3(M,M+1)
3085 V1=0
3090 FOR I=1 TO M
3100 FOR J=1 TO M
3110 D3(I,J)=C(I,J)
3130 V1=V1 MAX ABS(C(I,J))
3140 NEXT J
3150 NEXT I
3160 FOR I=1 TO M
3170 D3(I,M+1)=B(I)
3180 NEXT I
3190 D3=INV(D3)
3195 V2=0
3200 FOR I=1 TO M
3210 B(I)=D3(I,M+1)
3220 REM CHECK COND. NUMBER
```

```
3230 FOR J=1 TO M
3240 V2=V2 MAX ABS(D3(I,J))
3250 NEXT J
3260 NEXT I
3270 IF V1*V2>1.0E+10 THEN 3300
3280 DELETE D3
3290 RETURN
3300 PRINT "JTHE LINEAR SYSTEM HAS NO SOLUTION PROBABLY RANK(A).NE.MJ"
3310 PRI @D:"JTHE LINEAR SYSTEM HAS NO SOLUTION PROBABLY RANK(A).NE.MJ"
3330 GO TO 3730
3340 REM SUBROUTINE WEIGHTS
3350 W0=1
3360 DIM I5(N)
3370 FOR J=1 TO N
3380 I5(J)=J
3390 NEXT J
3400 FOR I=1 TO M
3410 M0=I5(I)
3420 I7=I
3430 J=I+1
3440 FOR L=J TO N
3450 IF NOT(F0(I5(L))<F0(I5(M0))) THEN 3480
3460 M0=I5(L)
3470 I7=L
3480 NEXT L
3490 I6=I5(I)
3500 I5(I)=M0
3510 I5(I7)=I6
3520 NEXT I
3530 B0=ABS(F0(I5(1)))
3540 Z=F0(I5(M+1))+B0
3550 K4=1
3560 IF Z=>1 THEN 3590
3570 K4=10
3580 B0=B0*K4
3590 W0(I5(1))=Z*K4
3600 I=M+1
3610 FOR J=2 TO M
3620 I=I-1
3630 Z=F0(I5(I))+B0
3640 IF Z=>1 THEN 3660
3650 Z=1
3660 W0(I5(J))=Z
3670 NEXT J
3680 DELETE I5,Z,B0
3690 RETURN
3700 END
```

APPENDIX C

FORTRAN Computer Program for BNR Algorithm

C.1 PROGRAM LISTING

```
      MAXIT=100                                                        RAN00010
      CALL RANDEQ(1,1,0,MAXIT)                                         RAN00020
      STOP                                                             RAN00030
      END                                                              RAN00040
      SUBROUTINE RANDEQ(IRP,IPR,IP1,MAXIT)                             RAN00050
C*******************************************************************  RAN00060
C              IDEAL SYSTEM EQUILIBRIUM CALCULATION                    RAN00070
C              USING RAND VARIATION OF BNR ALGORITHM                   RAN00080
C******* VERSION DATED DECEMBER 28, 1981, UNIVERSITY OF GUELPH ******* RAN00090
C                                                                      RAN00100
C     PROGRAM CAN HANDLE UP TO 150 SPECIES AND 15 ELEMENTS, BUT        RAN00110
C     DIMENSION STATEMENTS MAY BE MODIFIED TO ALTER THIS.              RAN00120
C     THIS PROGRAM IS DOUBLE-PRECISION VERSION FOR 32-BIT MACHINES.    RAN00130
C     MAXIT IS MAXIMUM ALLOWABLE NUMBER OF ITERATIONS IF CONVERGENCE NOT RAN00140
C     ACHIEVED.  USUALLY THE NUMBER OF ITERATIONS REQUIRED IS LESS THAN 20. RAN00150
C     UP TO TWO MULTI-SPECIES PHASES CAN BE HANDLED BY THE PRESENT VERSION RAN00160
C     (BUT THIS IS READILY MODIFIED).  FOR TREATMENT OF SINGLE-SPECIES RAN00170
C     PHASES, SEE USER'S GUIDE.                                        RAN00180
C     PHASE1 IS NOMINALLY A GAS, SINCE ALOG(P) IS ADDED TO THE STANDARD RAN00190
C     CHEMICAL POTENTIAL DATA.  THIS CAN BE OVER-RIDDEN BY SETTING P=1. RAN00200
C     PHASE2 IS NOMINALLY A LIQUID, OR ANY PHASE FOR WHICH THE STANDARD RAN00210
C     CHEMICAL POTENTIAL DATA ARE INDEPENDENT OF P.                    RAN00220
C     MULTI-SPECIES PHASE IS DEEMED TO BE ABSENT IF NT.LT.1.E-10.      RAN00230
C     LINEAR PROGRAMMING ROUTINE MUST BE PROVIDED FOR INITIAL ESTIMATE RAN00240
C     OF EQUILIBRIUM COMPOSITION FOLLOWING STATEMENT LABELLED 11705.   RAN00250
C*******************************************************************  RAN00260
      COMMON /RAND/BM(150,15),CT(150),YA(150),EA(15),E(15),TT,PT,YINT, RAN00270
     1SI(150),SP(3,150),NNCP,M,NCP,NL,IF,NE,ICE,IEST,NCP1,NNCP1,NP1,N1 RAN00280
      COMMON /FORCER/LIQ                                               RAN00290
      INTEGER SP,SP1,SP2,SP3                                           RAN00300
      INTEGER IW(303)                                                  RAN00310
      INTEGER OUT(2),VF(13),TEMP(3,150)                                RAN00320
      LOGICAL LIQ,ICONV                                                RAN00330
      REAL*8 S                                                         RAN00340
      REAL*8 C(150),X(150),Y(150),A(150,15),B(15),PI(40),CP(150),      RAN00350
     1R(20,20),DEL(150),T,P,YIN,RT,YL,YT,YLX,YTX,PAR,GF,DF,CX(150),    RAN00360
     2TMAX,TM,TEST,DA(150),B1(15),YLIN,Z1(15),Z2(15),S1,S2,T1,T2,SPAR, RAN00370
     3DSQRT,XN(15),ZZ,BB(15),DELB(15),PAR1,DFB,CL,OUT2,LIST,LOUT       RAN00380
      REAL*8 BBB(15),PSOL(150),DSOL(15),CCC(150),RW(623),AX(17,152)    RAN00390
      DIMENSION MM(15),CZ(150),AA(40),IR(150),INX(150),LIST(30),LIST1(14 RAN00400
     1),LOUT(10),TI(80)                                                RAN00410
      DATA OUT/'SI( ','I)  '/                                          RAN00420
      DATA OUT2/'SI(I)   '/                                            RAN00430
      DATA LIST1/'2','3','4','5','6','7','8','9','10','11','12','13','14 RAN00440
     *','15'/                                                          RAN00450
      DATA VF/'(1X,','3A4,',' 1X,',' ','F6.3',', I5',',T96',',D12',    RAN00460
     *'.5,6','X,D1','2.5,','D12.','5)  '/                              RAN00470
      DATA LIST/5H2I3 ,,5H42X ,,5H3I3 ,,5H39X ,,5H4I3 ,,5H36X ,,5H5I3 ,, RAN00480
     15H33X ,,5H6I3 ,,5H30X ,,5H7I3 ,,5H27X ,,5H8I3 ,,5H24X ,,5H9I3 ,, RAN00490
     25H21X ,,5H10I3,,5H18X ,,5H11I3,,5H15X ,,5H12I3,,5H12X ,,5H13I3,, RAN00500
     35H9X  ,,5H14I3,,5H6X  ,,5H15I3,,5H3X  ,,5H16I3,,5H1X  /          RAN00510
```

C.1 Program Listing

```
      DATA LOUT/5H(    ,5H    1X,5H,3A4,,5H       ,5H    ,5H 1PD1,5H 2.4,, RAN00520
     15H   2X,,5H D12.,5H4 )  /                                            RAN00530
C********************************                                          RAN00540
C**** INPUT FORMAT STATEMENTS ****                                         RAN00550
C********************************                                          RAN00560
  101 FORMAT(9I3)                                                          RAN00570
  102 FORMAT(3A4,1X,15F2.0,F1.0,F9.3)                                      RAN00580
 1022 FORMAT(3A4,1X,F1.0,F9.3)                                             RAN00590
 2022 FORMAT(15F5.3)                                                       RAN00600
  103 FORMAT(8E10.4)                                                       RAN00610
  901 FORMAT(80A1)                                                         RAN00620
C**************************                                                RAN00630
C**** OUTPUT FORMATS ******                                                RAN00640
C**************************                                                RAN00650
  902 FORMAT(//I5,8H SPECIES,I8,9H ELEMENTS,I16,11H COMPONENTS/            RAN00660
     1I5,15H PHASE1 SPECIES,I10,15H PHASE2 SPECIES,I8,22H SINGLE SPECIES   RAN00670
     2 PHASES//,9H PRESSURE,F22.3,4H ATM/12H TEMPERATURE,F19.3,2H K)       RAN00680
  903 FORMAT(14H PHASE2 INERTS,F17.3)                                      RAN00690
  904 FORMAT(14H PHASE1 INERTS,F17.3)                                      RAN00700
  905 FORMAT(/1X,51HMODIFIED LINEAR PROGRAMMING ESTIMATE OF EQUILIBRIUM)   RAN00710
  906 FORMAT(/1X,28HUSER ESTIMATE OF EQUILIBRIUM)                          RAN00720
  907 FORMAT( /8H SPECIES,5X,16H  FORMULA VECTOR,29X,16HSTAN. CHEM. POT.   RAN00730
     1,3X,16HEQUILIBRIUM EST./)                                            RAN00740
 9070 FORMAT( /8H SPECIES,5X,16H  FORMULA VECTOR,T96,16HSTAN. CHEM. POT.   RAN00750
     1,3X,16HEQUILIBRIUM EST./)                                            RAN00760
  911 FORMAT(/1X,20HELEMENTAL ABUNDANCES,20X,7HCORRECT,7X,13HFROM ESTIMA   RAN00770
     1TE//(26X,A2,1PD20.12,D20.12))                                        RAN00780
  912 FORMAT(15A2)                                                         RAN00790
  913 FORMAT(24H1RAND CALCULATION METHOD/)                                 RAN00800
  899 FORMAT(/1X,80A1)                                                     RAN00810
C********************************************                              RAN00820
C**** SET INITIAL VALUES, READ IN DATA ****                                RAN00830
C********************************************                              RAN00840
      IF(IRP.EQ.0)GO TO 671                                                RAN00850
      READ(5,101)NRUNS                                                     RAN00860
      NRU=0                                                                RAN00870
 9999 NRU=NRU+1                                                            RAN00880
      ICONV=.FALSE.                                                        RAN00890
      IF(NRU.GT.NRUNS)RETURN                                               RAN00900
      READ(5,101)NNCP,M,NCP,NL,IF,IEST,ICE,NE,NCP1                         RAN00910
      IF(NE.EQ.0)NE=M                                                      RAN00920
  671 CONTINUE                                                             RAN00930
      IF(IRP.EQ.0)GO TO 672                                                RAN00940
      IF(ICE.LT.0)GO TO 621                                                RAN00950
      READ(5,102)((SP(J,I),J=1,3),(BM(I,J),J=1,15),SI(I),CT(I),I=1,NNCP)   RAN00960
      GO TO 622                                                            RAN00970
  621 READ(5,1022)((SP(J,I),J=1,3),SI(I),CT(I),I=1,NNCP)                   RAN00980
      READ(5,2022)((BM(I,J),J=1,15),I=1,M)                                 RAN00990
  622 IF(IEST.LT.0)GO TO 293                                               RAN01000
      READ(5,103)(YA(I),I=1,NNCP)                                          RAN01010
  293 READ(5,103)(EA(I),I=1,NE)                                            RAN01020
  672 CONTINUE                                                             RAN01030
      DO 2931 I=1,NE                                                       RAN01040
 2931 B1(I)=EA(I)+0.D0                                                     RAN01050
```

```
      IF (IRP.EQ.0)GO TO 673                                   RAN01060
  295 READ (5,103)TT,PT,YINT,YLINT                              RAN01070
      READ (5,912) (E(I),I=1,NE)                                RAN01080
      READ (5,901)TI                                            RAN01090
      LOUT (4)=LIST (2*M-1)                                     RAN01100
      LOUT (5)=LIST (2*NE)                                      RAN01110
      VF (4)=LIST1 (M-1)                                        RAN01120
  673 CONTINUE                                                  RAN01130
C**************************************************************RAN01140
C**** CALCULATE INITIAL ELEMENTAL ABUNDANCES, IF NECESSARY **** RAN01150
C**************************************************************RAN01160
      IT=0                                                      RAN01170
      ITDG=0                                                    RAN01180
      NG=NNCP-NL-NCP                                            RAN01190
      N=NG+NL                                                   RAN01200
      NP1=N+1                                                   RAN01210
      M1=M+1                                                    RAN01220
      M2=M+2                                                    RAN01230
      NNCP1=N+NCP1                                              RAN01240
      JL=NNCP1                                                  RAN01250
      NT=NNCP                                                   RAN01260
      DO 2674 I=1,NT                                            RAN01270
 2674 INX (I)=I                                                 RAN01280
      N1=N                                                      RAN01290
      N7=NNCP-M                                                 RAN01300
      NNNCP=NNCP1                                               RAN01310
      NNE=NE                                                    RAN01320
      NR=NNNCP-M                                                RAN01330
      MR=NNNCP                                                  RAN01340
      LIQ=NL.GT.0                                               RAN01350
      IF (NL.GT.0)GO TO 62                                      RAN01360
      LX=M1+NCP1                                                RAN01370
      KP=M1+1                                                   RAN01380
      GO TO 63                                                  RAN01390
   62 LX=M2+NCP1                                                RAN01400
      KP=M2+1                                                   RAN01410
   63 CONTINUE                                                  RAN01420
      IF (IPR.EQ.0)GO TO 674                                    RAN01430
      WRITE (6,913)                                             RAN01440
      WRITE (6,899)TI                                           RAN01450
      WRITE (6,902)NNCP,NE,M,NG,NL,NCP,PT,TT                    RAN01460
      IF (NG.GT.0)WRITE (6,904)YINT                             RAN01470
      IF (LIQ)WRITE (6,903)YLINT                                RAN01480
  674 CONTINUE                                                  RAN01490
      DO 3358 J=1,NNCP                                          RAN01500
 3358 IR (J)=M+J                                                RAN01510
C**************************************************************RAN01520
C**** ADJUST FOR SPECIFIED TYPE OF STAN. CHEM. POT. DATA **** RAN01530
C**************************************************************RAN01540
      IF (IF)165,163,164                                        RAN01550
  165 TF=1000./1.9872/TT                                        RAN01560
  169 FORMAT (31H STAN. CHEM. POT. IN KCAL./MOLE/)               RAN01570
      GO TO 166                                                 RAN01580
  164 TF=1./(0.0083143*TT)                                      RAN01590
```

C.1 Program Listing

```
          GO TO 166
      168 FORMAT(29H STAN. CHEM. POT. IN KJ./MOLE/)
      163 TF=0.
      167 FORMAT(26H STAN. CHEM. POT. IS MU/RT/)
      166 DO 888 I=1,NNCP
          C(I)=CT(I)
          IF(TF.NE.0.)C(I)=CT(I)*TF
          IF(SI(I).EQ.1.)C(I)=C(I)+ALOG(PT)
      888 CZ(I)=C(I)
          IF(IEST.GE.0)GO TO 159
          M10=0
          M20=M
          IA=17
          DO 11703 I=1,M
    11703 BBB(I)=EA(I)
          DO 11704 J=1,NNCP
    11704 CCC(J)=-CT(J)
          DO 11705 I=1,NNCP
          DO 11705 J=1,M
    11705 AX(J,I)=BM(I,J)
C*********************************
C**** USE LIN. PROG. ROUTINE ****
C*********************************
C
C
C     CALL ROUTINE TO SOLVE MAX(CCC*PSOL) SUCH THAT AX*PSOL=BBB
C
C
          DO 11706 J=1,NNCP
          YA(J)=PSOL(J)
          IF(PSOL(J).EQ.0.AND.SI(J).NE.0.)YA(J)=1.0D-10
    11706 CONTINUE
      159 CONTINUE
C*********************************
C**** PUT SOLIDS AT END OF LIST ****
C*********************************
          KK=1
      666 DO 612 I=KK,JL
          IF(SI(I).EQ.0.)GO TO 613
      612 CONTINUE
          GO TO 459
      613 CALL SWTCH2(SP,I,JL)
          CALL SWITCH(SI,I,JL)
          CALL SWITCH(CT,I,JL)
          CALL SWDP(C,I,JL)
          CALL SWITCH(YA,I,JL)
          DO 614 J=1,NE
          TT1=BM(I,J)
          BM(I,J)=BM(JL,J)
      614 BM(JL,J)=TT1
          JL=JL-1
          IF(I.GE.JL)GO TO 459
          KK=I
          GO TO 666
```

```
C****************************************            RAN02140
C****  CONVERT DATA TO DOUBLE PRECISION FORM ****      RAN02150
C****************************************            RAN02160
      459 DO 113 I=1,NNCP1                             RAN02170
          Y(I)=YA(I)+0.D0                              RAN02180
          X(I)=Y(I)                                    RAN02190
          DO 113 J=1,NE                                RAN02200
      113 A(I,J)=DBLE(BM(I,J))                         RAN02210
          IF (NNCP1.EQ.NNCP)GO TO 9907                 RAN02220
          MXY=NNCP1+1                                  RAN02230
          DO 1131  I=MXY,NNCP                          RAN02240
          Y(I)=0.D0                                    RAN02250
          X(I)=Y(I)                                    RAN02260
          DO 1131 J=1,NE                               RAN02270
     1131 A(I,J)=DBLE(BM(I,J))                         RAN02280
     9907 CONTINUE                                     RAN02290
          T=TT+0.D0                                    RAN02300
          P=PT+0.D0                                    RAN02310
          YIN=YINT+0.D0                                RAN02320
          YLIN=YLINT+0.D0                              RAN02330
          DBMAX=0.D0                                   RAN02340
          DO 80 J=1,NE                                 RAN02350
          BB(J)=0.D0                                   RAN02360
          DO 82 I=1,NNCP                               RAN02370
       82 BB(J)=BB(J)+A(I,J)*Y(I)                      RAN02380
          DELB(J)=B1(J)-BB(J)                          RAN02390
       80 IF (DABS(DELB(J)).GT.DBMAX)DBMAX=DABS(DELB(J)) RAN02400
          IF (DBMAX.GT.1.D-12)GO TO 6176               RAN02410
          DBMAX=0.D0                                   RAN02420
          DO 6177 J=1,M                                RAN02430
     6177 DELB(J)=0.D0                                 RAN02440
     6176 CONTINUE                                     RAN02450
      908 WRITE(6,911) (E(I),B1(I),BB(I),I=1,NE)       RAN02460
          IF (IPR.EQ.0)GO TO 675                       RAN02470
C***********************                              RAN02480
C****  PRINT OUT DATA  ****                           RAN02490
C***********************                              RAN02500
          IF (IEST.LT.0)WRITE(6,905)                   RAN02510
          IF (IEST.GE.0)WRITE(6,906)                   RAN02520
          IF (IF)765,763,764                           RAN02530
      765 WRITE(6,169)                                 RAN02540
          GO TO 766                                    RAN02550
      763 WRITE(6,167)                                 RAN02560
          GO TO 766                                    RAN02570
      764 WRITE(6,168)                                 RAN02580
      766 CONTINUE                                     RAN02590
          IF (ICE.LT.0)GO TO 78                        RAN02600
          WRITE(6,907)                                 RAN02610
          WRITE(6,890) (E(I),I=1,NE),OUT               RAN02620
      890 FORMAT(' ',13X,17A3)                         RAN02630
          DO 77 I=1,NNCP                               RAN02640
          MN=SI(I)                                     RAN02650
          DO 859 J=1,NE                                RAN02660
      859 MM(J)=BM(I,J)                                RAN02670
```

C.1 Program Listing

```
        ZZ=CT(I)                                                    RAN02680
     77 WRITE (6,LOUT) (SP(J,I),J=1,3),(MM(J),J=1,NE),MN,ZZ,Y(I)     RAN02690
        GO TO 675                                                   RAN02700
     78 WRITE (6,9070)                                               RAN02710
        WRITE (6,891) (E(I),I=1,NE),OUT2                             RAN02720
    891 FORMAT(' ',13X,17(1X,A5))                                    RAN02730
        DO 7152 I=1,NNCP                                             RAN02740
        ZZ=CT(I)                                                     RAN02750
        MN=SI(I)                                                     RAN02760
   7152 WRITE (6,VF) (SP(J,I),J=1,3),(BM(I,J),J=1,NE),MN,ZZ,Y(I)     RAN02770
    675 CONTINUE                                                     RAN02780
C****************************************                           RAN02790
C**** EVALUATE TOTAL MOLES GAS, LIQUID ****                          RAN02800
C****************************************                           RAN02810
        YT=YIN                                                       RAN02820
        YL=YLIN                                                      RAN02830
        DO 1 I=1,N                                                   RAN02840
        IF(SI(I).EQ.1.)YT=YT+Y(I)                                    RAN02850
     01 IF(SI(I).EQ.2.)YL=YL+Y(I)                                    RAN02860
        CALL CHEMP(Y,C,SI,CP,YL,YT,NNCP)                             RAN02870
   1000 DO 6 I=1,N1                                                  RAN02880
     06 CP(I)=CP(I)*Y(I)                                             RAN02890
C****************************************                           RAN02900
C**** Q X Q SUB-MATRIX OF COEFFICIENTS ****                          RAN02910
C****************************************                           RAN02920
        DO 543 I=1,LX                                                RAN02930
        PI(I)=0.D0                                                   RAN02940
        DO 543 J=1,LX                                                RAN02950
    543 R(I,J)=0.D0                                                  RAN02960
        DO 11 J=1,M                                                  RAN02970
        DO 11 K=J,M                                                  RAN02980
        DO 13 I=1,N1                                                 RAN02990
     13 R(J,K)=R(J,K)+A(I,J)*A(I,K)*Y(I)                             RAN03000
     11 R(K,J)=R(J,K)                                                RAN03010
C*************************************************************RAN03020
C**** REMAINING PORTIONS OF COEFF. MATRIX AND ELEMENTS OF RHS VECTOR ***RAN03030
C*************************************************************RAN03040
        R(M1,M1)=-YIN                                                RAN03050
        DO 19 J=1,M                                                  RAN03060
        PI(J)=DELB(J)                                                RAN03070
        DO 16 I=1,NNCP1                                              RAN03080
        IF(SI(I)-1.)15,14,309                                        RAN03090
     15 PI(J)=PI(J)+A(I,J)*Y(I)                                      RAN03100
        GO TO 16                                                     RAN03110
    309 R(J,M2)=R(J,M2)+A(I,J)*Y(I)                                  RAN03120
        GO TO 882                                                    RAN03130
     14 R(J,M1)=R(J,M1)+A(I,J)*Y(I)                                  RAN03140
    882 PI(J)=PI(J)+A(I,J)*CP(I)                                     RAN03150
     16 CONTINUE                                                     RAN03160
        IF(LIQ)R(M2,J)=R(J,M2)                                       RAN03170
     19 R(M1,J)=R(J,M1)                                              RAN03180
        DO 20 I=1,N1                                                 RAN03190
        IF(SI(I).EQ.1.)PI(M1)=PI(M1)+CP(I)                           RAN03200
     20 IF(SI(I).EQ.2)PI(M2)=PI(M2)+CP(I)                            RAN03210
```

```
      IF (NCP1.EQ.0) GO TO 542                                        RAN03220
C*************************************************************RAN03230
C**** ELEMENTS OF COEFF. MATRIX CORRESPONDING TO PURE CONDENSED SPECIES RAN03240
C*************************************************************RAN03250
      I=N1                                                            RAN03260
      DO 540 K=KP,LX                                                  RAN03270
      I=I+1                                                           RAN03280
      PI(K)=C(I)                                                      RAN03290
      DO 540 J=1,M                                                    RAN03300
      R(J,K)=A(I,J)                                                   RAN03310
  540 R(K,J)=R(J,K)                                                   RAN03320
  542 CONTINUE                                                        RAN03330
      CALL DLEQU(R,20,LX,PI,1,IS)                                     RAN03340
      IF (IS.EQ.1) GO TO 452                                          RAN03350
      WRITE (6,454)                                                   RAN03360
  454 FORMAT (//)                                                     RAN03370
      WRITE (6,453)                                                   RAN03380
  453 FORMAT (33H **********METHOD FAILS**********/75H LINEAR DEPENDENCE,RAN03390
     1 PROBABLY FROM FORMULA VECTORS OF PURE CONDENSED SPECIES/27H CURRERAN03400
     2NT RESULTS AS FOLLOWS//)                                        RAN03410
      GO TO 4445                                                      RAN03420
  452 CONTINUE                                                        RAN03430
C*************************************************************       RAN03440
C**** COMPUTE MOLE NUMBER CHANGES FOR GASES AND LIQUIDS ****          RAN03450
C*************************************************************       RAN03460
      DO 24 I=1,N1                                                    RAN03470
      DEL(I)=0.D0                                                     RAN03480
      DO 25 J=1,M                                                     RAN03490
   25 DEL(I)=DEL(I)+PI(J)*A(I,J)                                      RAN03500
      IF (SI(I).EQ.1.) DEL(I)=(DEL(I)+PI(M1))*Y(I)-CP(I)              RAN03510
   24 IF (SI(I).EQ.2.) DEL(I)=(DEL(I)+PI(M2))*Y(I)-CP(I)              RAN03520
      IF (NCP1.EQ.0) GO TO 31                                         RAN03530
C*************************************************************       RAN03540
C****   COMPUTE MOLE NUMBER CHANGES FOR PURE CONDENSED SPECIES ****   RAN03550
C*************************************************************       RAN03560
      DO 33 I=NP1,NNCP1                                               RAN03570
      INN=LX-NNCP1+I                                                  RAN03580
   33 DEL(I)=PI(INN)-Y(I)                                             RAN03590
C*************************************************************RAN03600
C**** STEP-SIZE ADJUSTMENT FOR +VE MOLE NUMBERS FOR MULTI-SPECIE PHASES RAN03610
C*************************************************************RAN03620
   31 CONTINUE                                                        RAN03630
      PAR=.5D0                                                        RAN03640
      DO 318 I=1,N1                                                   RAN03650
  318 IF (PAR.LT.(-DEL(I)/Y(I))) PAR=-DEL(I)/Y(I)                     RAN03660
      PAR=1.D0/PAR                                                    RAN03670
      IF (PAR.LE.1.D0.AND.PAR.GT.0.D0) GO TO 29                       RAN03680
      PAR=1.D0                                                        RAN03690
      GO TO 30                                                        RAN03700
   29 IF (PAR.LT.0.1D0) GO TO 2919                                    RAN03710
      PAR=PAR*.99D0                                                   RAN03720
      GO TO 30                                                        RAN03730
 2919 PAR=PAR*.999D0                                                  RAN03740
C*************************************************************RAN03750
```

C.1 Program Listing

```
C****  COMPUTE TENTATIVE NEW MOLE NUMBERS, TOTAL GAS AND LIQUID MOLES ***RAN03760
C***********************************************************************RAN03770
   30 YTX=YIN                                                            RAN03780
      YLX=YLIN                                                           RAN03790
      DO 1130 I=1,NNCP1                                                  RAN03800
      X(I)=Y(I)+DEL(I)*PAR                                               RAN03810
      IF(SI(I).EQ.1.)YTX=YTX+X(I)                                        RAN03820
 1130 IF(SI(I).EQ.2.)YLX=YLX+X(I)                                        RAN03830
      CALL CHEMP(X,C,SI,CX,YLX,YTX,NNCP)                                 RAN03840
C*****************************                                           RAN03850
C****  CONVERGENCE FORCER ****                                           RAN03860
C*****************************                                           RAN03870
      IF(DBMAX.GT.0.D0)GO TO 4450                                        RAN03880
      DF=0.D0                                                            RAN03890
      DO 36 I=1,NNCP1                                                    RAN03900
   36 DF=DF+DEL(I)*CX(I)                                                 RAN03910
      IF(DF.LE.0.D0)GO TO 38                                             RAN03920
      GF=0.D0                                                            RAN03930
      DO 131 I=1,NNCP1                                                   RAN03940
      IF(SI(I).GT.0.)GF=GF+DEL(I)/Y(I)*CP(I)                             RAN03950
  131 IF(SI(I).EQ.0.)GF=GF+DEL(I)*C(I)                                   RAN03960
      IF(GF.LE.0.D0)GO TO 1132                                           RAN03970
      ITDG=ITDG+1                                                        RAN03980
      IF(ITDG.GT.3)GO TO 4449                                            RAN03990
      PAR=PAR*.1                                                         RAN04000
      GO TO 4450                                                         RAN04010
 1132 PAR=PAR*GF/(GF-DF)                                                 RAN04020
      IF(PAR.GT..95D0)GO TO 38                                           RAN04030
 4450 YTX=YT+PAR*(YTX-YT)                                                RAN04040
      IF(NL.GT.0)YLX=YL+PAR*(YLX-YL)                                     RAN04050
      DO 37 I=1,NNCP1                                                    RAN04060
   37 X(I)=Y(I)+PAR*DEL(I)                                               RAN04070
      CALL CHEMP(X,C,SI,CX,YLX,YTX,NNCP1)                                RAN04080
   38 YT=YTX                                                             RAN04090
      YL=YLX                                                             RAN04100
C***********************************************************************RAN04110
C****  RESET CHEMICAL POTENTIALS, CHECK FOR MICROSCOPIC SPECIES ****     RAN04120
C***********************************************************************RAN04130
      IF(YT.GT.1.D-10)GO TO 6671                                         RAN04140
      WRITE(6,397)                                                       RAN04150
      GO TO 4445                                                         RAN04160
  397 FORMAT(//59H***********PHASE1 ABSENT - LESS THAN 1.E-10 MOLES*****RAN04170
     1****/27H CURRENT RESULTS AS FOLLOWS//)                             RAN04180
 6671 IF(YL.GT.1.D-10.OR.NL.EQ.0)GO TO 6672                              RAN04190
      WRITE(6,398)                                                       RAN04200
      GO TO 4445                                                         RAN04210
  398 FORMAT(//59H***********PHASE2 ABSENT - LESS THAN 1.E-10 MOLES*****RAN04220
     1****/27H CURRENT RESULTS AS FOLLOWS//)                             RAN04230
 6672 CONTINUE                                                           RAN04240
      KK=1                                                               RAN04250
  145 DO 142 I=KK,NT                                                     RAN04260
      CP(I)=CX(I)                                                        RAN04270
      IF(X(I).LT.1.D-32)IF(I-N1)138,138,142                              RAN04280
  142 Y(I)=X(I)                                                          RAN04290
```

```
          GO TO 333                                            RAN04300
    138 IF (N1.GT.M)CALL DELETE (I,NT,C,CX,Y,A,X,INX)           RAN04310
          IF (I.GE.NT)GO TO 333                                 RAN04320
          KK=I                                                  RAN04330
          GO TO 145                                             RAN04340
    333 IT=IT+1                                                 RAN04350
          IF (IP1.EQ.0)GO TO 6730                               RAN04360
          GF=0.                                                 RAN04370
          IF (YT.GT.0.0.AND.YIN.GT.0.0)GF=GF+YIN*DLOG (YIN/YT)  RAN04380
          IF (LIQ.AND.YLIN.GT.0.0)GF=GF+YLIN*DLOG (YLIN/YL)     RAN04390
          DO 7166 I=1,NNCP1                                     RAN04400
   7166 IF (X(I).GT.1.E-30)GF=GF+CX (I)*X (I)                   RAN04410
          WRITE (6,7167)IT,GF                                   RAN04420
   7167 FORMAT (/10H ITERATION,I3,3X,8H G/RT = ,D14.7)          RAN04430
          WRITE (6,4729)LX,LX                                   RAN04440
   4729 FORMAT (I3,4H BY ,I3,19H COEFFICIENT MATRIX)            RAN04450
          DO 3829 I=1,LX                                        RAN04460
   3829 WRITE (6,3729) (R(I,J),J=1,LX)                          RAN04470
   3729 FORMAT (6D14.6)                                         RAN04480
          WRITE (6,4731)                                        RAN04490
   4721 FORMAT (29H SPECIES AND COMPUTED CHANGES)               RAN04500
          WRITE (6,4730) ((SP(J,I),J=1,3),X(I),DEL(I),I=1,NNCP1) RAN04510
   4730 FORMAT (' ',3A4,D14.6,2X,D14.6)                         RAN04520
   6730 CONTINUE                                                RAN04530
          IF (IT.LT.MAXIT) GO TO 4545                           RAN04540
          ICONV=.TRUE.                                          RAN04550
          GO TO 4445                                            RAN04560
   4545 DBMAX=0.D0                                              RAN04570
          DO 2381 J=1,M                                         RAN04580
          B (J)=0.D0                                            RAN04590
          DO 2382 K=1,NNCP                                      RAN04600
   2382 B (J)=B (J)+A (K,J)*Y (K)                               RAN04610
          DELB (J)=B1 (J)-B (J)                                 RAN04620
   2381 IF (DABS (DELB (J)).GT.DBMAX)DBMAX=DABS (DELB (J))      RAN04630
          IF (DBMAX.GT.1.D-12)GO TO 6178                        RAN04640
          DBMAX=0.D0                                            RAN04650
          DO 6179 J=1,M                                         RAN04660
   6179 DELB (J)=0.D0                                           RAN04670
   6178 CONTINUE                                                RAN04680
C***********************************                            RAN04690
C**** CHECK FOR 5-FIGURE ACCURACY ****                          RAN04700
C***********************************                            RAN04710
          DO 43 I=1,NNCP1                                       RAN04720
          IF (DABS (DEL(I)/Y(I)).GE..5D-05)GO TO 1000           RAN04730
      43 CONTINUE                                               RAN04740
   4449 CONTINUE                                                RAN04750
C*****************************************************          RAN04760
C**** CHECK FOR DELETED SPECIES .GT. 1.D-32 MOLES ****           RAN04770
C*****************************************************          RAN04780
      445 IF (NNCP.EQ.NT)GO TO 4446                             RAN04790
          MR1=NNCP+1                                            RAN04800
          IKL=MR1                                               RAN04810
   1968 DO 1713 I=IKL,NT                                        RAN04820
          IF (SI(I).EQ.1.)T=-C(I)+DLOG (YT)                     RAN04830
```

C.1 Program Listing

```
         IF (SI (I) .EQ.2.)T=-C (I)+DLOG (YL)              RAN04840
         DO 1717 J=1,M                                     RAN04850
1717     T=T+A (I,J)*PI (J)                               RAN04860
         IF (32.D0*DLOG (10.D0) .LT. (-T))GO TO 1713       RAN04870
         IF (T.GT. (-24.D0))GO TO 3728                    RAN04880
         CALL SWTCH2 (SP,MR1,I)                            RAN04890
         Y (MR1)=DEXP (T)                                  RAN04900
         DEL (MR1)=0.D0                                    RAN04910
         INXL=INX (I)                                      RAN04920
         INX (I)=INX (MR1)                                 RAN04930
         INX (MR1)=INXL                                    RAN04940
         NNCP=NNCP+1                                       RAN04950
         MR1=MR1+1                                         RAN04960
         GO TO 1713                                        RAN04970
3728     CONTINUE                                          RAN04980
         INXL=INX (MR1)                                    RAN04990
         SP1=SP (1,MR1)                                    RAN05000
         SP2=SP (2,MR1)                                    RAN05010
         SP3=SP (3,MR1)                                    RAN05020
         CL=C (MR1)                                        RAN05030
         SIL=SI (MR1)                                      RAN05040
         DO 3732 J=1,M                                     RAN05050
3732     AA (J)=A (MR1,J)                                  RAN05060
         NJ=MR1-N1                                         RAN05070
         DO 3730 JL=1,NJ                                   RAN05080
         J=MR1+1-JL                                        RAN05090
         INX (J)=INX (J-1)                                 RAN05100
         SP (1,J)=SP (1,J-1)                               RAN05110
         SP (2,J)=SP (2,J-1)                               RAN05120
         SP (3,J)=SP (3,J-1)                               RAN05130
         C (J)=C (J-1)                                     RAN05140
         CP (J)=CP (J-1)                                   RAN05150
         SI (J)=SI (J-1)                                   RAN05160
         Y (J)=Y (J-1)                                     RAN05170
         DO 3730 K=1,M                                     RAN05180
3730     A (J,K)=A (J-1,K)                                 RAN05190
         N1=N1+1                                           RAN05200
         NNCP1=NNCP1+1                                     RAN05210
         NNCP=NNCP+1                                       RAN05220
         NP1=NP1+1                                         RAN05230
         MR1=MR+1                                          RAN05240
         INX (N1)=INXL                                     RAN05250
         SP (1,N1)=SP1                                     RAN05260
         SP (2,N1)=SP2                                     RAN05270
         SP (3,N1)=SP3                                     RAN05280
         C (N1)=CL                                         RAN05290
         SI (N1)=SIL                                       RAN05300
         Y (N1)=1.D-10                                     RAN05310
         DO 3733 J=1,M                                     RAN05320
3733     A (N1,J)=AA (J)                                   RAN05330
         IF (SI (N1) .EQ.1.)CP (N1)=C (N1)+DLOG (Y (N1)/YTX) RAN05340
         IF (SI (N1) .EQ.2.)CP (N1)=C (N1)+DLOG (Y (N1)/YTL) RAN05350
         GO TO 1000                                        RAN05360
1713     CONTINUE                                          RAN05370
```

```
 4446 IF (NCP.EQ.NCP1)GO TO 4445
C****************************************************************
C**** CHECK TO SEE IF DOUBTFUL PURE CONDENSED SPECIES MUST BE INSERTED *
C****************************************************************
      LT=NCP-NCP1
      TMAX=0.D0
      DO 447 I=1,LT
      L=NNCP1+I
      T=C(L)
      DO 448 J=1,M
  448 T=T-A(L,J)*PI(J)
      IF (T.GE.TMAX)GO TO 447
      IMAX=L
      TMAX=T
  447 CONTINUE
      IF (TMAX.GE.0.D0)GO TO 4445
      LX=LX+1
      NNCP1=NNCP1+1
      NCP1=NCP1+1
      CALL SWTCH2(SP,IMAX,NNCP1)
      CALL SWDP(C,IMAX,NNCP1)
      DO 1453 J=1,NE
      TM=A(NNCP1,J)
      A(NNCP1,J)=A(IMAX,J)
 1453 A(IMAX,J)=TM
      GO TO 1000
 4445 CONTINUE
C************************************************
C**** CALCULATE FINAL ELEMENTAL ABUNDANCES ****
C************************************************
      DO 64 J=1,NE
      B(J)=0.D0
      DO 64 I=1,NNCP1
   64 B(J)=B(J)+A(I,J)*Y(I)
C************************************************
C**** CALCULATE FINAL G/RT, MOLE FRACTIONS ****
C************************************************
      GF=0.D0
      IF (YT.GT.0.0.AND.YIN.GT.0.0)GF=GF+YIN*DLOG(YIN/YT)
      IF (LIQ.AND.YLIN.GT.0.0)GF=GF+YLIN*DLOG(YLIN/YL)
      DO 67 I=1,NNCP1
      GF=GF+CP(I)*Y(I)
      IF (SI(I).EQ.1.)X(I)=Y(I)/YT
      IF (SI(I).EQ.2.)X(I)=Y(I)/YL
      IF (SI(I).EQ.0..AND.Y(I).LE.0.D0)X(I)=0.D0
   67 IF (SI(I).EQ.0..AND.Y(I).GT.0.D0)X(I)=1.D0
      IF (NCP.EQ.NCP1)GO TO 1121
      L=NNCP1+1
      DO 1010 I=L,NNCP
 1010 DEL(I)=0.D0
      IF (IPR.EQ.0)RETURN
C***********************
C**** PRINT RESULTS ****
C***********************
```

C.1 Program Listing

```
 1121 WRITE (6,117) IT
      IF (ITDG.GT.3) WRITE (6,5138)
 5138 FORMAT (/47H TERMINATION BECAUSE OF WRONG DIRECTION 4 TIMES/)
      IF (ICONV) WRITE (6,5139)
 5139 FORMAT (37H CONVERGENCE CRITERION NOT SATISFIED.)
      WRITE (6,120)
  120 FORMAT (/1X,7HSPECIES,11X,17HEQUILIBRIUM MOLES,3X,14H MOLE FRACTION
     1,4X,11HFINAL DELTA/)
  121 FORMAT (1X,3A4,7X,1PD14.7,5X,D14.7,4X,D11.4)
      WRITE (6,121) ((SP(J,I),J=1,3),Y(I),X(I),DEL(I),I=1,NNCP)
      IF (NNCP.EQ.NT) GO TO 9980
      WRITE (6,817)
      L=NNCP+1
      WRITE (6,818) ((SP(J,I),J=1,3),I=L,NT)
 9980 WRITE (6,803) GF,YT
      IF (NL.GT.0) WRITE (6,806) YL
      WRITE (6,8111) (E(I),B(I),I=1,NE)
 8111 FORMAT (/1X,20HELEMENTAL ABUNDANCES,5X,A2,1PD20.12/(26X,A2,D20.12))
      WRITE (6,119) (PI(I),I=1,M)
      NNCP=NT
      DO 367 I=1,NT
      L=INX(I)
      TEMP (1,L)=SP (1,I)
      TEMP (2,L)=SP (2,I)
  367 TEMP (3,L)=SP (3,I)
      DO 3367 I=1,NT
      SP (1,I)=TEMP (1,I)
      SP (2,I)=TEMP (2,I)
 3367 SP (3,I)=TEMP (3,I)
      DO 3368 I=1,NT
      L=INX(I)
 3368 YA (L)=Y (I)
      IF (IRP.NE.0) GO TO 9999
      RETURN
  117 FORMAT (///I4,11H ITERATIONS)
  817 FORMAT (//23H LESS THAN 1.E-32 MOLES/)
  818 FORMAT (1X,3A4)
  119 FORMAT (/39H FINAL LAGRANGE MULTIPLIERS  (LAMBDA/RT)//(1PD16.8))
  803 FORMAT (//8H G/RT = ,1PD14.7/21H TOTAL MOLES PHASE1 =,D11.4)
  806 FORMAT (21H TOTAL MOLES PHASE2 =,1PD11.4)
      END
      SUBROUTINE CHEMP (X,C,SI,CX,YLX,YTX,NNCP)
      DOUBLE PRECISION C(1),X(1),CX(1),YLX,YTX,DLOG
      DIMENSION SI(1)
      DO 1 I=1,NNCP
      IF (SI(I).EQ.0.) CX(I)=C(I)
      IF (SI(I).EQ.1.) CX(I)=C(I)+DLOG(X(I)/YTX)
   01 IF (SI(I).EQ.2.) CX(I)=C(I)+DLOG(X(I)/YLX)
      RETURN
      END
      SUBROUTINE SWITCH (X,I,J)
      DIMENSION X(1)
      T=X(I)
      X(I)=X(J)
```

```
      X(J)=T                                                              RAN06460
      RETURN                                                              RAN06470
      END                                                                 RAN06480
      SUBROUTINE SWDP(X,I,J)                                              RAN06490
      DOUBLE PRECISION X(1),T                                             RAN06500
      T=X(I)                                                              RAN06510
      X(I)=X(J)                                                           RAN06520
      X(J)=T                                                              RAN06530
      RETURN                                                              RAN06540
      END                                                                 RAN06550
      SUBROUTINE SWTCH2(X,I,J)                                            RAN06560
      DIMENSION X(3,1)                                                    RAN06570
      INTEGER X,T                                                         RAN06580
      T=X(1,I)                                                            RAN06590
      X(1,I)=X(1,J)                                                       RAN06600
      X(1,J)=T                                                            RAN06610
      T=X(2,I)                                                            RAN06620
      X(2,I)=X(2,J)                                                       RAN06630
      X(2,J)=T                                                            RAN06640
      T=X(3,I)                                                            RAN06650
      X(3,I)=X(3,J)                                                       RAN06660
      X(3,J)=T                                                            RAN06670
      RETURN                                                              RAN06680
      END                                                                 RAN06690
      SUBROUTINE DELETE(K,NT,C,CX,Y,A,X,INX)                              RAN06700
      COMMON /RAND/BM(150,15),CT(150),YA(150),EA(15),E(15),TT,PT,YINT,    RAN06710
     1SI(150),SP(3,150),NNCP,M,NCP,NL,IF,NE,ICE,IEST,NCP1,NNCP1,NP1,N1    RAN06720
      INTEGER SP                                                          RAN06730
      DOUBLE PRECISION C(1),CX(1),Y(1),A(150,15),X(1)                     RAN06740
      DIMENSION INX(1)                                                    RAN06750
      L=NT+1                                                              RAN06760
      INX(L)=INX(K)                                                       RAN06770
      SP(1,L)=SP(1,K)                                                     RAN06780
      SP(2,L)=SP(2,K)                                                     RAN06790
      SP(3,L)=SP(3,K)                                                     RAN06800
      C(L)=C(K)                                                           RAN06810
      CX(L)=CX(K)                                                         RAN06820
      SI(L)=SI(K)                                                         RAN06830
      X(L)=X(K)                                                           RAN06840
      DO 1 I=1,M                                                          RAN06850
    1 A(L,I)=A(K,I)                                                       RAN06860
      DO 2 I=K,NT                                                         RAN06870
      SP(1,I)=SP(1,I+1)                                                   RAN06880
      SP(2,I)=SP(2,I+1)                                                   RAN06890
      SP(3,I)=SP(3,I+1)                                                   RAN06900
      C(I)=C(I+1)                                                         RAN06910
      CX(I)=CX(I+1)                                                       RAN06920
      SI(I)=SI(I+1)                                                       RAN06930
      X(I)=X(I+1)                                                         RAN06940
      INX(I)=INX(I+1)                                                     RAN06950
      DO 2 J=1,M                                                          RAN06960
    2 A(I,J)=A(I+1,J)                                                     RAN06970
      N1=N1-1                                                             RAN06980
      NNCP1=NNCP1-1                                                       RAN06990
```

C.1 Program Listing

```
      NP1=NP1-1                                              RAN07000
      NNCP=NNCP-1                                            RAN07010
      RETURN                                                 RAN07020
      END                                                    RAN07030
      SUBROUTINE DLEQU(C,IDEM,N,B,M,IS)                      RAN07040
      DIMENSION C(IDEM,IDEM),B(IDEM,M)                       RAN07050
      DOUBLE PRECISION C,B,R                                 RAN07060
      DO 58 I=1,N                                            RAN07070
      IF(C(I,I).NE.0.D0)GO TO 51                             RAN07080
      DO 53 K=I,N                                            RAN07090
      IF(C(K,I).NE.0.D0)GO TO 54                             RAN07100
   53 CONTINUE                                               RAN07110
      WRITE(6,55)                                            RAN07120
   55 FORMAT(//40H LINEAR DEPENDENCE IN COEFFICIENT MATRIX//) RAN07130
      IS=0                                                   RAN07140
      RETURN                                                 RAN07150
   54 DO 57 J=1,N                                            RAN07160
   57 C(I,J)=C(I,J)+C(K,J)                                   RAN07170
      DO 67 J=1,M                                            RAN07180
   67 B(I,J)=B(I,J)+B(K,J)                                   RAN07190
   51 DO 58 L=1,N                                            RAN07200
      IF(L.EQ.I.OR.C(L,I).EQ.0.D0)GO TO 588                  RAN07210
      R=C(L,I)/C(I,I)                                        RAN07220
      DO 62 J=I,N                                            RAN07230
   62 C(L,J)=C(L,J)-C(I,J)*R                                 RAN07240
      DO 64 J=1,M                                            RAN07250
   64 B(L,J)=B(L,J)-B(I,J)*R                                 RAN07260
  588 CONTINUE                                               RAN07270
   58 CONTINUE                                               RAN07280
      DO 63 I=1,N                                            RAN07290
      DO 63 J=1,M                                            RAN07300
   63 B(I,J)=B(I,J)/C(I,I)                                   RAN07310
      IS=1                                                   RAN07320
      DET=1.                                                 RAN07330
      DO 65 I=1,N                                            RAN07340
   65 DET=DET*C(I,I)                                         RAN07350
      IF(ABS(DET).LT.1.E-15)IS=0                             RAN07360
      RETURN                                                 RAN07370
      END                                                    RAN07380
```

C.2 USER'S GUIDE

Data cards:

Card	Column	Variable Name	Input Format	Description
1	1-3	NRUNS	I3	Number of problems to be run
2	1-3	NNCP	I3	Number of species
	4-6	M	I3	Number of elements
	7-9	NCP	I3	Number of single-species phases (All must be placed at the end of the species data cards below. If NCP is greater than zero, a value for NCP1 is required-see below.)
	10-12	NL	I3	Number of phase2 species
	13-15	IF	I3	Type of chem. pot. data: -1 kcal/mole 0 MU/RT 1 kJ/mole
	16-18	IEST	I3	Initial estimate: 0 User estimate 1 Machine estimate
	19-21	ICE	I3	Type of formula vector: 0 integer entries -1 non-integer entries
	22-24	NE	I3	Number of components
	25-27	NCP1	I3	Number of single-species phases included in the initial computation (the remaining NCP-NCP1 must be at the end of the species list in the species data cards below).

Cards 3,4,...,NNCP+2 are data cards for each species, as follows:

	1-12	SP	3A4	Species name
	13			Blank

C.2 User's Guide

	14-43	BM	15F20	Formula vector
	44	SI	I1	Type of phase: 0 single-species
				1 multi-species(gas)
				2 multi-species(liquid)
	45-53	CT	F9.3	Chemical potential

If any formula vector has non-integer entries, the information on each of cards 3 to NNCP+2 must be coded on two cards as follows:

3a	1-12	SP	3A4	Species name
	13			Blank
	14	SI	I1	Type of phase
	15-23	CT	F9.3	Chemical potential
3b	1-5	BM(1)	F5.3	1st entry of formula vector
	6-10	BM(2)	F5.3	2nd entry of formula vector
				etc. for columns 11-75
NNCP+3	1-10	YA(1)	EN.4	Initial estimate for 1st species
	11-20	YA(2)	EN.4	Initial estimate for 2nd species
				etc. for columns 21-80
NNCP+4	1-10	EA(1)	E10.4	Elemental abundance for 1st element
	11-20	EA(2)	E10.4	Elemental abundance for 2nd element
				etc. for columns 21-80
NNCP+5	1-10	TT	E10.4	Temperature in K
	11-20	PT	E10.4	Pressure in atm
	21-30	YINT	E10.4	Total moles of inert gas
	31-40	YLINT	E10.4	Total moles of inert liquid
NNCP+6	1-2	E(1)	A2	Name of first element
	3-4	E(2)	A2	Name of second element
				etc. for columns 5-30

NNCP+7   1-80   TI   80A1   Arbitrary title

If the user makes his own initial estimate it must satisfy the equation   BM(I,J) * YA(I) = EA(J).

Subroutine arguments:

In the statement CALL RANDEQ(IRP,IPR,IP1,MAXIT) the first three arguments control the input and output. If IRP = 0, then no data is read; otherwise IRP = 1. If IPR = 0, then neither input data nor results are printed; otherwise IPR = 1. If IP1 = 1, then intermediate results are printed at each iteration; otherwise IP1 = 0. MAXIT is the maximum number of iterations allowed if convergence is not reached.

APPENDIX D

FORTRAN Computer Program for VCS Algorithm

Appendix D

D.1 PROGRAM LISTING

```
      IMPLICIT REAL*8 (A-H,O-Z)                                   VCS00010
      COMMON /STOICH/SC(15,149),FE(150),FF(150),W(150),DNG(150),DNL(150)VCS00020
     1,WT(150),DG(150),DS(150),FEL(150),GA(15),GAI(15),TG1,TL1,TG,TL,DTGVCS00030
     2,DTL,BM(150,15),T,P,TING,TINL,IF,IEST,DA(150),IND(150),SP(3,150),IVCS00040
     3C(150),IR(150),SI(150),M,N,NC,NR,MR,NE                       VCS00050
      INTEGER SI,SP                                                VCS00060
      MAXIT=500                                                    VCS00070
      CALL VCS(1,1,0,MAXIT)                                        VCS00080
      STOP                                                         VCS00090
      END                                                          VCS00100
      SUBROUTINE VCS(IRP,IPR,IP1,MAXIT)                            VCS00110
C*****************************************************************VCS00120
C   IDEAL SYSTEM STOICHIOMETRIC EQUILIBRIUM ALGORITHM USING VCS METHOD. VCS00130
C******* VERSION DATED DECEMBER 28, 1981, UNIVERSITY OF GUELPH *******VCS00140
C                                                                  VCS00150
C   PROGRAM CAN HANDLE UP TO 150 SPECIES AND 15 ELEMENTS, BUT      VCS00160
C   DIMENSION STATEMENTS MAY BE MODIFIED TO ALTER THIS.            VCS00170
C   THIS PROGRAM IS DOUBLE-PRECISION VERSION FOR 32-BIT MACHINES.  VCS00180
C   MAXIT IS MAXIMUM NUMBER OF ALLOWABLE ITERATIONS IF CONVERGENCE NOT  VCS00190
C   ACHIEVED.  USUALLY THE NUMBER OF ITERATIONS REQUIRED IS LESS THAN   VCS00200
C   100, BUT VERY NEAR A MULTISPECIES PHASE BOUNDARY THOUSANDS OF ITERA-VCS00210
C   TIONS MAY BE REQUIRED.                                         VCS00220
C   ANY NUMBER OF SINGLE-SPECIES PHASES AND TWO MULTI-SPECIES PHASES   VCS00230
C   CAN BE HANDLED BY THE PRESENT VERSION (THE LATTER IS READILY MODIFIED)VCS00240
C   PHASE1 IS NOMINALLY A GAS, SINCE ALOG(P) IS ADDED TO THE STANDARD  VCS00250
C   CHEMICAL POTENTIAL DATA.  THIS CAN BE OVER-RIDDEN BY SETTING P=1.  VCS00260
C   PHASE2 IS NOMINALLY A LIQUID, OR ANY PHASE FOR WHICH THE STANDARD  VCS00270
C   CHEMICAL POTENTIAL DATA ARE INDEPENDENT OF P.                  VCS00280
C   MULTI-SPECIES PHASE IS DEEMED TO BE ABSENT IF NT.LT.1.E-10.  IF  VCS00290
C   MULTI-SPECIES PHASE ABSENT AT EQUILIBRIUM,  DG/RT VALUE REFERS TO  VCS00300
C   1 - SIGMA(X(I)), WHERE X(I) ARE VIRTUAL MOLE FRACTIONS AT THE  VCS00310
C   CURRENT EQUILIBRIUM.                                           VCS00320
C   LINEAR PROGRAMMING ROUTINE MUST BE PROVIDED FOR INITIAL ESTIMATE  VCS00330
C   OF EQUILIBRIUM COMPOSITION OPTION (SEE SUBROUTINE INEST)       VCS00340
C*****************************************************************VCS00350
      IMPLICIT REAL*8 (A-H,O-Z)                                    VCS00360
      REAL*8 LOUT,LIST                                             VCS00370
      COMMON /STOICH/SC(15,149),FE(150),FF(150),W(150),DNG(150),DNL(150)VCS00380
     1,WT(150),DG(150),DS(150),FEL(150),GA(15),GAI(15),TG1,TL1,TG,TL,DTGVCS00390
     2,DTL,BM(150,15),T,P,TING,TINL,IF,IEST,DA(150),IND(150),SP(3,150),IVCS00400
     3C(150),IR(150),SI(150),M,N,NC,NR,MR,NE                       VCS00410
      INTEGER SI,SP,XM,STATE,STAT1,STAT2                           VCS00420
      INTEGER OUT(2),VF(13)                                        VCS00430
      LOGICAL LINDEP,SOLDEL,IFIRST,CONV,ICONV                      VCS00440
      LOGICAL DAT,LIQ,LEC,ITL,ILT,LBO,IM,MACH,FORCED               VCS00450
      COMMON /FORCER/LIQ                                           VCS00460
      DIMENSION PS(150),WX(150),E(15),MJ(15),EA(15),TI(80),FFO(150), VCS00470
     1LIST(30),LIST1(14)                                           VCS00480
      DIMENSION GAMM(25),DB(10),V(10),PHY(10)                      VCS00490
      DIMENSION LOUT(10)                                           VCS00500
      DIMENSION SA(15),SS(15),SM(15,15)                            VCS00510
```

D.1 Program Listing

```
      DIMENSION YY(15,1)                                           VCS00520
      DIMENSION XY(150)                                            VCS00530
      DIMENSION X(16),Z(16),B(16)                                  VCS00540
      DIMENSION AW(150),AA(15,15)                                  VCS00550
      DIMENSION BB(15),PSOL(150),DSOL(15),CC(150),RW(623),IW(303)  VCS00560
      DIMENSION AX(17,152)                                         VCS00570
      EQUIVALENCE (X,YY)                                           VCS00580
      EQUIVALENCE (FFO,FF),(PS,WX,MJ,DS)                           VCS00590
      EXP(X)=DEXP(X)                                               VCS00600
      ALOG(X)=DLOG(X)                                              VCS00610
      ABS(X)=DABS(X)                                               VCS00620
      DATA OUT/'SI( ','I)  '/                                      VCS00630
      DATA OUT2/'SI(I)  '/                                         VCS00640
      DATA LIST1/'2','3','4','5','6','7','8','9','10','11','12','13','14VCS00650
     *','15'/                                                      VCS00660
      DATA VF/'(1X,','3A4,',' 1X,',' ',  ','F6.3',',  I5',',T96',',D12', VCS00670
     *'.5,6','X,D1','2.5,','D12.','5)  '/                          VCS00680
      DATA LIST/5H2I3 ,,5H1I2 ,,5H3I3 ,,5H39X ,,5H4I3 ,,5H36X ,,5H5I3 ,,VCS00690
     15H33X ,,5H6I3 ,,5H30X ,,5H7I3 ,,5H27X ,,5H8I3 ,,5H24X ,,5H9I3 ,,  VCS00700
     25H21X ,,5H10I3,,5H18X ,,5H11I3,,5H15X ,,5H12I3,,5H12X ,,5H13I3,,  VCS00710
     35H9X  ,,5H14I3,,5H6X  ,,5H15I3,,5H3X  ,,5H16I3,,5H1X  /          VCS00720
      DATA LOUT/5H(    ,5H   1X,5H,3A4,,5H      ,5H     ,5H 1PD1,5H 2.5,,VCS00730
     15H  2X,,5H D12.,,5H5  )  /                                   VCS00740
      DO 9191 J=1,150                                              VCS00750
      FE(J)=0.                                                     VCS00760
      DA(J)=0.                                                     VCS00770
      IF(IRP.NE.0)W(J)=1.D0                                        VCS00780
      FEL(J)=0.                                                    VCS00790
 9191 WT(J)=0.D0                                                   VCS00800
C*******************************                                   VCS00810
C**** INPUT FORMAT STATEMENTS ****                                 VCS00820
C*******************************                                   VCS00830
  101 FORMAT(7I3)                                                  VCS00840
  102 FORMAT(3A4,1X,15F2.0,I1,F9.3)                                VCS00850
 1022 FORMAT(3A4,1X,I1,F9.3)                                       VCS00860
 2022 FORMAT(15F5.3)                                               VCS00870
  103 FORMAT(8E10.4)                                               VCS00880
  901 FORMAT(80A1)                                                 VCS00890
C**************************                                        VCS00900
C**** OUTPUT FORMATS ******                                        VCS00910
C**************************                                        VCS00920
  902 FORMAT(//I5,8H SPECIES,I8,9H ELEMENTS,I16,11H COMPONENTS/    VCS00930
     1I5,15H PHASE1 SPECIES,I10,15H PHASE2 SPECIES,I8,22H SINGLE SPECIESVCS00940
     2 PHASES//,9H PRESSURE,F22.3,4H ATM/12H TEMPERATURE,F19.3,2H K) VCS00950
  903 FORMAT(14H PHASE2 INERTS,F17.3)                              VCS00960
  904 FORMAT(14H PHASE1 INERTS,F17.3)                              VCS00970
  905 FORMAT(/1X,51HMODIFIED LINEAR PROGRAMMING ESTIMATE OF EQUILIBRIUM)VCS00980
  906 FORMAT(/1X,28HUSER ESTIMATE OF EQUILIBRIUM)                  VCS00990
  907 FORMAT( /8H SPECIES,5X,16H  FORMULA VECTOR,29X,16HSTAN. CHEM. POT.VCS01000
     1,3X,16HEQUILIBRIUM EST./)                                    VCS01010
 9070 FORMAT( /8H SPECIES,5X,16H  FORMULA VECTOR,T96,16HSTAN. CHEM. POT.VCS01020
     1,3X,16HEQUILIBRIUM EST./)                                    VCS01030
 9111 FORMAT(/21H ELEMENTAL ABUNDANCES,13X,7HCORRECT,10X,13HFROM ESTIMATVCS01040
     1E//(26X,A2,1PD20.12,D20.12))                                 VCS01050
```

```
      911 FORMAT (/21H ELEMENTAL ABUNDANCES,5X,A2,1PD20.12/(26X,A2,D20.12))   VCS01060
      912 FORMAT (15A2)                                                        VCS01070
      913 FORMAT (23H1VCS CALCULATION METHOD/)                                 VCS01080
      914 FORMAT (/1X,80A1)                                                    VCS01090
C***********************************************************                   VCS01100
C**** SET INITIAL VALUES, READ IN DATA, PRINT TITLES ****                      VCS01110
C***********************************************************                   VCS01120
          IF (IRP.EQ.0)NRUNS=1                                                 VCS01130
          NID=0                                                                VCS01140
          IF (IRP.EQ.0)GO TO 620                                               VCS01150
          READ (5,101) NRUNS                                                   VCS01160
      620 DO 9999 IJ=1,NRUNS                                                   VCS01170
          ICONV=.FALSE.                                                        VCS01180
          IF (IRP.EQ.0)GO TO 623                                               VCS01190
          READ (5,101)M, NE,NS1,NL1,IF,IEST,ICE                                VCS01200
          NC=NE                                                                VCS01210
          IF (ICE.LT.0)GO TO 621                                               VCS01220
          READ (5,102) ((SP(J,I),J=1,3), (BM(I,J),J=1,15),SI(I),FFO(I),I=1,M)  VCS01230
          GO TO 622                                                            VCS01240
      621 READ (5,1022) ((SP(J,I),J=1,3),SI(I),FFO(I),I=1,M)                   VCS01250
          READ (5,2022) ((BM(I,J),J=1,15),I=1,M)                               VCS01260
      622 IF (IEST.LT.0)GO TO 293                                              VCS01270
          READ (5,103) (W(I),I=1,M)                                            VCS01280
      293 READ (5,103) (GAI(I),I=1,NE)                                         VCS01290
          READ (5,103)T,P,TING,TINL                                            VCS01300
          READ (5,912) (E(I),I=1,NE)                                           VCS01310
          READ (5,901)TI                                                       VCS01320
      623 IT=0                                                                 VCS01330
          NOPT=0                                                               VCS01340
          IM=.FALSE.                                                           VCS01350
C********************************                                              VCS01360
C**** INIT. VARIABLE FORMATS ****                                              VCS01370
C********************************                                              VCS01380
          LOUT(4)=LIST(2*NE-1)                                                 VCS01390
          LOUT(5)=LIST(2*NE)                                                   VCS01400
          VF(4)=LIST1(NE-1)                                                    VCS01410
C************************                                                      VCS01420
C**** PRINT TITLES *****                                                       VCS01430
C************************                                                      VCS01440
          WRITE (6,913)                                                        VCS01450
          WRITE (6,914)TI                                                      VCS01460
C**************************************************                            VCS01470
C***** COMPUTE NUMBER AND TYPE OF PHASES *****                                 VCS01480
C**************************************************                            VCS01490
          NS=0                                                                 VCS01500
          NL=0                                                                 VCS01510
          DO 855 I=1,M                                                         VCS01520
          IND(I)=I                                                             VCS01530
          IF (SI(I).EQ.0)NS=NS+1                                               VCS01540
      855 IF (SI(I).EQ.2)NL=NL+1                                               VCS01550
          IF (NS.NE.NS1)WRITE (6,105)                                          VCS01560
          IF (NL.NE.NL1)WRITE (6,106)                                          VCS01570
      105 FORMAT (57H *** NUMBER OF SINGLE-SPECIES PHASES DOES NOT COMPUTE **VCS01580
         1*)                                                                   VCS01590
```

D.1 Program Listing

```
  106 FORMAT(46H NUMBER OF PHASE2 SPECIES DOES NOT COMPUTE ***)    VCS01600
      NG=M-NL-NS                                                    VCS01610
      IF(NG.NE.1)GO TO 852                                          VCS01620
      NS=NS+1                                                       VCS01630
      NG=0                                                          VCS01640
      DO 856 I=1,M                                                  VCS01650
      IND(I)=I                                                      VCS01660
      IF(SI(I).NE.1)GO TO 856                                       VCS01670
      IF(IF)8561,8562,8563                                          VCS01680
 8561 FFO(I)=FFO(I)+1.9872D0*T*ALOG(P)                              VCS01690
      GO TO 856                                                     VCS01700
 8562 FFO(I)=FFO(I)+ALOG(P)                                         VCS01710
      GO TO 856                                                     VCS01720
 8563 FFO(I)=FFO(I)+.0083143D0*T*ALOG(P)                            VCS01730
      SI(I)=0                                                       VCS01740
      WRITE(6,107)(SP(J,I),J=1,3)                                   VCS01750
  856 CONTINUE                                                      VCS01760
  852 IF(NL.NE.1)GO TO 854                                          VCS01770
      NS=NS+1                                                       VCS01780
      NL=0                                                          VCS01790
      DO 858 I=1,M                                                  VCS01800
      IF(SI(I).NE.2)GO TO 858                                       VCS01810
      SI(I)=0                                                       VCS01820
      WRITE(6,108)(SP(J,I),J=1,3)                                   VCS01830
  858 CONTINUE                                                      VCS01840
  107 FORMAT(14H THIS SPECIES:,3A4,58H IS THE ONLY GAS. IT WILL THEREFORVCS01850
     1E BE TREATED AS A SOLID. )                                    VCS01860
  108 FORMAT(14H THIS SPECIES:,3A4,61H IS THE ONLY LIQUID. IT WILL THEREVCS01870
     1FORE BE TREATED AS A SOLID. )                                 VCS01880
  854 N=M-NC                                                        VCS01890
      LIQ=NL.GT.0                                                   VCS01900
      NR=N                                                          VCS01910
      MR=M                                                          VCS01920
      NC1=NC+1                                                      VCS01930
      DO 889 I=1,M                                                  VCS01940
      IR(I)=NC+I                                                    VCS01950
  889 CONTINUE                                                      VCS01960
      TEST=-1.D-10                                                  VCS01970
      IF(IEST.GE.0)GO TO 30001                                      VCS01980
      DO 30000 I=1,MR                                               VCS01990
30000 W(I)=-FFO(I)                                                  VCS02000
      TEST=-1.E20                                                   VCS02010
30001 CONTINUE                                                      VCS02020
C****************************************                          VCS02030
C****  DETERMINE NUMBER OF COMPONENTS.*****                         VCS02040
C****************************************                          VCS02050
      CALL BASOPT(.TRUE.,NOPT,AW,SA,SM,SS,AA,TEST,IT,CONV)           VCS02060
      IF(IPR.NE.0)WRITE(6,902)M,NE,NC,NG,NL,NS,P,T                  VCS02070
      NC1=NC+1                                                      VCS02080
      IF(IPR.NE.0.AND.NG.GT.0)WRITE(6,904)TING                      VCS02090
      IF(IPR.NE.0.AND.LIQ)WRITE(6,903)TINL                          VCS02100
C**************************************************************    VCS02110
C**** ADJUST FOR SPECIFIED TYPE OF STAN. CHEM. POT. DATA ****       VCS02120
C**************************************************************    VCS02130
```

```
      910 RT=1.9872D0*T                                              VCS02140
          IF(IF)165,163,164                                          VCS02150
      165 TF=1000.D0/RT                                              VCS02160
      169 FORMAT (31H STAN. CHEM. POT. IN KCAL./MOLE)                VCS02170
          GO TO 166                                                  VCS02180
      164 TF=1.D0/(0.0083143D0*T)                                    VCS02190
      168 FORMAT (29H STAN. CHEM. POT. IN KJ./MOLE)                  VCS02200
          GO TO 166                                                  VCS02210
      163 TF=0.                                                      VCS02220
      167 FORMAT (26H STAN. CHEM. POT. IS MU/RT)                     VCS02230
      166 DAT=TF.NE.0.                                               VCS02240
          DO 888 I=1,M                                               VCS02250
          IR(I)=NC+I                                                 VCS02260
          DA(I)=FFO(I)                                               VCS02270
          IF(DAT)FFO(I)=FFO(I)*TF                                    VCS02280
          IF(SI(I).NE.1)GO TO 8866                                   VCS02290
          FFO(I)=FFO(I)+ALOG(P)                                      VCS02300
          IF(NG.EQ.1)SI(I)=0                                         VCS02310
     8866 IF(SI(I).EQ.0)FE(I)=FF(I)                                  VCS02320
      888 IF(SI(I).EQ.2)SI(I)=-1                                     VCS02330
C*******************************************                        VCS02340
C**** EVALUATE EQUILIBRIUM ESTIMATE, IF REQUIRED ****                VCS02350
C*******************************************                        VCS02360
          IF(IEST.GE.0)GO TO 159                                     VCS02370
          CALL INEST(L,.FALSE.,NOPT,AW,SA,SM,SS,AA,TEST,IT,NC1)      VCS02380
      494 CALL ELAB                                                  VCS02390
    11591 CALL ELCORR                                                VCS02400
C*******************************************                        VCS02410
C**** PRINT DATA, EQUILIBRIUM ESTIMATE ****                         VCS02420
C*******************************************                        VCS02430
      159 CONTINUE                                                   VCS02440
          IF(IEST.GE.0)CALL ELAB                                     VCS02450
          IF(IPR.EQ.0)GO TO 1955                                     VCS02460
          WRITE(6,9111) (E(I),GAI(I),GA(I),I=1,NE)                   VCS02470
          IF(IEST.LT.0)WRITE(6,905)                                  VCS02480
          IF(IEST.GE.0)WRITE(6,906)                                  VCS02490
          IF(IF.LT.0)WRITE(6,169)                                    VCS02500
          IF(IF.EQ.0)WRITE(6,167)                                    VCS02510
          IF(IF.GT.0)WRITE(6,168)                                    VCS02520
          IF(ICE.LT.0)GO TO 78                                       VCS02530
          WRITE(6,907)                                               VCS02540
          WRITE(6,890) (E(I),I=1,NE),OUT                             VCS02550
      890 FORMAT(' ',13X,17A3)                                       VCS02560
          DO 865 I=1,M                                               VCS02570
          IXA=SI(I)                                                  VCS02580
          IF(SI(I).EQ.-1)IXA=2                                       VCS02590
          DO 860 J=1,NE                                              VCS02600
      860 MJ(J)=BM(I,J)                                              VCS02610
      865 WRITE(6,LOUT) (SP(J,I),J=1,3),(MJ(J),J=1,NE),IXA,DA(I),W(I) VCS02620
          GO TO 1955                                                 VCS02630
       78 CONTINUE                                                   VCS02640
          WRITE(6,9070)                                              VCS02650
          WRITE(6,891) (E(I),I=1,NE),OUT2                            VCS02660
      891 FORMAT(' ',13X,17(1X,A5))                                  VCS02670
```

D.1 Program Listing

```
          DO 77 I=1,M                                          VCS02680
          IXA=SI(I)                                            VCS02690
          IF(SI(I).EQ.-1)IXA=2                                 VCS02700
       77 WRITE(6,VF) (SP(J,I),J=1,3),(BM(I,J),J=1,NE),IXA,DA(I),W(I)  VCS02710
C******************************************                   VCS02720
C**** EVALUATE TOTAL MOLES GAS, LIQUID ****                   VCS02730
C******************************************                   VCS02740
     1955 TG=TING                                              VCS02750
          TL=TINL                                              VCS02760
          DO 1 I=1,MR                                          VCS02770
          IF(SI(I))3,1,2                                       VCS02780
       03 TL=TL+W(I)                                           VCS02790
          GO TO 1                                              VCS02800
       02 TG=TG+W(I)                                           VCS02810
       01 CONTINUE                                             VCS02820
C******************************************                   VCS02830
C**** EVALUATE ALL CHEMICAL POTENTIALS ****                   VCS02840
C******************************************                   VCS02850
          CALL DFE(W,0,0)                                      VCS02860
          IF(IT.GT.0)GO TO 430                                 VCS02870
          IF(IEST.LT.0)GO TO 441                               VCS02880
C**************************************************           VCS02890
C**** DETERMINE BASIS SPECIES, EVALUATE STOICHIOMETRY ****    VCS02900
C**************************************************           VCS02910
      429 CALL BASOPT(.FALSE.,NOPT,AW,SA,SM,SS,AA,TEST,IT,CONV) VCS02920
          IF(.NOT.CONV)GO TO 24192                             VCS02930
     6714 WRITE(6,6713)                                        VCS02940
          GO TO 857                                            VCS02950
     6713 FORMAT(53H CONVERGENCE TO NUMBER OF POSITIVE N(I) LESS THAN C. VCS02960
         1/35H CHECK RESULTS TO FOLLOW CAREFULLY.  //)         VCS02970
    24192 IT1=1                                                VCS02980
          ILT=.FALSE.                                          VCS02990
C******************************************                   VCS03000
C**** EVALUATE INITIAL MAJOR-MINOR VECTOR ****                VCS03010
C******************************************                   VCS03020
      323 JM=0                                                 VCS03030
          DO 140 I=1,NR                                        VCS03040
          L=IR(I)                                              VCS03050
          IF(W(L).GT.0.)GO TO 400                              VCS03060
          JM=JM+1                                              VCS03070
          IC(I)=-1                                             VCS03080
          GO TO 140                                            VCS03090
      400 IC(I)=1                                              VCS03100
      140 CONTINUE                                             VCS03110
          LEC=.FALSE.                                          VCS03120
          LBO=.TRUE.                                           VCS03130
          IF(IT.GT.0)GO TO 96                                  VCS03140
C******************************************                   VCS03150
C**** EVALUATE ALL REACTION FREE ENERGY CHANGES ****          VCS03160
C******************************************                   VCS03170
      430 CALL DELTAG(0,1,NR)                                  VCS03180
          IF(.NOT.LBO)GO TO 212                                VCS03190
     1430 LBO=.FALSE.                                          VCS03200
          IF(.NOT.IM)GO TO 93                                  VCS03210
```

```
      9090 ITI=0                                              VCS03220
           GO TO 90                                           VCS03230
C*******************************************                 VCS03240
C**** SET INITIAL VALUES FOR ITERATION ****                  VCS03250
C**** EVALUATE REACTION ADJUSTMENTS      ****                VCS03260
C*******************************************                 VCS03270
        93 ITI=4*(IT1/4)-IT1                                  VCS03280
           IF(ITI.NE.0)GO TO 90                               VCS03290
           CALL DFE(W,0,1)                                    VCS03300
           CALL DELTAG(1,1,NR)                                VCS03310
        90 DO 1707 I=1,MR                                     VCS03320
      1707 FEL(I)=FE(I)                                       VCS03330
           IF(ILT.OR.IM)GO TO 11608                           VCS03340
           CALL ST2(SOLDEL,&429)                              VCS03350
     11608 ITL=.FALSE.                                        VCS03360
           LEC=.FALSE.                                        VCS03370
           DTG=0.                                             VCS03380
           IF(LIQ)DTL=0.                                      VCS03390
           DO 1092 I=1,NC                                     VCS03400
      1092 DS(I)=0.                                           VCS03410
C***************************************************************  VCS03420
C**** MAIN LOOP IN CALCULATION FROM HERE TO ST. NO. 33 ****   VCS03430
C***************************************************************  VCS03440
           I=1                                                VCS03450
      9058 L=IR(I)                                            VCS03460
      9059 IF(IC(I))134,34,22                                 VCS03470
       134 IF(DG(I).GE.0.)GO TO 3331                          VCS03480
           JM=JM-1                                            VCS03490
           IC(I)=1                                            VCS03500
           IM=.FALSE.                                         VCS03510
           ILT=.FALSE.                                        VCS03520
           GO TO 3331                                         VCS03530
C***********************                                      VCS03540
C**** MINOR SPECIES ****                                      VCS03550
C***********************                                      VCS03560
        34 IF(ITI.NE.0)GO TO 333                              VCS03570
           DGG=DG(I)                                          VCS03580
           WW=W(L)                                            VCS03590
           IF(DGG.LT.82.)GO TO 36                             VCS03600
      1373 IF(WW.LT.1.E-26)GO TO 1374                         VCS03610
           WT(L)=WW*1.E-6                                     VCS03620
           GO TO 1135                                         VCS03630
        36 IF(ABS(DGG).LE.1.E-5)GO TO 3331                    VCS03640
           C=ALOG(WW)-DGG                                     VCS03650
           IF(C.LE.(-73.6))GO TO 1373                         VCS03660
           WT(L)=EXP(C)                                       VCS03670
           IF(WT(L).LE.W(1))GO TO 1135                        VCS03680
           WT(L)=WW*1.D+6                                     VCS03690
      1135 DX=WT(L)-WW                                        VCS03700
           DS(L)=DX                                           VCS03710
           IF(WW.LE.(-DX))ITL=.TRUE.                          VCS03720
           GO TO 18                                           VCS03730
        22 IF(ABS(DG(I)).LE.1.E-6)GO TO 3331                  VCS03740
           DX=PS(L)                                           VCS03750
```

D.1 Program Listing

```
        WT(L)=W(L)+DX                                      VCS03760
C***************************************************      VCS03770
C**** CHECK FOR NON-POSITIVE MOLES OF MAJOR SPECIES ****   VCS03780
C***************************************************      VCS03790
        IF(WT(L).LE.0.)GO TO 85                            VCS03800
C*********************************************************  VCS03810
C**** CALCULATE MOLE NUMBER CHANGES FOR GAS, LIQUID, BASIS ****  VCS03820
C*********************************************************  VCS03830
     18 DO 14 K=1,NC                                       VCS03840
     14 DS(K)=DS(K)+SC(K,I)*DX                             VCS03850
        DTG=DTG+DNG(I)*DX                                  VCS03860
        IF(LIQ)DTL=DTL+DNL(I)*DX                           VCS03870
        GO TO 33                                           VCS03880
   3331 WT(L)=W(L)                                         VCS03890
    333 DS(L)=0.                                           VCS03900
     33 I=I+1                                              VCS03910
   1232 IF(I.LE.NR)GO TO 9058                              VCS03920
C***************************************************      VCS03930
C**** LIMIT REDUCTION OF BASIS SPECIES TO 99 PERCENT ****  VCS03940
C***************************************************      VCS03950
   1587 PAR=.5                                             VCS03960
        DO 1086 I=1,NC                                     VCS03970
        XX=-DS(I)/W(I)                                     VCS03980
   1086 IF(PAR.LT.XX)PAR=XX                                VCS03990
        PAR=1./PAR                                         VCS04000
        IF(PAR.LE.1.01.AND.PAR.GT.0.)GO TO 29              VCS04010
        PAR=1.                                             VCS04020
        LK=NC                                              VCS04030
        GO TO 1589                                         VCS04040
C*****************************************                 VCS04050
C**** REDUCTION IN BASIS TOO LARGE ****                    VCS04060
C**** REDUCE OVERALL STEP SIZE      ****                   VCS04070
C*****************************************                 VCS04080
     29 PAR=.99*PAR                                        VCS04090
        ITL=.FALSE.                                        VCS04100
        DO 1088 I=1,MR                                     VCS04110
   1088 IF(DS(I).NE.0.)DS(I)=DS(I)*PAR                     VCS04120
        DTG=DTG*PAR                                        VCS04130
        IF(LIQ)DTL=DTL*PAR                                 VCS04140
        ITL=.FALSE.                                        VCS04150
        LK=MR                                              VCS04160
   1589 DO 1591 I=1,LK                                     VCS04170
   1591 WT(I)=W(I)+DS(I)                                   VCS04180
        TG1=TG+DTG                                         VCS04190
        IF(LIQ)TL1=TL+DTL                                  VCS04200
        GO TO 95                                           VCS04210
C*****************************************                 VCS04220
C**** NON-POSITIVE MOLES OF MAJOR SPECIES ****             VCS04230
C*****************************************                 VCS04240
     85 IF(SI(L).NE.0)GO TO 86                             VCS04250
C*****************************************                 VCS04260
C**** SET PURE PHASE SPECIES TO ZERO ****                  VCS04270
C*****************************************                 VCS04280
        DX=-W(L)                                           VCS04290
```

```
 1123 DO 1119 J=1,NC                                          VCS04300
      WT(J)=W(J)+SC(J,I)*DX                                   VCS04310
      IF(WT(J).LE.0.)GO TO 1120                               VCS04320
 1119 CONTINUE                                                VCS04330
      DO 1124 J=1,NC                                          VCS04340
 1124 W(J)=WT(J)                                              VCS04350
      GO TO 1125                                              VCS04360
 1120 DX=DX/2.                                                VCS04370
      GO TO 1123                                              VCS04380
 1125 W(L)=W(L)+DX                                            VCS04390
      TG=TG+DNG(I)*DX                                         VCS04400
      IF(LIQ)TL=TL+DNL(I)*DX                                  VCS04410
      IF(W(L).GT.0.)GO TO 3331                                VCS04420
      IC(I)=-1                                                VCS04430
      CALL DFE(W,0,ITI)                                       VCS04440
      CALL DELTAG(ITI,1,NR)                                   VCS04450
      DO 5111 LL=1,MR                                         VCS04460
 5111 FEL(LL)=FE(LL)                                          VCS04470
      GO TO 1218                                              VCS04480
C*****************************                                VCS04490
C**** CHANGE MAJOR TO MINOR ****                              VCS04500
C*****************************                                VCS04510
 1118 IC(I)=0                                                 VCS04520
 1218 JM=JM+1                                                 VCS04530
      IM=JM.EQ.NR                                             VCS04540
      IF(IM.AND.ITI.NE.0)GO TO 227                            VCS04550
      IF(SI(L).EQ.0)GO TO 3331                                VCS04560
      GO TO 18                                                VCS04570
C**********************************************************  VCS04580
C**** DELETE MINOR SPECIES LESS THAN 1.E-32 MOLES *****       VCS04590
C**********************************************************  VCS04600
 1374 MM=0                                                    VCS04610
      CALL DELETE(L,0,&1366)                                  VCS04620
11375 JM=JM-1                                                 VCS04630
      GO TO 1232                                              VCS04640
C*********************************************               VCS04650
C**** CUT REACTION ADJUSTMENT TO 1/3      ****                VCS04660
C**** FOR POSITIVE MOLES OF MAJOR SPECIE ****                 VCS04670
C*********************************************               VCS04680
   86 DX=-.9*W(L)                                             VCS04690
      DS(L)=DX                                                VCS04700
      WT(L)=W(L)+DX                                           VCS04710
      IF(W(L).LT..01*W(NC))GO TO 1118                         VCS04720
      GO TO 18                                                VCS04730
C*****************************************                    VCS04740
C**** TENTATIVE CHEMICAL POTENTIALS ****                      VCS04750
C*****************************************                    VCS04760
   95 CALL DFE(WT,1,ITI)                                      VCS04770
C***********************************                          VCS04780
C**** PRINT INTERMEDIATE RESULTS ****                         VCS04790
C***********************************                          VCS04800
      IF(IP1.EQ.0)GO TO 28616                                 VCS04810
      L1=IT1-1                                                VCS04820
      WRITE(6,7167)IT,L1                                      VCS04830
```

D.1 Program Listing

```
      7167 FORMAT(/ 10H ITERATION,I3,52H,   ITERATIONS SINCE LAST EVALUATION O VCS04840
          1F STOICHIOMETRY,I3)                                                 VCS04850
           WRITE(6,4731)                                                       VCS04860
      4731 FORMAT(8H SPECIES,19X,13HINITIAL MOLES,6X,11HFINAL MOLES, 8X,5HMU/  VCS04870
          1RT,16X,10HDELTA G/RT)                                               VCS04880
           WRITE(6,802)((SP(J,I),J=1,3),W(I),WT(I),FE(I),I=1,NC)               VCS04890
      8055 FORMAT(1X,3A4,13X,1PD14.7,3D19.7)                                   VCS04900
           DO 4732 I=NC1,MR                                                    VCS04910
           L1=I-NC                                                             VCS04920
      4732 WRITE(6,8055)(SP(J,I),J=1,3),W(I),WT(I),FE(I),DG(L1)                VCS04930
     28616 CONTINUE                                                            VCS04940
           FORCED=.FALSE.                                                      VCS04950
           IF(IM.OR.ILT.OR.ITL)GO TO 18617                                     VCS04960
     C*****************************                                            VCS04970
     C**** CONVERGENCE FORCER ****                                             VCS04980
     C*****************************                                            VCS04990
           CALL FORCE(ITI,FORCED)                                              VCS05000
     C****************************************************                    VCS05010
     C**** RESET VALUES AT END OF ITERATION         ****                       VCS05020
     C**** CALCULATE NEW REACTION FREE ENERGY CHANGES ****                     VCS05030
     C****************************************************                    VCS05040
     18617 IT=IT+1                                                             VCS05050
           ITI=ITI+1                                                           VCS05060
       142 IF(ITI.EQ.0)CALL DELTAG(0,1,NR)                                     VCS05070
           IF(ITI.NE.0)CALL DELTAG(-1,1,NR)                                    VCS05080
           IF(FORCED)GO TO 4414                                                VCS05090
           TG=TG1                                                              VCS05100
           IF(LIQ)TL=TL1                                                       VCS05110
           DO 10 I=1,MR                                                        VCS05120
        10 IF(DS(I).NE.0.)W(I)=WT(I)                                           VCS05130
     C*********************************************************************   VCS05140
     C**** SET MICROSCOPIC MULTISPECIES PHASE (NT.LT.1.E-10) TO ZERO ****      VCS05150
     C*********************************************************************   VCS05160
      4414 CONTINUE                                                            VCS05170
           XM=0                                                                VCS05180
           IF(NG.EQ.0.OR.TG.EQ.0..OR.TG.GT.1.E-10)GO TO 9951                   VCS05190
           DO 9952 I=1,MR                                                      VCS05200
           IF(SI(I).NE.1)GO TO 9952                                            VCS05210
           W(I)=0.                                                             VCS05220
           WT(I)=0.                                                            VCS05230
           DS(I)=0.                                                            VCS05240
      9952 CONTINUE                                                            VCS05250
           TG=0.                                                               VCS05260
           TG1=0.                                                              VCS05270
           DTG=0.                                                              VCS05280
           XM=1                                                                VCS05290
      9951 IF(NL.EQ.0.OR.TL.EQ.0..OR.TL.GT.1.E-10)GO TO 9554                   VCS05300
           DO 9953 I=1,MR                                                      VCS05310
           IF(SI(I).NE.(-1))GO TO 9953                                         VCS05320
           W(I)=0.                                                             VCS05330
           WT(I)=0.                                                            VCS05340
           DS(I)=0.                                                            VCS05350
      9953 CONTINUE                                                            VCS05360
           TL=0.                                                               VCS05370
```

```
              TL1=0.                                               VCS05380
              DTL=0.                                               VCS05390
              XM=1                                                 VCS05400
         9554 IF(XM.EQ.0)GO TO 441                                 VCS05410
              DO 9956 I=1,NR                                       VCS05420
              L=IR(I)                                              VCS05430
         9956 IF(W(L).EQ.0.)IC(I)=-1                               VCS05440
              CALL BASOPT(.FALSE.,NOPT,AW,SA,SM,SS,AA,TEST,IT,CONV) VCS05450
              CALL DFE(W,0,0)                                      VCS05460
              CALL DELTAG(0,1,NR)                                  VCS05470
              IF(.NOT.CONV)GO TO 441                               VCS05480
              WRITE(6,9993)                                        VCS05490
              GO TO 6714                                           VCS05500
         9993 FORMAT(/32H DELETION OF MULTISPECIES PHASE. )        VCS05510
      C********************************                            VCS05520
      C**** CHECK FOR OPTIMUM BASIS ****                           VCS05530
      C********************************                            VCS05540
          441 CONTINUE                                             VCS05550
              IF(NC.EQ.1)GO TO 4442                                VCS05560
              DO 4441 I=2,NC                                       VCS05570
              IF(W(I-1).LT.W(I))GO TO 4442                         VCS05580
         4441 CONTINUE                                             VCS05590
              DO 426 I=1,NR                                        VCS05600
              L=IR(I)                                              VCS05610
              J=NC                                                 VCS05620
          427 IF(W(L).LE.W(J))GO TO 426                            VCS05630
              IF(SC(J,I).NE.0.)GO TO 429                           VCS05640
              J=J-1                                                VCS05650
              GO TO 427                                            VCS05660
          426 CONTINUE                                             VCS05670
              GO TO 1241                                           VCS05680
         4442 DO 4443 I=1,NR                                       VCS05690
              L=IR(I)                                              VCS05700
              DO 4443 J=1,NC                                       VCS05710
              IF(W(L).LE.W(J))GO TO 4443                           VCS05720
              IF(SC(J,I).NE.0.)GO TO 429                           VCS05730
         4443 CONTINUE                                             VCS05740
         1241 IF(IT.EQ.0)GO TO 24192                               VCS05750
      C*********************************************************   VCS05760
      C**** RE-EVALUATE MAJOR-MINOR VECTOR, IF NECESSARY ****      VCS05770
      C*********************************************************   VCS05780
              IF(ITI.NE.0)GO TO 212                                VCS05790
          221 JM=0                                                 VCS05800
              DO 3731 I=1,NC                                       VCS05810
         3731 WX(I)=.01*W(I)                                       VCS05820
              I=1                                                  VCS05830
         1920 L=IR(I)                                              VCS05840
              IF(W(L).GT.0.)GO TO 211                              VCS05850
              IF(DG(I))20,20,1191                                  VCS05860
          211 IF(SI(L).EQ.0)GO TO 13                               VCS05870
              DO 11 K=1,NC                                         VCS05880
              IF(SI(K).NE.0.AND.SC(K,I).NE.0..AND.W(L).GE.WX(K))GO TO 13  VCS05890
           11 CONTINUE                                             VCS05900
              IF(SI(L).EQ.1.AND.W(L).GT..10*TG)GO TO 13            VCS05910
```

D.1 Program Listing

```
      IF(SI(L).EQ.(-1).AND.W(L).GT..10*TL)GO TO 13          VCS05920
      IC(I)=0                                                VCS05930
      GO TO 119                                              VCS05940
 1191 IC(I)=-1                                               VCS05950
  119 JM=JM+1                                                VCS05960
      GO TO 19                                               VCS05970
   13 IF(IC(I).GT.0)GO TO 19                                 VCS05980
   20 IC(I)=1                                                VCS05990
      IF(ITI.EQ.0)GO TO 19                                   VCS06000
C*********************************************               VCS06010
C**** CHEM. POT., DELTA G FOR NEW MAJOR ****                 VCS06020
C*********************************************               VCS06030
      IF(W(L).GT.0.)GO TO 1216                               VCS06040
      FE(L)=FF(L)                                            VCS06050
      GO TO 1919                                             VCS06060
 1216 IF(SI(L))1213,1919,1215                                VCS06070
 1213 FE(L)=FF(L)+ALOG(W(L)/TL)                              VCS06080
      GO TO 1919                                             VCS06090
 1215 FE(L)=FF(L)+ALOG(W(L)/TG)                              VCS06100
 1919 CALL DELTAG(0,I,I)                                     VCS06110
   19 I=I+1                                                  VCS06120
      IF(I.LE.NR)GO TO 1920                                  VCS06130
      IM=JM.EQ.NR                                            VCS06140
  212 IF(IM)GO TO 227                                        VCS06150
C*********************************************               VCS06160
C**** EQUILIBRIUM CHECK FOR MAJOR SPECIES ****               VCS06170
C*********************************************               VCS06180
  222 DO 15 I=1,NR                                           VCS06190
      IF(IC(I).LE.0.OR.ABS(DG(I)).LE.1.E-06)GO TO 15         VCS06200
      IF(IT.LT.MAXIT)GO TO 3857                              VCS06210
      ICONV=.TRUE.                                           VCS06220
      GO TO 857                                              VCS06230
 3857 ILT=.FALSE.                                            VCS06240
      GO TO 93                                               VCS06250
   15 CONTINUE                                               VCS06260
C*********************************************               VCS06270
C**** EQUILIBRIUM CHECK FOR MINOR SPECIES ****               VCS06280
C*********************************************               VCS06290
      IF(JM.EQ.0)GO TO 96                                    VCS06300
  227 IF(ITI.EQ.0)GO TO 228                                  VCS06310
      CALL DFE(W,0,1)                                        VCS06320
  228 CALL DELTAG(1,1,NR)                                    VCS06330
      DO 315 I=1,NR                                          VCS06340
      IF(IC(I).NE.0.OR.ABS(DG(I)).LE.1.E-05)GO TO 315        VCS06350
      IF(IT.LT.MAXIT)GO TO 2857                              VCS06360
      ICONV=.TRUE.                                           VCS06370
      GO TO 857                                              VCS06380
C*********************************************               VCS06390
C**** CONVERGENCE AMONG MAJORS BUT NOT MINORS ****           VCS06400
C*********************************************               VCS06410
 2857 ILT=.TRUE.                                             VCS06420
      IF(ITI.EQ.0)GO TO 90                                   VCS06430
      GO TO 9090                                             VCS06440
  315 CONTINUE                                               VCS06450
```

```
C*************************************************              VCS06460
C****  EVALUATE FINAL OR INTERMEDIATE VALUES  ****               VCS06470
C****  OF ELEMENTAL ABUNDANCES                ****               VCS06480
C*************************************************              VCS06490
   96 CALL ELAB                                                  VCS06500
      IF(.NOT.LEC)GO TO 1368                                     VCS06510
C*************************************************              VCS06520
C****  ELEMENTAL ABUNDANCE ACCURACY CHECK  ****                  VCS06530
C*************************************************              VCS06540
      DO 1362 I=1,NC                                             VCS06550
      IF(GAI(I).EQ.0.0D0)GO TO 1367                              VCS06560
      IF(ABS(GA(I)-GAI(I)).GT..5D-08*GAI(I))GO TO 1363           VCS06570
      GO TO 1362                                                 VCS06580
 1367 IF(ABS(GA(I)).GT.1.D-10)GO TO 1363                         VCS06590
 1362 CONTINUE                                                   VCS06600
      GO TO 1365                                                 VCS06610
 1368 LEC=.TRUE.                                                 VCS06620
C*****************************************                       VCS06630
C****  CORRECT ELEMENTAL ABUNDANCES  ****                        VCS06640
C*****************************************                       VCS06650
 1363 CALL ELCORR                                                VCS06660
      GO TO 1955                                                 VCS06670
 1365 IF(MR.EQ.M)GO TO 857                                       VCS06680
C***********************************                             VCS06690
C****  RE-CHECK DELETED SPECIES  ****                            VCS06700
C***********************************                             VCS06710
 1366 MR1=MR+1                                                   VCS06720
      NR1=NR+1                                                   VCS06730
      DO 849 I=MR1,M                                             VCS06740
  849 FE(I)=FF(I)                                                VCS06750
      CALL DELTAG(0,NR1,N)                                       VCS06760
      I=NR1                                                      VCS06770
      NPB=0                                                      VCS06780
      XT1=0.                                                     VCS06790
      XT2=0.                                                     VCS06800
      IF(TG.GT.0.)XT1=ALOG(1.E+32*TG)                            VCS06810
      IF(LIQ.AND.TL.GT.0.)XT2=ALOG(1.E+32*TL)                    VCS06820
 1968 L=IR(I)                                                    VCS06830
      IF(SI(L))850,8501,8501                                     VCS06840
 8501 IF(TG.GT.0..AND.DG(I).LT.XT1.OR.TG.EQ.0..AND.DG(I).LT.0.)GO TO 853VCS06850
      GO TO 18531                                                VCS06860
  850 IF(TL.GT.0..AND.DG(I).LT.XT2.OR.TL.EQ.0..AND.DG(I).LT.0.)GO TO 853VCS06870
18531 I=I+1                                                      VCS06880
      IF(I.LE.N)GO TO 1968                                       VCS06890
      IF(NPB)857,857,1969                                        VCS06900
C**************************************                          VCS06910
C****  RE-INSERT DELETED SPECIES  ****                           VCS06920
C**************************************                          VCS06930
  853 W(L)=1.E-20                                                VCS06940
      IF(SI(L))8531,8533,8532                                    VCS06950
 8531 TL=TL+W(MR)                                                VCS06960
      GO TO 8533                                                 VCS06970
 8532 TG=TG+W(MR)                                                VCS06980
 8533 IC(I)=0                                                    VCS06990
```

D.1 Program Listing

```
            JM=JM+1                                                VCS07000
            NR=NR+1                                                VCS07010
            MR=MR+1                                                VCS07020
            NPB=1                                                  VCS07030
            CALL DELETE (L,1,&1366)                                VCS07040
            GO TO 18531                                            VCS07050
       1969 ILT=.TRUE.                                             VCS07060
            CALL DFE (W,0,1)                                       VCS07070
            CALL DELTAG (0,NR1,NR)                                 VCS07080
            GO TO 9090                                             VCS07090
      C*****************************************                  VCS07100
      C**** FINAL DELTA G/RT VALUES ********************           VCS07110
      C*****************************************                  VCS07120
        857 J=MR+1                                                 VCS07130
            K=NR+1                                                 VCS07140
            DO 847 I=1,NR                                          VCS07150
            J=J-1                                                  VCS07160
            K=K-1                                                  VCS07170
        847 DG(J)=DG(K)                                            VCS07180
      C*****************************************                  VCS07190
      C**** EVALUATE FINAL MOLE FRACTIONS ****                     VCS07200
      C*****************************************                  VCS07210
            DO 1810 I=1,MR                                         VCS07220
            IF (SI(I).NE.0) GO TO 7153                             VCS07230
            IF (W(I).EQ.0.) WT(I)=0.                               VCS07240
            IF (W(I).GT.0.) WT(I)=1.                               VCS07250
       7153 IF (SI(I).EQ.1.AND.W(I).NE.0.) WT(I)=W(I)/TG           VCS07260
       1010 IF (SI(I).LT.0.AND.W(I).NE.0.) WT(I)=W(I)/TL           VCS07270
      C*****************************************                  VCS07280
      C**** SORT DEPENDENT SPECIES IN DECREASING ORDER ****        VCS07290
      C*****************************************                  VCS07300
            DO 8475 I=NC1,MR                                       VCS07310
            IR(I)=I                                                VCS07320
       8475 XY(I)=W(I)                                             VCS07330
            DO 8476 L=NC1,MR                                       VCS07340
            CALL AMAX (XY,K,L,MR)                                  VCS07350
            IF (K.EQ.L) GO TO 8476                                 VCS07360
            CALL DSW (XY,K,L)                                      VCS07370
            CALL SWITCH (IR,K,L)                                   VCS07380
       8476 CONTINUE                                               VCS07390
            IF (IPR.EQ.0) RETURN                                   VCS07400
      C*****************************************                  VCS07410
      C**** OUTPUT FORMAT STATEMENTS ****                          VCS07420
      C*****************************************                  VCS07430
        801 FORMAT (14H ITERATIONS = ,I5/32H EVALUATIONS OF STOICHIOMETRY =   VCS07440
           1,I4//1X,7HSPECIES,17X,17HEQUILIBRIUM MOLES,4X,13HMOLE FRACTION,4X,VCS07450
           214HDG/RT REACTION/)                                    VCS07460
        802 FORMAT (1X,3A4,13X,1PD14.7,5X,D14.7,5X,D11.4)          VCS07470
        803 FORMAT (//8H G/RT = ,1PD15.7/22H TOTAL PHASE1 MOLES = ,D11.4)  VCS07480
        804 FORMAT (1X,3A4)                                        VCS07490
        805 FORMAT (1X,3A4,13X,1PD14.7,5X,D14.7)                   VCS07500
        806 FORMAT (22H TOTAL PHASE2 MOLES =    ,1PD11.4)           VCS07510
      C*****************************************                  VCS07520
      C**** PRINT OUT RESULTS ****                                 VCS07530
```

```
C****************************                              VCS07540
      WRITE (6,5138)                                        VCS07550
 5138 FORMAT (///)                                          VCS07560
      IF (ICONV) WRITE (6,5139)                             VCS07570
      WRITE (6,801) IT,NOPT                                 VCS07580
 5139 FORMAT (37H CONVERGENCE CRITERION NOT SATISFIED.)     VCS07590
 1809 WRITE (6,805) ((SP(J,I),J=1,3),W(I),WT(I),I=1,NC)     VCS07600
      DO 18990 I=NC1,MR                                     VCS07610
      L=IR(I)                                               VCS07620
18990 WRITE (6,802) (SP(J,L),J=1,3),W(L),WT(L),DG(L)        VCS07630
      IF (MR.EQ.M) GO TO 1873                               VCS07640
      WRITE (6,9074)                                        VCS07650
 9074 FORMAT (/23H LESS THAN 1.E-32 MOLES/)                 VCS07660
      MR1=MR+1                                              VCS07670
      WRITE (6,804) ((SP(J,I),J=1,3),I=MR1,M)               VCS07680
 1873 G=0.                                                  VCS07690
      IF (TG.GT.0.0.AND.TING.GT.0.0) G=G+TING*ALOG(TING/TG) VCS07700
      IF (LIQ.AND.TINL.GT.0.0) G=G+TINL*ALOG(TINL/TL)       VCS07710
      DO 1847 I=1,MR                                        VCS07720
 1847 G=G+W(I)*FE(I)                                        VCS07730
  374 WRITE (6,803) G,TG                                    VCS07740
      IF (NL.GT.0) WRITE (6,806) TL                         VCS07750
      WRITE (6,911) (E(I),GA(I),I=1,NE)                     VCS07760
      IF (NRUNS.NE.1) RETURN                                VCS07770
C********************************                          VCS07780
C****  RE-ARRANGE INPUT DATA ****                          VCS07790
C********************************                          VCS07800
      DO 1375 I=1,M                                         VCS07810
      DO 1378 J=I,M                                         VCS07820
      L=IND(J)                                              VCS07830
      K1=J                                                  VCS07840
      IF (L.EQ.I) GO TO 1379                                VCS07850
 1378 CONTINUE                                              VCS07860
 1379 CALL DSW (W,I,K1)                                     VCS07870
      CALL DSW (DA,I,K1)                                    VCS07880
      CALL DSW (WT,I,K1)                                    VCS07890
      CALL DSW2 (SP,I,K1)                                   VCS07900
      CALL SWITCH (SI,I,K1)                                 VCS07910
      CALL SWITCH (IND,I,K1)                                VCS07920
      DO 1376 J=1,NE                                        VCS07930
 1376 CALL DSW (BM(1,J),I,K1)                               VCS07940
 1375 CONTINUE                                              VCS07950
      DO 1377 I=1,M                                         VCS07960
      IF (SI(I).EQ.(-1)) SI(I)=2                            VCS07970
 1377 FF(I)=DA(I)                                           VCS07980
 9999 CONTINUE                                              VCS07990
      RETURN                                                VCS08000
      END                                                   VCS08010
C*****************************************************     VCS08020
C****  REMOVES OR ADDS A SPECIES TO CALCULATIONS ****      VCS08030
C*****************************************************     VCS08040
      SUBROUTINE DELETE (L,MM,*)                            VCS08050
      IMPLICIT REAL*8 (A-H,O-Z)                             VCS08060
      COMMON /STOICH/SC(15,149),FE(150),FF(150),W(150),DNG(150),DNL(150) VCS08070
```

D.1 Program Listing

```
      1,WT(150),DG(150),DS(150),FEL(150),GA(15),GAI(15),TG1,TL1,TG,TL,DTG    VCS08080
      2,DTL,BM(150,15),T,P,TING,TINL,IF,IEST,DA(150),IND(150),SP(3,150),IV   VCS08090
      3C(150),IR(150),SI(150),M,N,NC,NR,MR,NE                                VCS08100
       INTEGER SP,SI                                                         VCS08110
       LOGICAL LIQ                                                           VCS08120
16700  NNL=L-NC                                                              VCS08130
       IF(L.EQ.MR)GO TO 16796                                                VCS08140
C******************************************                                  VCS08150
C**** RE-ARRANGE DATA WHEN SPECIE ADDED OR REMOVED ****                      VCS08160
C******************************************                                  VCS08170
       CALL DSW(WT,MR,L)                                                     VCS08180
       CALL DSW(W,MR,L)                                                      VCS08190
       CALL DSW2(SP,MR,L)                                                    VCS08200
       CALL DSW(FF,MR,L)                                                     VCS08210
       IF(SI(MR).EQ.0)CALL DSW(FE,MR,L)                                      VCS08220
       CALL SWITCH(SI,MR,L)                                                  VCS08230
       CALL DSW(FEL,MR,L)                                                    VCS08240
       CALL SWITCH(IC,NR,NNL)                                                VCS08250
       CALL DSW(DA,MR,L)                                                     VCS08260
       CALL SWITCH(IND,MR,L)                                                 VCS08270
       DO 16701 K=1,NC                                                       VCS08280
       T1=SC(K,NR)                                                           VCS08290
       SC(K,NR)=SC(K,NNL)                                                    VCS08300
16701  SC(K,NNL)=T1                                                          VCS08310
       CALL DSW(DNG,NR,NNL)                                                  VCS08320
       IF(LIQ)CALL DSW(DNL,NR,NNL)                                           VCS08330
       DO 3030 J=1,NE                                                        VCS08340
       CALL DSW(BM(1,J),MR,L)                                                VCS08350
3030   CONTINUE                                                              VCS08360
16796  IF(MM.NE.0)RETURN                                                     VCS08370
C******************************************                                  VCS08380
C**** EXTRA PROCEDURES WHEN REMOVING A SPECIES ****                          VCS08390
C******************************************                                  VCS08400
       CALL DSW(DG,NR,NNL)                                                   VCS08410
       CALL DSW(DS,MR,L)                                                     VCS08420
C******************************************                                  VCS08430
C**** SPECIAL PROCEDURES FOR DELETION OF PHASE ****                          VCS08440
C******************************************                                  VCS08450
       IF(SI(MR))16798,16797,16799                                           VCS08460
16798  TL=TL-W(MR)                                                           VCS08470
       GO TO 16797                                                           VCS08480
16799  TG=TG-W(MR)                                                           VCS08490
16797  NR=NR-1                                                               VCS08500
       MR=MR-1                                                               VCS08510
       IF(NR.EQ.0)RETURN1                                                    VCS08520
       RETURN                                                                VCS08530
       END                                                                   VCS08540
       SUBROUTINE FORCE(ITI,FORCED)                                          VCS08550
C*****************************                                               VCS08560
C**** CONVERGENCE FORCER ****                                                VCS08570
C*****************************                                               VCS08580
       IMPLICIT REAL*8 (A-H,O-Z)                                             VCS08590
       COMMON /STOICH/SC(15,149),FE(150),FF(150),W(150),DNG(150),DNL(150)    VCS08600
      1,WT(150),DG(150),DS(150),FEL(150),GA(15),GAI(15),TG1,TL1,TG,TL,DTG    VCS08610
```

```
      2,DTL,BM(150,15),T,P,TING,TINL,IF,IEST,DA(150),IND(150),SP(3,150),IVCS08620
      3C(150),IR(150),SI(150),M,N,NC,NR,MR,NE                            VCS08630
      INTEGER SI,SP                                                       VCS08640
      COMMON /FORCER/LIQ                                                  VCS08650
      LOGICAL LIQ,FORCED                                                  VCS08660
C*****************************************                               VCS08670
C**** CALCULATE SLOPE AT END OF STEP ****                                VCS08680
C*****************************************                               VCS08690
13700 S2=0.                                                               VCS08700
      DO 13702 I=1,MR                                                     VCS08710
13702 IF(DS(I).NE.0.)S2=S2+FE(I)*DS(I)                                    VCS08720
      IF(S2.LE.0.)RETURN                                                  VCS08730
C***********************************                                      VCS08740
C**** CALCULATE ORIGINAL SLOPE ****                                       VCS08750
C***********************************                                      VCS08760
      S1=0.                                                               VCS08770
      DO 13704 I=1,MR                                                     VCS08780
13704 IF(DS(I).NE.0.)S1=S1+FEL(I)*DS(I)                                   VCS08790
      IF(S1.GE.0.)RETURN                                                  VCS08800
C**********************                                                   VCS08810
C**** FIT PARABOLA ****                                                   VCS08820
C**********************                                                   VCS08830
      AL=S1/(S1-S2)                                                       VCS08840
      IF(AL.GE..95)RETURN                                                 VCS08850
C*********************************************                           VCS08860
C**** ADJUST MOLE NUMBERS, CHEM. POT. ****                                VCS08870
C*********************************************                           VCS08880
      DO 13705 I=1,MR                                                     VCS08890
13705 IF(DS(I).NE.0.)W(I)=W(I)+AL*DS(I)                                   VCS08900
      TG=TG+AL*DTG                                                        VCS08910
      IF(LIQ)TL=TL+AL*DTL                                                 VCS08920
      CALL DFE(W,0,ITI)                                                   VCS08930
      FORCED=.TRUE.                                                       VCS08940
      RETURN                                                              VCS08950
      END                                                                 VCS08960
      SUBROUTINE ST2(SOLDEL,*)                                            VCS08970
C*********************************************                           VCS08980
C**** CALCULATES REACTION ADJUSTMENTS ****                                VCS08990
C*********************************************                           VCS09000
      IMPLICIT REAL*8 (A-H,O-Z)                                           VCS09010
      COMMON /STOICH/SC(15,149),FE(150),FF(150),W(150),DNG(150),DNL(150)VCS09020
     1,WT(150),DG(150),DS(150),FEL(150),GA(15),GAI(15),TG1,TL1,TG,TL,DTGVCS09030
     2,DTL,BM(150,15),T,P,TING,TINL,IF,IEST,DA(150),IND(150),SP(3,150),IVCS09040
     3C(150),IR(150),SI(150),M,N,NC,NR,MR,NE                              VCS09050
      INTEGER SI,SP                                                       VCS09060
      COMMON /FORCER/LIQ                                                  VCS09070
      LOGICAL LIQ,SOLDEL                                                  VCS09080
      ABS(X)=DABS(X)                                                      VCS09090
14700 SOLDEL=.FALSE.                                                      VCS09100
      DO 14701 I=1,NR                                                     VCS09110
      L=IR(I)                                                             VCS09120
      IF(W(L).EQ.0..AND.SI(L).NE.0)GO TO 47101                            VCS09130
      IF(ABS(DG(I)).LE.1.D-6)GO TO 14701                                  VCS09140
      IF(IC(I).LE.0.AND.DG(I).GE.0.)GO TO 14701                           VCS09150
```

D.1 Program Listing

```
           IF(SI(L).NE.0)GO TO 14703                    VCS09160
           S=0.                                         VCS09170
           GO TO 14711                                  VCS09180
14703      S=1./W(L)                                    VCS09190
14711      DO 14704 J=1,NC                              VCS09200
14704      IF(SI(J).NE.0)S=S+SC(J,I)**2/W(J)            VCS09210
           IF(TG.GT.0.)S=S-DNG(I)**2/TG                 VCS09220
           IF(TL.GT.0.)S=S-DNL(I)**2/TL                 VCS09230
           IF(S.EQ.0.)GO TO 47103                       VCS09240
           DS(L)=-DG(I)/S                               VCS09250
           GO TO 14701                                  VCS09260
C****************************************************  VCS09270
C**** REACTION ENTIRELY AMONG CONDENSED PHASES ****    VCS09280
C**** DELETE ONE SOLID AND RECOMPUTE BASIS     ****    VCS09290
C****************************************************  VCS09300
47103      CONTINUE                                     VCS09310
           IF(DG(I).LE.0.)GO TO 47104                   VCS09320
           DSS=W(L)                                     VCS09330
           K=L                                          VCS09340
           DO 47106 J=1,NC                              VCS09350
           IF(SC(J,I).LE.0.)GO TO 47106                 VCS09360
           XX=W(J)/SC(J,I)                              VCS09370
           IF(XX.GE.DSS)GO TO 47106                     VCS09380
           DSS=XX                                       VCS09390
           K=J                                          VCS09400
47106      CONTINUE                                     VCS09410
           DSS=-DSS                                     VCS09420
           GO TO 47109                                  VCS09430
47104      DSS=1.E10                                    VCS09440
           DO 47105 J=1,NC                              VCS09450
           IF(SC(J,I).GE.0.)GO TO 47105                 VCS09460
           XX=-W(J)/SC(J,I)                             VCS09470
           IF(XX.GE.DSS)GO TO 47105                     VCS09480
           DSS=XX                                       VCS09490
           K=J                                          VCS09500
47105      CONTINUE                                     VCS09510
47109      CONTINUE                                     VCS09520
           IF(DSS.EQ.0.)GO TO 14701                     VCS09530
           W(L)=W(L)+DSS                                VCS09540
           DO 47108 J=1,NC                              VCS09550
47108      W(J)=W(J)+DSS*SC(J,I)                        VCS09560
           W(K)=0.                                      VCS09570
           IF(K.NE.L)SOLDEL=.TRUE.                      VCS09580
           GO TO 14701                                  VCS09590
C******************************************            VCS09600
C**** MULTISPECIES PHASE WITH NT=0 *****                VCS09610
C******************************************            VCS09620
47101      IF(DG(I).GE.-1.E-4)GO TO 47102               VCS09630
           DS(L)=1.E-10                                 VCS09640
           IC(I)=1                                      VCS09650
           GO TO 14701                                  VCS09660
47102      DS(L)=0.                                     VCS09670
14701      CONTINUE                                     VCS09680
           IF(SOLDEL)RETURN1                            VCS09690
```

```
              RETURN                                              VCS09700
              END                                                 VCS09710
C****************************************                        VCS09720
C****  SUBROUTINE TO             *****                            VCS09730
C****  CORRECT   ELEMENTAL ABUNDANCES ****                        VCS09740
C****************************************                        VCS09750
              SUBROUTINE ELCORR                                   VCS09760
              IMPLICIT REAL*8 (A-H,O-Z)                           VCS09770
              COMMON /STOICH/SC(15,149),FE(150),FF(150),W(150),DNG(150),DNL(150)VCS09780
             1,WT(150),DG(150),DS(150),FEL(150),GA(15),GAI(15),TG1,TL1,TG,TL,DTGVCS09790
             2,DTL,BM(150,15),T,P,TING,TINL,IF,IEST,DA(150),IND(150),SP(3,150),IVCS09800
             3C(150),IR(150),SI(150),M,N,NC,NR,MR,NE               VCS09810
              INTEGER SI,SP                                       VCS09820
              DIMENSION X(16),XY(150),AA(15,15),YY(15,1)           VCS09830
              EQUIVALENCE (X,YY)                                  VCS09840
        18700 DO 18701 I=1,NC                                     VCS09850
              X(I)=GA(I)-GAI(I)                                   VCS09860
              CALL UNPACK (BM,I,NC,XY)                            VCS09870
              DO 18701 J=1,NC                                     VCS09880
        18701 AA(J,I)=XY(J)                                       VCS09890
              CALL MLEQU (AA,15,NC,YY,1)                          VCS09900
              PAR=.5                                              VCS09910
              DO 18702 I=1,NC                                     VCS09920
              XX=-X(I)/W(I)                                       VCS09930
        18702 IF (PAR.LT.XX) PAR=XX                                VCS09940
              PAR=1./PAR                                          VCS09950
              IF (PAR.LE.1.01.AND.PAR.GT.0.)GO TO 18729            VCS09960
              PAR=1.                                              VCS09970
              GO TO 18759                                         VCS09980
        18729 PAR=.99*PAR                                         VCS09990
        18759 DO 18704 I=1,NC                                     VCS10000
        18704 W(I)=W(I)+PAR*X(I)                                  VCS10010
              RETURN                                              VCS10020
              END                                                 VCS10030
              SUBROUTINE DELTAG (L,J,KP)                          VCS10040
C****************************************************            VCS10050
C****  CALCULATES REACTION FREE ENERGY CHANGES ****               VCS10060
C****************************************************            VCS10070
              IMPLICIT REAL*8 (A-H,O-Z)                           VCS10080
              COMMON /STOICH/SC(15,149),FE(150),FF(150),W(150),DNG(150),DNL(150)VCS10090
             1,WT(150),DG(150),DS(150),FEL(150),GA(15),GAI(15),TG1,TL1,TG,TL,DTGVCS10100
             2,DTL,BM(150,15),T,P,TING,TINL,IF,IEST,DA(150),IND(150),SP(3,150),IVCS10110
             3C(150),IR(150),SI(150),M,N,NC,NR,MR,NE               VCS10120
              INTEGER SI,SP                                       VCS10130
              EXP(X)=DEXP(X)                                      VCS10140
        15700 IF(L)15710,15711,15712                              VCS10150
C*********************                                            VCS10160
C****  MAJORS ONLY ****                                           VCS10170
C*********************                                            VCS10180
        15710 DO 15701 K=1,NR                                     VCS10190
              IF(IC(K).EQ.0)GO TO 15701                           VCS10200
              LL=IR(K)                                            VCS10210
              DG(K)=FE(LL)                                        VCS10220
              DO 15713 LL=1,NC                                    VCS10230
```

D.1 Program Listing

```
15713 DG(K)=DG(K)+SC(LL,K)*FE(LL)                              VCS10240
15701 CONTINUE                                                 VCS10250
      GO TO 15715                                              VCS10260
C***********************                                       VCS10270
C**** ALL REACTIONS ****                                       VCS10280
C***********************                                       VCS10290
15711 DO 15702 K=J,KP                                          VCS10300
      LL=IR(K)                                                 VCS10310
      DG(K)=FE(LL)                                             VCS10320
      DO 15702 LL=1,NC                                         VCS10330
15702 DG(K)=DG(K)+SC(LL,K)*FE(LL)                              VCS10340
      GO TO 15715                                              VCS10350
C*********************                                         VCS10360
C**** MINORS ONLY ****                                         VCS10370
C*********************                                         VCS10380
15712 DO 15703 K=1,NR                                          VCS10390
      IF(IC(K).GT.0)GO TO 15703                                VCS10400
      LL=IR(K)                                                 VCS10410
      DG(K)=FE(LL)                                             VCS10420
      DO 15714 LL=1,NC                                         VCS10430
15714 DG(K)=DG(K)+SC(LL,K)*FE(LL)                              VCS10440
15703 CONTINUE                                                 VCS10450
C*******************************************                   VCS10460
C**** MULTISPECIES PHASES WITH NT=0 *****                      VCS10470
C*******************************************                   VCS10480
15715 CONTINUE                                                 VCS10490
      SDEL1=0.                                                 VCS10500
      SDEL2=0.                                                 VCS10510
      DO 15716 I=1,NR                                          VCS10520
      LL=IR(I)                                                 VCS10530
      IF(W(LL).NE.0.)GO TO 15716                               VCS10540
      IF(DG(I).GT.50.)DG(I)=50.                                VCS10550
      IF(DG(I).LT.-50.)DG(I)=-50.                              VCS10560
      IF(SI(LL).EQ.1)SDEL1=SDEL1+EXP(-DG(I))                   VCS10570
      IF(SI(LL).EQ.-1)SDEL2=SDEL2+EXP(-DG(I))                  VCS10580
15716 CONTINUE                                                 VCS10590
      DO 15717 I=1,NR                                          VCS10600
      LL=IR(I)                                                 VCS10610
      IF(W(LL).NE.0.)GO TO 15717                               VCS10620
      IF(SI(LL).EQ.1)DG(I)=1.-SDEL1                            VCS10630
      IF(SI(LL).EQ.-1)DG(I)=1.-SDEL2                           VCS10640
15717 CONTINUE                                                 VCS10650
      RETURN                                                   VCS10660
      END                                                      VCS10670
      SUBROUTINE BASOPT(IFIRST,NOPT,AW,SA,SM,SS,AA,TEST,IT,CONV) VCS10680
C***************************************************************VCS10690
C**** CHOOSES OPTIMUM BASIS, CALCULATES STOICHIOMETRY ****      VCS10700
C***************************************************************VCS10710
      IMPLICIT REAL*8 (A-H,O-Z)                                VCS10720
      INTEGER SP,SI,XM                                         VCS10730
      COMMON /STOICH/SC(15,149),FE(150),FF(150),W(150),DNG(150),DNL(150)VCS10740
     1,WT(150),DG(150),DS(150),FEL(150),GA(15),GAI(15),TG1,TL1,TG,TL,DTGVCS10750
     2,DTL,BM(150,15),T,P,TING,TINL,IF,IEST,DA(150),IND(150),SP(3,150),IVCS10760
     3C(150),IR(150),SI(150),M,N,NC,NR,MR,NE                   VCS10770
```

```
      LOGICAL IFIRST,LINDEP                                VCS10780
      LOGICAL CONV                                         VCS10790
      DIMENSION AW(150),SA(15),SM(15,15),SS(15),AA(15,15)  VCS10800
      ABS(X)=DABS(X)                                       VCS10810
      CONV=.FALSE.                                         VCS10820
12700 NOPT=NOPT+1                                          VCS10830
      DO 12701 I=1,MR                                      VCS10840
12701 AW(I)=W(I)                                           VCS10850
      JR=0                                                 VCS10860
12750 JR=JR+1                                              VCS10870
C***************************************************      VCS10880
C****  DETERMINE LARGEST REMAINING MOLE NUMBER ****        VCS10890
C***************************************************      VCS10900
12796 CALL AMAX(AW,K,JR,MR)                                VCS10910
      IF(AW(K).EQ.0)CONV=.TRUE.                            VCS10920
      IF(CONV) RETURN                                      VCS10930
        IF(AW(K).NE.TEST)GO TO 12759                       VCS10940
        NC=JR-1                                            VCS10950
        N=M-NC                                             VCS10960
        NR=N                                               VCS10970
        DO12760 I=1,M                                      VCS10980
12760   IR(I)=NC+I                                         VCS10990
        GO TO 12758                                        VCS11000
12759   CONTINUE                                           VCS11010
        AW(K)=TEST                                         VCS11020
C***********************************************************   VCS11030
C****  CHECK LINEAR INDEPENDENCE WITH PREVIOUS SPECIES ****     VCS11040
C***********************************************************   VCS11050
C************************************************************************VCS11060
C                                                          VCS11070
C     LOGICAL FUNCTION LINDEP(BM,JR,K,NC)                  VCS11080
C                                                          VCS11090
      JL=JR-1                                              VCS11100
      SA(JR)=0.                                            VCS11110
      CALL UNPACK(BM,K,NC,DG)                              VCS11120
      DO 13748 J=1,NC                                      VCS11130
13748 SM(J,JR)=DG(J)                                       VCS11140
      IF(JL.EQ.0)GO TO 13731                               VCS11150
      DO 13758 J=1,JL                                      VCS11160
      SS(J)=0.                                             VCS11170
      DO 13749 I=1,NC                                      VCS11180
13749 SS(J)=SS(J)+SM(I,JR)*SM(I,J)                         VCS11190
13758 SS(J)=SS(J)/SA(J)                                    VCS11200
      DO 13746 J=1,JL                                      VCS11210
      DO 13746 L=1,NC                                      VCS11220
13746 SM(L,JR)=SM(L,JR)-SS(J)*SM(L,J)                      VCS11230
13731 DO 13747 ML=1,NC                                     VCS11240
13747 IF(ABS(SM(ML,JR)).GT.1.E-17)SA(JR)=SA(JR)+SM(ML,JR)**2  VCS11250
C***************************************************      VCS11260
C****  IF NORM OF NEW ROW .LT. 10**-3, REJECT ****         VCS11270
C***************************************************      VCS11280
      IF(SA(JR).LT.1.E-06)GO TO 13790                      VCS11290
      LINDEP=.FALSE.                                       VCS11300
      GO TO 12791                                          VCS11310
```

D.1 Program Listing

```
13790       LINDEP=.TRUE.                                    VCS11320
      C                                                      VCS11330
      C     END OF LINDEP                                    VCS11340
      C                                                      VCS11350
      C**********************************************************VCS11360
12791 IF (LINDEP) GO TO 12796                                VCS11370
      C*************************                             VCS11380
      C**** RE-ARRANGE DATA ****                             VCS11390
      C*************************                             VCS11400
            CALL DSW (W,JR,K)                                VCS11410
            CALL DSW (WT,JR,K)                               VCS11420
            CALL DSW2 (SP,JR,K)                              VCS11430
            CALL DSW (FF,JR,K)                               VCS11440
            CALL DSW (FE,JR,K)                               VCS11450
            CALL DSW (AW,JR,K)                               VCS11460
            CALL SWITCH (SI,JR,K)                            VCS11470
            CALL DSW (DA,JR,K)                               VCS11480
            CALL DSW (FEL,JR,K)                              VCS11490
            CALL SWITCH (IND,JR,K)                           VCS11500
            DO 2929 J=1,NE                                   VCS11510
            CALL DSW (BM (1,J),JR,K)                         VCS11520
 2929 CONTINUE                                               VCS11530
12758 IF (JR.LT.NC) GO TO 12750                              VCS11540
            IF (IFIRST) RETURN                               VCS11550
      C********************************                      VCS11560
      C**** EVALUATE STOICHIOMETRY ****                      VCS11570
      C********************************                      VCS11580
            DO 12754 J=1,NC                                  VCS11590
            CALL UNPACK (BM,J,NC,DG)                         VCS11600
            DO 12754 I=1,NC                                  VCS11610
12754 AA (I,J) =DG (I)                                       VCS11620
            DO 12753 I=1,N                                   VCS11630
            CALL UNPACK (BM,IR (I),NC,DG)                    VCS11640
            DO 12753 J=1,NC                                  VCS11650
12753 SC (J,I) =DG (J)                                       VCS11660
            CALL MLEQU (AA,15,NC,SC,N)                       VCS11670
            DO 12755 I=1,N                                   VCS11680
            K=IR (I)                                         VCS11690
      C********************************                      VCS11700
      C**** EVALUATE DELTA N VALUES ****                     VCS11710
      C********************************                      VCS11720
            IF (SI (K)) 12763,12762,12761                    VCS11730
12762 DNG (I) =0.                                            VCS11740
            GO TO 12765                                      VCS11750
12763 DNL (I) =1.                                            VCS11760
            DNG (I) =0.                                      VCS11770
            GO TO 12764                                      VCS11780
12761 DNG (I) =1.                                            VCS11790
12765 DNL (I) =0.                                            VCS11800
12764 DO 12755 J=1,NC                                        VCS11810
            IF (ABS (SC (J,I)).LE.1.E-06) SC (J,I) =0.       VCS11820
            IF (SI (J)) 12756,12755,12757                    VCS11830
12756 DNL (I) =DNL (I) +SC (J,I)                             VCS11840
            GO TO 12755                                      VCS11850
```

```
12757 DNG(I)=DNG(I)+SC(J,I)                                          VCS11860
12755 CONTINUE                                                       VCS11870
      RETURN                                                         VCS11880
      END                                                            VCS11890
      SUBROUTINE ELAB                                                VCS11900
C****************************************                           VCS11910
C**** COMPUTES ELEMENTAL ABUNDANCES ****                             VCS11920
C****************************************                           VCS11930
      IMPLICIT REAL*8 (A-H,O-Z)                                      VCS11940
      COMMON /STOICH/SC(15,149),FE(150),FF(150),W(150),DNG(150),DNL(150)VCS11950
     1,WT(150),DG(150),DS(150),FEL(150),GA(15),GAI(15),TG1,TL1,TG,TL,DTGVCS11960
     2,DTL,BM(150,15),T,P,TING,TINL,IF,IEST,DA(150),IND(150),SP(3,150),IVCS11970
     3C(150),IR(150),SI(150),M,N,NC,NR,MR,NE                         VCS11980
      INTEGER SI,XM,SP                                               VCS11990
17700 DO 17702 I=1,NE                                                VCS12000
17702 GA(I)=0.D0                                                     VCS12010
      DO 17703 J=1,NE                                                VCS12020
      DO 17703 I=1,M                                                 VCS12030
17703 GA(J)=GA(J)+BM(I,J)*W(I)                                       VCS12040
      RETURN                                                         VCS12050
      END                                                            VCS12060
C********************************************************           VCS12070
C**** SUBROUTINE TO EVALUATE CHEMICAL POTENTIALS *****               VCS12080
C********************************************************           VCS12090
      SUBROUTINE DFE(Z,KK,LL)                                        VCS12100
      IMPLICIT REAL*8 (A-H,O-Z)                                      VCS12110
      COMMON /STOICH/SC(15,149),FE(150),FF(150),W(150),DNG(150),DNL(150)VCS12120
     1,WT(150),DG(150),DS(150),FEL(150),GA(15),GAI(15),TG1,TL1,TG,TL,DTGVCS12130
     2,DTL,BM(150,15),T,P,TING,TINL,IF,IEST,DA(150),IND(150),SP(3,150),IVCS12140
     3C(150),IR(150),SI(150),M,N,NC,NR,MR,NE                         VCS12150
      INTEGER SI,SP                                                  VCS12160
      COMMON /FORCER/LIQ                                             VCS12170
      LOGICAL LIQ                                                    VCS12180
      DIMENSION Z(1)                                                 VCS12190
      ALOG(X)=DLOG(X)                                                VCS12200
      IF(KK.GT.0)GO TO 402                                           VCS12210
      X=TG                                                           VCS12220
      IF(LIQ)Y=TL                                                    VCS12230
      GO TO 403                                                      VCS12240
  402 X=TG1                                                          VCS12250
      IF(LIQ)Y=TL1                                                   VCS12260
  403 IF(X.GT.0.)X=ALOG(X)                                           VCS12270
      IF(.NOT.LIQ)GO TO 404                                          VCS12280
      IF(Y.GT.0.)Y=ALOG(Y)                                           VCS12290
  404 IF(LL.EQ.0)GO TO 55                                            VCS12300
      L1=NC                                                          VCS12310
      GO TO 56                                                       VCS12320
   55 L1=MR                                                          VCS12330
C************************************                                VCS12340
C**** ALL SPECIES, OR COMPONENTS ****                                VCS12350
C************************************                                VCS12360
   56 DO 1 I=1,L1                                                    VCS12370
      IF(Z(I).NE.0.)GO TO 1111                                       VCS12380
      FE(I)=FF(I)                                                    VCS12390
```

D.1 Program Listing

```
         GO TO 1                                VCS12400
   1111  IF (SI (I)) 11,1,14                    VCS12410
     11  FE (I)=FF (I)-Y+ALOG (Z (I))           VCS12420
         GO TO 1                                VCS12430
     14  FE (I)=FF (I)+ALOG (Z (I))-X           VCS12440
     01  CONTINUE                               VCS12450
         IF (LL) 10,99,12                       VCS12460
C*********************                          VCS12470
C**** MAJORS ONLY ****                          VCS12480
C*********************                          VCS12490
     10  DO 2 I=1,NR                            VCS12500
         IF (IC (I).EQ.0) GO TO 2               VCS12510
         L=IR (I)                               VCS12520
         IF (Z (L).NE.0.) GO TO  1112           VCS12530
         FE (L)=FF (L)                          VCS12540
         GO TO 2                                VCS12550
   1112  IF (SI (L)) 21,2,23                    VCS12560
     21  FE (L)=FF (L)-Y+ALOG (Z (L))           VCS12570
         GO TO 2                                VCS12580
     23  FE (L)=FF (L)+ALOG (Z (L))-X           VCS12590
     02  CONTINUE                               VCS12600
         RETURN                                 VCS12610
C*********************                          VCS12620
C**** MINORS ONLY ****                          VCS12630
C*********************                          VCS12640
     12  DO 3 I=1,NR                            VCS12650
         IF (IC (I).NE.0) GO TO 3               VCS12660
         L=IR (I)                               VCS12670
         IF (Z (L).NE.0.) GO TO 1113            VCS12680
         FE (L)=FF (L)                          VCS12690
         GO TO 3                                VCS12700
   1113  IF (SI (L)) 31,3,32                    VCS12710
     31  FE (L)=FF (L)-Y+ALOG (Z (L))           VCS12720
         GO TO 3                                VCS12730
     32  FE (L)=FF (L)+ALOG (Z (L))-X           VCS12740
     03  CONTINUE                               VCS12750
     99  RETURN                                 VCS12760
         END                                    VCS12770
         SUBROUTINE UNPACK (BM,K,NC,X)          VCS12780
         DIMENSION X (15),BM (150,15)           VCS12790
         DOUBLE PRECISION X,BM                  VCS12800
         DO 3 L=1,NC                            VCS12810
      3  X (L)=BM (K,L)                         VCS12820
         RETURN                                 VCS12830
         END                                    VCS12840
         SUBROUTINE MLEQU (C,IDEM,N,B,M)        VCS12850
         IMPLICIT REAL*8 (A-H,O-Z)              VCS12860
         DIMENSION C (IDEM,IDEM),B (IDEM,M)     VCS12870
         DO 58 I=1,N                            VCS12880
         IF (C (I,I).NE.0.) GO TO 51            VCS12890
         IP1=I+1                                VCS12900
         DO 53 K=IP1,N                          VCS12910
         IF (C (K,I).NE.0.) GO TO 54            VCS12920
     53  CONTINUE                               VCS12930
```

```
      WRITE (6,55)
   55 FORMAT (19H NO UNIQUE SOLUTION)
      CALL EXIT
   54 DO 57 J=I,N
   57 C(I,J)=C(I,J)+C(K,J)
      DO 67 J=1,M
   67 B(I,J)=B(I,J)+B(K,J)
   51 DO 58 L=1,N
      IF (L.EQ.I.OR.C(L,I).EQ.0.) GO TO 58
      R=C(L,I)/C(I,I)
      DO 62 J=I,N
   62 C(L,J)=C(L,J)-C(I,J)*R
      DO 64 J=1,M
   64 B(L,J)=B(L,J)-B(I,J)*R
   58 CONTINUE
      DO 63 I=1,N
      DO 63 J=1,M
   63 B(I,J)=-B(I,J)/C(I,I)
      RETURN
      END
C*****************************************************
C****  SUBROUTINE TO RE-ARRANGE VECTOR ELEMENTS  *****
C*****************************************************
      SUBROUTINE SWITCH (X,J1,J2)
      INTEGER X(1),T
      T=X(J1)
      X(J1)=X(J2)
      X(J2)=T
      RETURN
      END
      SUBROUTINE AMAX (X,K,J,N)
      DIMENSION X(1)
      DOUBLE PRECISION X,T
      K=J
      BIG=X(J)
      DO 1 I=J,N
      IF (X(I).LE.BIG) GO TO 1
      K=I
      BIG=X(I)
   01 CONTINUE
      RETURN
      END
      SUBROUTINE DSW (X,J1,J2)
      DOUBLE PRECISION X,T
      DIMENSION X(1)
      T=X(J1)
      X(J1)=X(J2)
      X(J2)=T
      RETURN
      END
      SUBROUTINE DSW2 (X,J1,J2)
      INTEGER X,T
      DIMENSION X(3,1)
      T=X(1,J1)
```

```
      X(1,J1)=X(1,J2)                                        VCS13480
      X(1,J2)=T                                              VCS13490
      T=X(2,J1)                                              VCS13500
      X(2,J1)=X(2,J2)                                        VCS13510
      X(2,J2)=T                                              VCS13520
      T=X(3,J1)                                              VCS13530
      X(3,J1)=X(3,J2)                                        VCS13540
      X(3,J2)=T                                              VCS13550
      RETURN                                                 VCS13560
      END                                                    VCS13570
      SUBROUTINE INEST (L,IFIRST,NOPT,AW,SA,SM,SS,AA,TEST,IT,NCVCS13580
     11)                                                     VCS13590
C*******************************************                 VCS13600
C**** ESTIMATES EQUILIBRIUM COMPOSITION ****                 VCS13610
C*******************************************                 VCS13620
      IMPLICIT REAL*8 (A-H,O-Z)                              VCS13630
      COMMON /STOICH/SC(15,149),FE(150),FF(150),W(150),DNG(150),DNL(150)VCS13640
     1,WT(150),DG(150),DS(150),FEL(150),GA(15),GAI(15),TG1,TL1,TG,TL,DTGVCS13650
     2,DTL,BM(150,15),T,P,TING,TINL,IF,IEST,DA(150),IND(150),SP(3,150),IVCS13660
     3C(150),IR(150),SI(150),M,N,NC,NR,MR,NE                 VCS13670
      INTEGER SP,SI,XM                                       VCS13680
      COMMON/PORCEP/LIO                                      VCS13690
      LOGICAL LINDEP,IFIRST,LIQ,CONV                         VCS13700
      DIMENSION BB(15),PSOL(150),DSOL(150),CC(150),RW(623),IW(303),AX(L/VCS13710
     1,152),AW(150),CA(15),SM(15,15),SS(15),AA(15,15),FFO(150) VCS13720
      ALOG(X)=DLOG(X)                                        VCS13730
      EXP(X)=DEXP(X)                                         VCS13740
      ABS(X)=DABS(X)                                         VCS13750
11700 IKL=0                                                  VCS13760
      LT=0                                                   VCS13770
      TL1=0.                                                 VCS13780
C*********************************************               VCS13790
C**** SET UP DATA FOR LIN. PROG. ROUTINE ****                VCS13800
C*********************************************               VCS13810
      M1=0                                                   VCS13820
      M2=NE                                                  VCS13830
      IA=17                                                  VCS13840
      DO 11703 I=1,NE                                        VCS13850
11703 BB(I)=GAI(I)                                           VCS13860
      DO 11701 J=1,M                                         VCS13870
      CALL UNPACK (BM,J,NE,DG)                               VCS13880
      DO 11701 I=1,NE                                        VCS13890
11701 AX(I,J)=DG(I)                                          VCS13900
      DO 11702 J=1,M                                         VCS13910
11702 CC(J)=-FF(J)                                           VCS13920
C*********************************                           VCS13930
C**** USE LIN. PROG. ROUTINE ****                            VCS13940
C*********************************                           VCS13950
C                                                            VCS13960
C                                                            VCS13970
C     CALL ROUTINE TO SOLVE MAX(CC*PSOL) SUCH THAT AX*PSOL = BB VCS13980
C                                                            VCS13990
C                                                            VCS14000
      DO 11706 I=1,M                                         VCS14010
```

```
            DS(I)=0.                                            VCS14020
            W(I)=PSOL(I)                                        VCS14030
            IF(PSOL(I).EQ.0.)W(I)=1.D-10                        VCS14040
11706   CONTINUE                                                VCS14050
            DO 11707 I=1,M                                      VCS14060
11707   WT(I)=W(I)                                              VCS14070
            CALL BASOPT(.FALSE.,NOPT,AW,SA,SM,SS,AA,TEST,IT,CONV) VCS14080
C****************************************************           VCS14090
C****   CALCULATE TOTAL GASEOUS AND LIQUID MOLES, ****           VCS14100
C****   CHEMICAL POTENTIALS OF BASIS                ****        VCS14110
C****************************************************           VCS14120
11717   TG1=TING                                                VCS14130
            IF(LIQ)TL1=TINL                                     VCS14140
            DO 11711 I=1,NC                                     VCS14150
            IF(SI(I).GT.0)TG1=TG1+WT(I)                         VCS14160
11711   IF(SI(I).LT.0)TL1=TL1+WT(I)                             VCS14170
            DO 11712 I=1,NC                                     VCS14180
            IF(SI(I).EQ.1)FE(I)=FF(I)+ALOG(WT(I)/TG1)           VCS14190
11712   IF(SI(I).LT.0)FE(I)=FF(I)+ALOG(WT(I)/TL1)               VCS14200
            DO 11725 I=NC1,M                                    VCS14210
11725   FE(I)=FF(I)                                             VCS14220
            CALL DELTAG(0,1,N)                                  VCS14230
C*****************************************                      VCS14240
C****   ESTIMATE REACTION ADJUSTMENTS ****                      VCS14250
C*****************************************                      VCS14260
11726   DTG=0.                                                  VCS14270
            IF(LIQ)DTL=0.                                       VCS14280
            IF(TG1.EQ.0.)GO TO 11753                            VCS14290
            XT1=ALOG(1.E+32*TG1)                                VCS14300
            XT2=ALOG(1.E-32*TG1)                                VCS14310
11753   IF(TL1.EQ.0.)GO TO 11754                                VCS14320
            XT3=ALOG(1.E+32*TL1)                                VCS14330
            XT4=ALOG(1.E-32*TL1)                                VCS14340
11754   DO 11718 I=1,NR                                         VCS14350
            L=IR(I)                                             VCS14360
            IF(SI(L))11721,11718,11720                          VCS14370
11720   IF(DG(I).GE.XT1.OR.DG(I).LE.XT2)GO TO 11718             VCS14380
            DS(L)=TG1*EXP(-DG(I))                               VCS14390
            GO TO 11733                                         VCS14400
11721   IF(DG(I).GE.XT3.OR.DG(I).LE.XT4)GO TO 11718             VCS14410
            DS(L)=TL1*EXP(-DG(I))                               VCS14420
11733   DO 11724 K=1,NC                                         VCS14430
11724   DS(K)=DS(K)+SC(K,I)*DS(L)                               VCS14440
            DTG=DTG+DNG(I)*DS(L)                                VCS14450
            IF(LIQ)DTL=DTL+DNL(I)*DS(L)                         VCS14460
11718   CONTINUE                                                VCS14470
C*****************************************                      VCS14480
C****   KEEP BASIS SPECIES POSITIVE ****                        VCS14490
C*****************************************                      VCS14500
            PAR=.5                                              VCS14510
            DO 11786 I=1,NC                                     VCS14520
11786   IF(PAR.LT.(-DS(I)/WT(I)))PAR=-DS(I)/WT(I)               VCS14530
            PAR=1./PAR                                          VCS14540
            IF(PAR.LE.1..AND.PAR.GT.0.)GO TO 11729              VCS14550
```

D.1 Program Listing

```
            PAR=1.
            GO TO 11788
11729 PAR=.8*PAR
C******************************
C**** CALCULATE NEW MOLE NUMBERS ****
C******************************
11788 DO 11794 I=1,NC
11794 W(I)=WT(I)+PAR*DS(I)
      DO 11791 I=NC1,M
11791 IF(DS(I).NE.0.)W(I)=DS(I)*PAR
      TG=TG1+DTG*PAR
      IF(LIQ)TL=TL1+DTL*PAR
      IF(LT.GT.0)RETURN
C******************************
C**** CONVERGENCE-FORCING SECTION ****
C******************************
      CALL DFE(W,0,0)
      S=0.
      DO 11731 I=1,MR
11731 IF(DS(I).NE.0.)S=S+DS(I)*FE(I)
      IF(S)11773,494,11735
11773 IF(IKL)494,494,11738
11735 IF(IKL.GT.0)GO TO 11738
C******************************
C**** TRY HALF STEP SIZE ****
C******************************
      S1=S
      PAR=.5*PAR
      IKL=1
      GO TO 11788
C******************************
C**** FIT PARABOLA THRO HALF AND FULL STEPS ****
C******************************
11738 XL=.5*(1.-S/(S1-S))
      IF(XL.GE.0.)GO TO 17138
C******************************
C**** POOR DIRECTION, REDUCE STEP SIZE TO .2 ****
C******************************
      PAR=PAR*.2
      GO TO 11740
17138 IF(XL.LE.1.)GO TO 11742
C******************************
C**** TOO BIG A STEP, TAKE ORIGINAL FULL STEP ****
C******************************
      PAR=PAR*2.
      GO TO 11740
C******************************
C**** ACCEPT RESULT OF FORCER ****
C******************************
11742 PAR=PAR*2.*XL
11740 LT=1
      GO TO 11788
  494 RETURN
      END
```

D.2 USER'S GUIDE

Data cards:

Card	Column	Variable Name	Input Format	Description
1	1-3	NRUNS	I3	Number of problems to be run
2	1-3	M	I3	Number of species
	4-6	NE	I3	Number of elements
	7-9	NS1	I3	Number of single-species phases
	10-12	NL1	I3	Number of phase2 species
	13-15	IF	I3	Type of chem. pot. data: −1 Kcal/mole
				0 MU/RT
				1 kJ/mole
	16-18	IEST	I3	Initial estimate: 0 User est.
				1 Machine est.
	19-21	ICE	I3	Type of formula vectors: 0 integer entries
				−1 non-integer entries

Cards 3,4,...,M+2 are data cards for each species, as follows:

	Column	Variable Name	Input Format	Description
	1-12	SP	3A4	Species name
	13			Blank
	14-43	BM	15F2.0	Formula vector
	44	SI	I1	Type of phase: 0 single-species
				1 multi-species(gas)
				2 multi-species(liquid)
	45-53	FFO	F9.3	Chemical potential

If the formula vector has non-integer entries, the information on each of cards 3 to M+2 must be coded on two cards as follows:

Card	Column	Variable	Format	Description
3a	1-12	SP	3A4	Species name

D.2 User's Guide

	13			Blank
	14	SI	I1	Type of phase
	15-23	FF0	F9.3	Chemical potential
3b	1-5	BM(1)	F5.3	1st entry of formula vector
	6-10	BM(2)	F5.3	2nd entry of formula vector
				etc. for columns 11-75
M+3	1-10	W(1)	E10.4	Initial estimate for 1st species
	11-20	W(2)	E10.4	Initial estimate for 2nd species
				etc. for columns 21-80
M+4	1-10	GAI(1)	E10.4	Elemental abundance for 1st species
	11-20	GAI(2)	E10.4	Elemental abundance for 2nd species
				etc. for columns 21-80
M+5	1-10	T	E10.3	Temperature in K
	11-20	P	E10.3	Pressure in atm
	21-30	TING	E10.3	Total moles of inert gas
	31-40	TINL	E10.3	Total moles of inert liquid
M+6	1-2	E(1)	A2	Name of first element
	3-4	E(2)	A2	Name of second element
				etc. for columns 5-30
M+7	1-80	TI	80A1	Arbitrary title

When ion problems are run and an electronic charge is read as one of the elements, it must be placed last in the formula vector. If the user makes his own initial estimate it must satisfy the equation $\Sigma\, BM(I,J) * W(I) = GAI(J)$.

Subroutine arguments:

In the statement CALL VCS(IRP,IPR,IP1,MAXIT) the first three arguments control the input and output. If IRP = 0, then no data are read; otherwise IRP = 1. If IPR = 0, then neither input data nor results are printed; otherwise IPR = 1. If IP1 = 1, then intermediate results are printed at each iteration; otherwise IP1 = 0. MAXIT is the maximum number of iterations allowed if convergence is not reached.

LIST OF SYMBOLS*

LATIN LETTERS

a	parameter in Redlich-Kwong equation of state
a_i	activity of species i
a_{ki}	subscript to kth element in molecular formula of species i
a_{li}	subscript to isotope l of an element in molecular formula of species i (Section 9.8)
$\mathbf{a}_i$	formula vector of species i; entry k is a_{ki}
atm	atmosphere(s), unit of pressure
A	Helmholtz function $(U - TS)$
A	accessible
A_i	symbolic designation of species i, including molecular formula, possibly together with structural and phase designations
$\mathbf{A}$	formula matrix: the $(M \times N)$ matrix whose columns are the N formula vectors; that is, $\mathbf{A} = (\mathbf{a}_1, \mathbf{a}_2, \ldots, \mathbf{a}_N)$; entry (k, i) of $\mathbf{A}$ is a_{ki}
$\mathbf{A}_1$	part of $\mathbf{A}$ referring to multispecies phase(s)
$\mathbf{A}_2$	part of $\mathbf{A}$ referring to single-species phase(s)
$\mathbf{A}^*$	unit matrix form of $\mathbf{A}$ or compatible formula matrix for $\mathbf{N}$
$\mathbf{A}'$	modified formula matrix
b	parameter in Redlich-Kwong equation of state
b_k	moles of element k in basis amount of chemical system
$\mathbf{b}$	element-abundance vector with entries b_k
$\mathbf{b}^*$	element-abundance vector corresponding to $\mathbf{A}^*$
$\mathbf{b}'$	modified element-abundance vector
B	second virial coefficient
BNR	Brinkley-NASA-RAND (algorithm)
$\bar{c}_{Pi}$	partial molar heat capacity of species i
C	rank ($\mathbf{A}$) (number of components); third virial coefficient
C_i	molarity of species i

*Symbols used only once, particularly in Chapter 7, are not included in this list.

Latin Letters

C_P	heat capacity at constant pressure
C_V	heat capacity at constant volume
ΔC_P^*	standard heat capacity change/of reaction
D	diagonal matrix
e	mole of electrons
emf	electromotive force
E	emf of chemical cell
$E°$	standard emf of cell or electrode process
E_k	symbolic designation of element k
EIE	equilibrium isotope effect (Section 9.8)
f	fugacity; fractional conversion (of reactant); scalar function
f	vector function
F	Faraday constant, 96,487 coulombs (mole electrons)$^{-1}$
F_s	number of stoichiometric degrees of freedom
g	gas phase (as phase designation)
g^E	molar excess free energy
g	vector function
G	Gibbs function or free energy ($H - TS$)
$G_T°$	standard Gibbs function at T (function of T)
ΔG	free-energy change/of reaction, $\Sigma_i \nu_i \mu_i$
$\Delta G°$	standard free-energy change/of reaction (function of T)
ΔG^*	standard free-energy change/of reaction (function of T and P)
$\Delta G_f°$	standard free energy of formation
h	molar enthalpy; number of species in a multispecies phase
$\bar{h}_i$	partial molar enthalpy of species i
h^E	molar excess enthalpy
h	vector function
H	enthalpy function ($U + PV$)
$H_T°$	standard enthalpy at T (function of T)
ΔH	enthalpy change/of reaction, $\Sigma_i \nu_i \bar{h}_i$
$\Delta H_c°$	standard enthalpy of combustion (function of T)
$\Delta H_f°$	standard enthalpy of formation (function of T)
$\Delta H_s°$	standard enthalpy of solution (function of T)
$\Delta H_T°$	standard enthalpy change/of reaction at T (function of T)
ΔH^*	standard enthalpy change/of reaction (function of T and P)
H	Hessian matrix; the Hessian of the scalar function $f(\mathbf{x})$ is the matrix with elements $\partial^2 f / \partial x_i \, \partial x_j$

I	ionic strength; integration constant; substance capable of existence in two or more isomeric forms
I	unit (identity) matrix
IO	inner-outer (iterations)
J	joules; kJ, kilojoules
J	Jacobian matrix; the Jacobian of the vector function $\mathbf{f}(\mathbf{x})$ is the matrix with elements $\partial f_i/\partial x_j$; $\mathbf{J}(\mathbf{x}) = (\partial \mathbf{f}^T/\partial \mathbf{x})^T$
K	equilibrium constant
K_C	equilibrium constant in terms of molarities
K_m	equilibrium constant in terms of molalities
K_p	equilibrium constant in terms of partial pressures
K_x	equilibrium constant in terms of mole fractions
ℓ	liquid phase (as phase designation)
ln	logarithm to base e
$\log_{10}$	logarithm to base 10
L	number of parameters
L_k	number of isotopes of element E_k (Section 9.8)
$\mathcal{L}$	Lagrangian (function)
m_i	molality of species i
m_I	number of accessible isomeric forms of substance I
m_{lk}	mass number of lth isotope of element E_k (Section 9.8)
m	molality vector with entries m_i
max	maximize
min	minimize
M	number of elements; number of constraints; molecular weight
n_i	number of moles of species i
n_I	number of moles of substance I capable of existing in two or more isomeric forms
n_t	total number of moles
$n_{t\alpha}$	total number of moles in phase α
$n_{t'}$	total number of moles in a phase, excluding inert species
$n_{t'}^o$	defined by equation 8.4-33
n	species-abundance vector with entries n_i
n_z	number of moles of inert species
$n_{z\alpha}$	number of moles of inert species in phase α
$\mathbf{n}_z$	inert-species mole-number vector with entries $n_{z\alpha}$
$\delta \mathbf{n}$	species-abundance-change vector
N	number of species; number of variables

Latin Letters

N'	number of species excluding inert species
$\mathbf{N}$	complete stoichiometric matrix: the $(N \times R)$ matrix whose columns are the R stoichiometric vectors; that is, $\mathbf{N} = (\boldsymbol{\nu}_1, \boldsymbol{\nu}_2, \ldots, \boldsymbol{\nu}_R)$; entry (i, j) of $\mathbf{N}$ is ν_{ij}
$\mathbf{N}^*$	unit matrix form of $\mathbf{N}$
$\mathbf{N}'$	modified form of $\mathbf{N}$
N_I	number of isotopic molecular forms of a species (Section 9.8)
NA	nonaccessible
p	protonic charge; parameter
p_i	partial pressure of species i
p_{kl}	relative molar (atomic) abundance of lth isotope of element k
p^*	vapor pressure
$\mathbf{p}$	parameter vector
P	pressure
$\mathbf{P}$	projection matrix, defined by equation 5.2-22
q_j	defined by equation 6.3-43
Q	quadratic form (equation 3.9-1)
$\mathbf{Q}$	null matrix of order $\pi \times \pi$, except that $Q_{11} = -n_z$
$Q(\mathbf{x})$	quadratic function (equation 5.2-7)
r	number of special stoichiometric restrictions
r_k	b_k/b_1; $k = 2, 3, \ldots, M$
r_1	n_z/b_1
R	maximum number of linearly independent chemical equations; gas constant, 8.3143 J mole^{-1} K^{-1}
R^L, R^M, R^N	vector space of L-tuples, M-tuples, and N-tuples of real numbers
s	solid phase (as phase designation)
$\bar{s}_i$	partial molar entropy of species i
S	entropy function; number of stoichiometric equations in Section 9.4.3, $\geq R$
$S°$	conventional standard absolute entropy (function of T)
ΔS	entropy change/of reaction
$\Delta S°$	standard entropy change/of reaction (function of T)
t	auxiliary parameter (Chapter 5)
T	temperature
u	variable defined by equation 6.3-25
U	internal energy function
v	molar volume; variable defined by equation 6.3-38

$\bar{v}_i$	partial molar volume of species i
V	volume
VCS	Villars-Cruise-Smith (algorithm)
ΔV	volume change/of reaction, $\Sigma_i \nu_i \bar{v}_i$
ΔV^*	standard volume change/of reaction (function of T and P)
w	work; statistical weight
w'	work other than pressure-volume work
W	set defined by footnote to Table 9.1
x_i	mole fraction of species i; variable in general
$\mathbf{x}$	mole-fraction vector with entries n_i; N vector in general
X	designation of inert species in list of elements
y_i	$\ln n_i$
$\mathbf{y}$	arbitrary vector
z	moles of electrons associated with chemical cell or electrode process; $\exp(\lambda/RT)$; compressibility factor; ion charge
$\mathbf{Z}$	$(C \times R)$ matrix (equation 2.3-11)

GREEK LETTERS

α	parameter in equation 7.3-4
$\boldsymbol{\alpha}$	auxiliary parameters (Chapter 5)
α_{li}	defined by equations 4.4-36 and 4.4-37
β_{ki}	defined by equations 4.4-29 and 4.4-30
$\boldsymbol{\beta}$	matrix defined by equation 9.2-1
γ_i	activity coefficient of species i
δ	change in (a quantity) or small amount of (path-dependent quantity)
δ_{ij}	Kronecker delta function; $\delta_{ij} = 1$, if $i = j$; $\delta_{ij} = 0$, if $i \neq j$
Δ	change in (a quantity)
ε	small, positive real number
η	a thermodynamic function (Section 9.9)
$\bar{\eta}_i$	partial molar η of species i
θ	defined by equations 4.4-34 and 4.4-35
λ	Lagrange multiplier
$\boldsymbol{\lambda}$	Lagrange multiplier vector
$\boldsymbol{\Lambda}$	diagonal matrix with entries $n_1, n_2, \ldots, n_h$
μ_i	chemical potential of species i
$\boldsymbol{\mu}$	chemical-potential vector with entries μ_i
μ_i°	standard chemical potential of species i (function of T)

Superscripts

$\boldsymbol{\mu}^\circ$	standard-chemical-potential vector with entries μ_i°
μ_i^*	standard chemical potential of species i (function of T and P)
$\boldsymbol{\mu}^*$	standard-chemical-potential vector with entries μ_i^*
ν_i	stoichiometric coefficient of species i
ν_{ij}	stoichiometric coefficient of species i in stoichiometric vector (equation) j
$\boldsymbol{\nu}$	stoichiometric vector with entries ν_i
$\boldsymbol{\nu}_j$	stoictiometric vector for stoichiometric equation j with entries ν_{ij}
ν_+, ν_-	subscript to cation, anion in molecular formula of electrolyte
ν	$\nu_+ + \nu_-$
$\bar{\nu}$	sum of stoichiometric coefficients in a stoichiometric equation
$\bar{\nu}_j$	sum of stoichiometric coefficients in stoichiometric equation j defined by equation 4.3-6 or 6.4-9
ξ	extent of reaction
$\boldsymbol{\xi}$	extent-of-reaction vector
π	number of phases
π_m	number of multispecies phases
π_s	number of single-species phases
Π	product of quantities
σ_i	$\exp(-\mu_i^*/RT)$
Σ	sum of quantities
ρ	density
ϕ	scalar function
ϕ_i	fugacity coefficient of species i
χ	a thermodynamic function (Section 9.9)
$\bar{\chi}_i$	partial molar χ of species i
ψ_k	λ_k/RT; in Brinkley variation of BNR algorithm (Section 6.3.2.2), chemical potential of component species k; see equation 6.3-51
ω	step-size parameter; acentric factor
Ω	constraint set

SUPERSCRIPTS

E	excess (function)
id	ideal (solution)

LP	linear programming (value)
m	iteration index ($m = 0, 1, 2, \ldots$)
$^\circ$	standard or reference state (function of T only); initial state; particular solution (with n)
T	transposed vector or matrix
0, 1	particular states (of a system) or sets (of parameters)
$(0), (1), (2), \ldots$	zeroth, first, second, etc., iteration/estimate
$(1), (2)$	compositional states of closed chemical system
$*$	standard state (function of T and P); vapor pressure (with p); unit matrix form
$'$	modified matrix or vector (Sections 2.4.3 and 2.4.4); first derivative (of a function)
-1	matrix inverse

SUBSCRIPTS

a	in terms of activities (equilibrium constant)
ad	adiabatic
C	critical; Henry convention (for standard state of solution) on a molarity basis; relating to a $(C \times C)$ matrix
f	of formation
g	of gas
H	Henry convention (for standard state of solution) on a mole fraction basis
i	index (often refers to species); in Section 9.7, refers to isomer
I	of substance I, capable of existing in two or more isomeric forms (Section 9.7); of isotopes (Section 9.8)
j	index (often refers to stoichiometric equation)
k	index (often refers to element)
l	index (refers to isotope in Section 9.8)
m	index; Henry convention (for standard state of solution) on a molality basis
n	at fixed mole numbers
P	at fixed pressure
R	reduced; relating to an $(R \times R)$ matrix
rev	reversible
s	of solvent; stoichiometric

Others

t	total
T	at fixed temperature
V	at fixed volume
z	relating to inert species
α	phase index
λ	at fixed Lagrange multipliers
0	0 K; evaluated at state denoted by super 0; reference species or hypothetical fluid
$+$	cation
$-$	anion
$\pm$	mean ion

OTHERS

$\in$	is a member of
∞	infinity
$\nabla \mathbf{f}$	gradient vector with entries $\partial f / \partial x_i$
$\|x\|$	absolute value of x
$\{\ \}$	a set of

In general, a boldface capital letter denotes a matrix, and a boldface lowercase letter denotes a vector.

References

Apse, J. I. (1965), B.A.Sc. Thesis, University of Toronto.
Apse, J. I., and R. W. Missen (1967), *J. Chem. Educ.*, **44**, 30.
Aris, R., and R. H. S. Mah (1963), *Ind. Eng. Chem. Fundam.*, **2**, 90.
Bakemeier, H., P. R. Laurer, and W. Schroder (1970), in G. A. Danner, Ed., *Methanol Technology and Economics*, Chemical Engineering Progress Symposium Series, No. 98, Vol. 66, pp. 1–10.
Balzhiser, R. E., M. R. Samuels, and J. D. Eliassen (1972), *Chemical Engineering Thermodynamics*, Prentice-Hall, Englewood Cliffs, NJ.
Bard, Y. (1974), *Nonlinear Parameter Estimation*, Academic, New York.
Barin, I., and O. Knacke (1973), *Thermochemical Properties of Inorganic Substances*, Springer-Verlag, Berlin.
Barner, H. E., and R. V. Scheuerman (1978), *Handbook of Thermochemical Data for Compounds and Aqueous Species*, Wiley-Interscience, New York.
Barnhard, P., and A. W. Hawkins (1963), in G. S. Bahn, Ed., *Kinetics, Equilibria, and Performance of High-Temperature Systems, Proceedings of Second Conference*, Gordon and Breach, New York, p. 235.
Bazaraa, M. S., and C. M. Shetty (1979), *Nonlinear Programming-Theory and Algorithms*, Wiley, New York.
Benedict, M., G. B. Webb, and L. C. Rubin (1940), *J. Chem. Phys.*, **8**, 334.
Benedict, M., G. B. Webb, and L. C. Rubin (1942), *J. Chem. Phys.*, **10**, 747.
Benedict, M., G. B. Webb, and L. C. Rubin (1951), *Chem. Eng. Progr.*, **47**, 419, 449, 571, 609.
Berthelot, M., and Péan de St.-Gilles (1862), *Ann. Chim. Phys.* (3), **65**, 385.
Bigelow, J. H. (1970), *Operations Research House*, Technical Report 70-3, Stanford University.
Bigelow, J. H., and N. Z. Shapiro (1971), *Sensitivity Analysis in Chemical Thermodynamics*, RAND Corporation Report P-4628, Santa Monica, California.
Björnbom, P. H. (1975), *Ind. Eng. Chem. Fundam.*, **14**, 102.
Björnbom, P. H. (1977), *A. I. Ch. E. J.*, **23**, 285.
Björnbom, P. H. (1981), *Ind. Eng. Chem. Fundam.*, **20**, 161.
Blum, E. H., and R. Luus (1964), *Chem. Eng. Sci.*, **19**, 322.
Boll, R. H. (1961), *J. Chem. Phys.*, **34**, 1108.
Bos, M., and H. Q. J. Meerschoek (1972), *Anal. Chim. Acta*, **61**, 185.
Boublik, T., V. Fried, and E. Hala (1973), *The Vapour Pressures of Pure Substances*, Elsevier, Amsterdam.
Boynton, F. P. (1960), *J. Chem. Phys.*, **32**, 1880.
Boynton, F. P. (1963), in G. S. Bahn, Ed., *Kinetics, Equilibria, and Performance of High-Temperature Systems, Proceedings of Second Conference*, Gordon and Breach, New York, p. 187.
Braun, F. (1887), *Z. Physik. Chem.*, **1**, 269.
Braun, F. (1888), *Ann. Physik.* **33**, 337.
Brinkley, S. R., Jr. (1946), *J. Chem. Phys.*, **14**, 563; erratum, p. 686.
Brinkley, S. R., Jr. (1947), *J. Chem. Phys.*, **15**, 107.
Brinkley, S. R., Jr. (1951), *Ind. Eng. Chem.*, **43**, 2471.
Brinkley, S. R., Jr. (1956), *High Speed Aerodynamics and Jet Propulsion-Combustion Processes*, Vol. II, Princeton University Press, Princeton, NJ, p. 64.

References

Brinkley, S. R., Jr. (1960), in G. S. Bahn and E. E. Zukoski, Ed., *Kinetics, Equilibria, and Performance of High Temperature Systems, Proceedings of First Conference*, Butterworths, London, p. 74.

Brinkley, S. R., Jr. (1966), paper presented at 16th Canadian Chemical Engineering Conference, Windsor, Ontario.

Bromley, L. A. (1973), *A. I. Ch. E. J.*, **19**, 313.

Browne, H. N., M. M. Williams, and D. R. Cruise (1960), NOTS TP 2434, U.S. Naval Ordnance Test Station, China Lake, California.

Callen, H. B. (1960), *Thermodynamics*, Wiley, New York.

Castillo, J., and I. E. Grossman (1979), paper presented at 86th National Meeting of American Institute of Chemical Engineers, Houston.

Ceram, H. S., and L. E. Scriven (1976), *Chem. Eng. Sci.*, **31**, 163.

Charlot, G. (1958), *Tables of Constants and Numerical Data*, No. 8, *Selected Constants: Oxydo-Reduction Potentials*, Pergamon, London (for IUPAC).

Chaston, S. H. H. (1975), *J. Chem. Educ.*, **52**, 206.

Core, T. C., S. G. Saunders, and P. S. McKittrick (1963), in G. S. Bahn, Ed., *Kinetics, Equilibria, and Performance of High-Temperature Systems, Proceedings of Second Conference*, Gordon and Breach, New York, p. 243.

Cowperthwaite, M., and W. H. Zwisler (1973), *TIGER Computer Program Documentation*, Stanford Research Institute (SRI) Publication No. Z106, January.

Cruise, D. R. (1964), *J. Phys. Chem.*, **68**, 3797.

Cumme, G. A. (1973), *Talanta*, **20**, 1009.

Damköhler, G., and R. Edse (1943), *Z. Elektrochem.*, **49**, 178.

Dantzig, G. B., and J. C. De Haven (1962), *J. Chem. Phys.*, **36**, 2620.

Davies, C. W. (1962), *Ion Association*, Butterworths, London.

De Donder, Th. (1936), *L'Affinité*, revised by P. Van Rysselberghe, Gauthier-Villars, Paris.

de Heer, J. (1957), *J. Chem. Educ.*, **34**, 375.

Dembo, R. S. (1976), in D. J. Phillips and C. S. Beightler, *Applied Geometric Programming*, Wiley, New York, pp. 425–573.

Deming, H. D. (1930), *J. Chem. Educ.*, **7**, 591.

Denbigh, K. G. (1981), *The Principles of Chemical Equilibrium*, 4th ed., Cambridge University Press.

Din, F., Ed. (1956, 1961), *Thermodynamic Functions of Gases*, Vols. 1 and 2 (1956), Vol. 3 (1961), Butterworths, London.

Dinkel, J. J., and R. Lakshmanan (1977), *Computers Chem. Eng.*, **1**, 41.

Dodge, B. F. (1944), *Chemical Engineering Thermodynamics*, McGraw-Hill, New York.

Duff, R. E., and S. H. Bauer (1962), *J. Chem. Phys.*, **36**, 1754.

Duffin, R. J., E. L. Peterson, and C. Zener (1967), *Geometric Programming*, Wiley, New York.

Dunsmore, H. S., and D. Midgley (1974), *Anal. Chim. Acta*, **72**, 121.

Elliott, R. P. (1965), *Constitution of Binary Alloys, First Supplement*, McGraw-Hill, New York.

Engelder, C. J. (1942), *Calculations of Qualitative Analysis*, 2nd ed., Wiley, New York.

Erickson, W. D., J. T. Kemper, and D. O. Allison (1966), NASA Tech. Note D-3488.

Eriksson, G. (1971), *Acta Chem. Scand.*, **25**, 2651.

Eriksson, G. (1975), *Chem. Scripta*, **8**, 100.

Eriksson, G. (1979), *Anal. Chim. Acta*, **112**, 375.

Eriksson, G., and E. Rosen (1973), *Chem. Scripta*, **4**, 193.

Ermenc, E. D. (1970), *Chem. Eng.*, **77** (11), 193.

Feenstra, T. P. (1979), *J. Chem. Educ.*, **56**, 104.

Fickett, W. (1963), *Phys. Fluids*, **6**, 997.

Fickett, W. (1976), *The MES Code: Chemical-Equilibrium Detonation-Product States of Condensed Explosives*, Los Alamos Scientific Laboratory, Report LA-6250, December.

Fletcher, R. (1980), *Practical Methods of Optimization*, Vol. 1, *Unconstrained Optimization*, Wiley-Interscience, New York.

Folkman, J., and N. Z. Shapiro (1968), *SIAM J. Appl. Math.*, **16**, 993.

Gautam, R., and W. D. Seider, (1979), *A. I. Ch. E. J.*, **25**, 991, 999, 1006.
George, B., L. P. Brown, C. H. Farmer, P. Buthod, and F. S. Manning (1976), *Ind. Eng. Chem. Process Des. Dev.*, **15**, 372.
Gibbs, J. W. (1876), in *The Scientific Papers of J. Willard Gibbs*, Vol. I, Thermodynamics, Dover, New York, 1961, p. 96 et seq.
Giggenbach, W. F. (1980), *Geochim. Cosmochim. Acta*, **44**, 2021.
Gordon, S., and B. J. McBride (1971, 1976), NASA Special Publication-273 and interim revision, Washington, DC.
Gordon, S., F. J. Zeleznik, and V. N. Huff (1959), NASA Technical Note, D-132, Washington, DC.
Guggenheim, E. A. (1956), *J. Chem. Educ.*, **33**, 544.
Guldberg, C. M., and P. Waage (1864), *Forh. Vid. Selsk. Christiania*, **35**, 92, 111.
Hadley, G. (1964), *Nonlinear and Dynamic Programming*, Addison-Wesley, Reading, MA.
Hala, E., J. Pick, V. Fried, and O. Vilim (1967), *Vapour-Liquid Equilibrium*, 2nd English ed., translated by G. Standart, Pergamon, Oxford. (Contains list of systems and bibliography.)
Hala, E., I. Wichterle, J. Polak, and T. Boublik (1968), *Vapour-Liquid Equilibrium Data at Normal Pressures*, Elsevier, Amsterdam.
Hamer, W. J. (1968), *Theoretical Mean Activity Coefficients of Strong Electrolytes in Aqueous Solution from 0 to 100°C*, National Bureau of Standards, NSRDS-NBS 24, Washington DC, December.
Hancock, J. H., and T. S. Motzkin (1960), in G. S. Bahn and E. E. Zukoski, Eds., *Kinetics, Equilibria and Performance of High-Temperature Systems, Proceedings of First Conference*, Butterworths, London, p. 82.
Hansen, M., and K. Anderko (1958), *Constitution of Binary Alloys*, 2nd ed., McGraw-Hill, New York.
Harned, H. S., and B. B. Owen (1958), *The Physical Chemistry of Electrolytic Solutions*, 3rd ed., Reinhold, New York.
Harned, H. S., and R. A. Robinson (1968), *Multicomponent Electrolyte Solutions*, Pergamon, Oxford.
Harvey, K. B., and G. B. Porter (1963), *Introduction to Physical Inorganic Chemistry*, Addison-Wesley, Reading, MA.
Heidemann, R. A. (1978), *Chem. Eng. Sci.*, **33**, 1517.
Helgeson, H. C., T. H. Brown, and R. H. Leeper (1969), *Handbook of Theoretical Activity Diagrams Depicting Chemical Equilibria in Geologic Systems Involving An Aqueous Phase at 1 atm and 0 to 300°C*, Freeman, San Francisco.
Hildebrand, J. H., J. M. Prausnitz, and R. L. Scott (1970), *Regular and Related Solutions*, Van Nostrand Reinhold, New York.
Hochstim, A. R., and B. Adams (1962), in J. F. Masi and D. H. Tsai, Eds., *Progress in International Research on Thermodynamic and Transport Properties*, Princeton University Press, Princeton, NJ, p. 228.
Holman, K. L. (1980), University of Idaho, private communication.
Holub, R., and P. Vonka (1976), *The Chemical Equilibrium of Gaseous Systems*, D. Reidel, Dordrecht.
Hougen, O. A., K. M. Watson, and R. A. Ragatz (1959), *Chemical Process Principles*, Part II, Thermodynamics, 2nd ed., Wiley, New York.
Huff, V. N., S. Gordon, and V. E. Morrell (1951), *Natl. Advis. Comm. Aeronaut.*, Report 1037, Washington.
Hutchison, H. P. (1962), *Chem. Eng. Sci.*, **17**, 703.
JANAF (1971), *JANAF Thermochemical Tables*, 2nd ed., D. R. Stull and H. Prophet (project directors), National Bureau of Standards, NSRDS-NBS 37, Washington, DC, June. Supplements by M. W. Chase, J. L. Curnutt, A. T. Hu, H. Prophet, A. N. Syverud, L. C. Walker, and R. A. McDonald, *J. Phys. Chem. Ref. Data*, **3**, 311 (1974); **4**, 1 (1975); **7**, 793 (1978). In Examples 4.2 and 4.5, data have inadvertently been used from earlier JANAF tables.
Johansen, E. S. (1967), *Acta Chem. Scand.*, **21**, 2273.

References

Jones, L. H., and R. S. McDowell (1959), *J. Mol. Spectrosc.*, **3**, 632.
Kandiner, H. J., and S. R. Brinkley, Jr. (1950a), *Ind. Eng. Chem.*, **42**, 850.
Kandiner, H. J., and S. R. Brinkley, Jr. (1950b), *Ind. Eng. Chem.*, **42**, 1526.
Karapet'yants, M. Kh., and M. L. Karapet'yants (1970), *Thermodynamic Constants of Inorganic and Organic Compounds*, translated by J. Schmorak, Ann Arbor-Humphrey, Ann Arbor, MI.
Kaskan, W. E., and G. L. Schott (1962), *Combust. Flame*, **6**, 73.
Kay, W. B. (1936), *Ind. Eng. Chem.*, **28**, 1014.
Kelley, K. K. (1960), *Contributions to Data on Theoretical Metallurgy. XIII. High-Temperature Heat Content, Heat Capacity and Entropy Data for Elements and Inorganic Compounds*. U.S. Bureau Mines, Bulletin 584.
Kelley, K. K., and E. G. King (1961), *Contributions to the Data on Theoretical Metallurgy. XIV. Entropies of the Elements and Inorganic Compounds*. U.S. Bureau of Mines, Bulletin 592.
Klein, M. (1971), in H. Eyring, D. Henderson, and W. Yost, Eds., *Physical Chemistry, An Advanced Treatise*, Vol. 1, Academic, New York, p. 489.
Kobe, K. A., and T. W. Leland (1954), *The Calculation of Chemical Equilibrium in a Complex System*, University of Texas Bureau of Engineering Research, Special Publication No. 26, Austin, Texas.
Kobe, K. A., and R. E. Lynn, Jr. (1953), *Chem. Rev.*, **52**, 117.
Kramer, J. R. (1967), in R. F. Gould, Ed., *Equilibrium Concepts in Natural Water Systems*, Advances in Chemistry Series, No. 67, ACS Publications, Washington.
Kubaschewski, O., and C. B. Alcock (1979), *Metallurgical Thermochemistry*, 5th ed., Pergamon, Oxford.
Kubert, B. R., and S. E. Stephanou (1960), in G. S. Bahn and E. E. Zukoski, Eds., *Kinetics, Equilibria, and Performance of High Temperature Systems, Proceedings of First Conference*, Butterworths, London, p. 166.
Lahiri, A. K. (1979), *Fluid Phase Equilibria*, **3**, 113.
Laitenen, H. A. (1960), *Chemical Analysis*, McGraw-Hill, New York.
Larsen, A. H., C. S. Lu, and C. J. Pings (1968), *Chem. Eng. Sci.*, **23**, 289.
Latimer, W. M. (1952), *The Oxidation States of the Elements and Their Potentials in Aqueous Solutions*, 2nd ed., Prentice-Hall, Englewood Cliffs, NJ.
Le Chatelier, H. (1884), *Compt. Rend.*, **99**, 786.
Le Chatelier, H. (1888), *Ann. Mines*, **13**, 157.
Leland, T. W., P. S. Chappelear and B. W. Gamson (1962), *A. I. Ch. E. J.*, **8**, 482.
Leland, T. W., J. S. Rowlinson, and G. A. Sather (1968), *Trans. Faraday Soc.*, **64**, 1447.
Levenspiel, O. (1972), *Chemical Reaction Engineering*, 2nd ed., Wiley, New York.
Levin, E. M., C. R. Robbins, and H. F. McMurdie (1964), *Phase Diagrams for Ceramists*, The American Ceramic Society, Columbus, Ohio; 1969 supplement; 1975 supplement.
Levine, H. B. (1962), *J. Chem. Phys.*, **36**, 3049.
Lewis, G. N., and M. Randall (1961), *Thermodynamics*, 2nd ed., revised by K. S. Pitzer and L. Brewer, McGraw-Hill, New York.
Lindauer, M. W. (1962), *J. Chem. Educ.*, **39**, 384.
Linke, W. F. (1958), *Solubilities of Inorganic and Metal-Organic Compounds*, 4th ed., Vol. I, Van Nostrand, Princeton, NJ.
Linke, W. F. (1965), *Solubilities of Inorganic and Metal-Organic Compounds*, 4th ed., Vol. II, American Chemical Society, Washington, DC.
Lu, C. S. (1967), *A Thermodynamic Study of Multiple Reaction Systems At and Near Equilibrium*, Dissertation California Institute of Technology 67-6160, University Microfilms Inc., Ann Arbor, Michigan.
Lund, E. W. (1965), *J. Chem. Educ.*, **42**, 548.
Ma, Y. H. and C. W. Shipman (1972), *A. I. Ch. E. J.*, **18**, 299.
Madeley, W. D., and J. M. Toguri (1973a), *Ind. Eng. Chem. Fundam.*, **12**, 261.
Madeley, W. D., and J. M. Toguri (1973b), *Can. Metall. Quart.*, **12**, 71.
Mahan, B. H. (1975), *University Chemistry*, 3rd ed., Addison-Wesley, Reading, MA.
Margrave, J. L., and R. B. Polansky (1962), *J. Chem. Educ.*, **39**, 335.

Margules, H. (1895), *Sitzungber. Wien Akad.*, **104**, 1243.
Marquardt, D. W. (1963), *SIAM J. Appl. Math.*, **11**, 431.
Martell, A. E., and R. M. Smith (1974–1977), *Critical Stability Constants*, Vols. 1–4, Plenum, New York.
Martin, F. J., and M. Yachter (1951), *Ind. Eng. Chem.*, **43**, 2446.
Mash, C. J., and R. C. Pemberton (1980), *Activity Coefficients at Very Low Concentrations for Organic Solutes in Water Determined by an Automatic Chromatographic Method*, NPL Report CHEM 111.
Mason, E. A., and T. H. Spurling (1969), *The Virial Equation of State*, Pergamon, Oxford.
Mathewson, C. H., Ed. (1959), *Zinc: The Science and Technology of the Metal, Its Alloys and Compounds*, Reinhold, New York.
McKay, H. A. C. (1971), *Principles of Radiochemistry*, Butterworths, London.
McKeown, J. J. (1980), in L. C. W. Dixon, E. Spedicato, and G. P. Szegö, Eds., *Nonlinear Optimization, Theory and Algorithms*, Birkhäuser, Boston, p. 387.
Mehrotra, A. K., P. R. Bishnoi, and W. Y. Svrcek (1979), *Can. J. Chem. Eng.*, **57**, 225.
Meissner, H. P., C. L. Kusik, and W. H. Dalzell (1969), *Ind. Eng. Chem. Fundam.*, **8**, 659.
Meissner, H. P., and C. L. Kusik (1972), *A. I. Ch. E. J.*, **18**, 294.
Meites, L., and P. Zuman (1974), *Electrochemical Data*, Part I. *Organic, Organometallic and Biochemical Substances*, Vol. A, Wiley-Interscience, New York.
Mellon, E. K. (1979), *J. Chem. Educ.*, **56**, 380.
Mentzer, R. A., R. A. Greenkorn, and K. C. Chao (1980), *Separation Sci. Technol.*, **15**, 1613.
Michels, H. H., and S. B. Schneiderman (1963), in G. S. Bahn, Ed., *Kinetics, Equilibria, and Performance of High-Temperature Systems, Proceedings of Second Conference*, Gordon and Breach, New York, p. 205.
Mihail, R., C. V. Radu, and I. M. Belcea (1978), *Revue Roumaine de Chimie*, **23**, 681.
Milazzo, G., and S. Caroli (1978), *Tables of Standard Electrode Potentials*, Wiley-Interscience, New York.
Missen, R. W. (1963), *Trans. Eng. Inst. Canada*, **6**, Paper EIC-63-CHEM-11, December.
Missen, R. W. (1969), *Ind. Eng. Chem. Fundam.*, **8**, 81.
Moffatt, W. G. (1977, et seq.), *The Handbook of Binary Phase Diagrams*, Vol. 1–3, General Electric Company, Schenectady, NY.
Moore, W. G. (1972), *Physical Chemistry*, 4th ed., Prentice-Hall, Englewood Cliffs, NJ.
Mysels, K. J. (1956), *J. Chem. Educ.*, **33**, 178.
Naphtali, L. M. (1959), *J. Chem. Phys.*, **31**, 263.
Naphtali, L. M. (1960), in G. S. Bahn and E. E. Zukoski, Eds., *Kinetics, Equilibria and Performance of High Temperature Systems, Proceedings of First Conference*, Butterworths, London, p. 181.
Naphtali, L. M. (1961), *Ind. Eng. Chem.*, **53**, 387.
Nernst, W. (1916), *Theoretical Chemistry*, 4th English ed., translated by H. T. Tizard, Macmillan, London.
Neumann, Klaus-Kurt (1962), in J. F. Masi and D. H. Tsai, Ed., *Progress in International Research on Thermodynamic and Transport Properties*, Princeton University Press, Princeton, NJ, p. 209.
Neumann, Klaus-Kurt (1963), *Ber. Bunsenges Physik. Chem.*, **67**, 373.
Noble, B. (1969), *Applied Linear Algebra*, Prentice-Hall, Englewood Cliffs, NJ.
Nordstrom, D. K., et al. (1979), in E. A. Jenne, Ed., *Chemical Modeling in Aqueous Systems*, ACS Symposium Series No. 93, *American Chemical Society*, Washington, p. 857; this paper contains an extensive bibliography.
Oliver, P. (1980), *Internatl. J. Chem. Kin.*, **12**, 509.
Oliver, R. C., S. E. Stephanou, and R. W. Baier (1962), *Chem. Eng.*, **69** (4), 121.
Othmer, H. G. (1976), *Chem. Eng. Sci.*, **31**, 993.
Otterstedt, J.-E. A., and R. W. Missen (1962), *Can. J. Chem. Eng.*, **40**, 12.
Partington, J. R. (1964), *A History of Chemistry*, Vol. 4, Macmillan, London.
Passey, U., and D. J. Wilde (1968), *SIAM J. Appl. Math.*, **16**, 363.

References

Pedley, J. B. (1972), Ed., *Computer Analysis of Thermochemical Data (CATCH Tables)*, Tables 1–4, School of Molecular Sciences, University of Sussex, Brighton.
Pedley, J. B., and J. Rylance (1977), *Sussex-NPL Computer Analyzed Thermochemical Data: Organic and Organometallic Compounds*, University of Sussex, Brighton.
Pings, C. J. (1961), *Chem. Eng. Sci.*, **16**, 181.
Pings, C. J. (1963), *Chem. Eng. Sci.*, **18**, 671.
Pings, C. J. (1964), *A. I. Ch. E. J.*, **10**, 934.
Pings, C. J. (1966), *Chem. Eng. Sci.*, **21**, 693.
Pitzer, K. S. (1939), *J. Chem. Phys.*, **7**, 583.
Pitzer, K. S. (1955), *J. Am. Chem. Soc.*, (1955), **77**, 3427.
Pitzer, K. S., and J. J. Kim (1973), *J. Phys. Chem.*, **77**, 268.
Pitzer, K. S., D. Z. Lippman, R. F. Curl, Jr., C. M. Huggins, and D. E. Petersen (1955), *J. Am. Chem. Soc.*, **77**, 3433.
Powell, M. J. D. (1970), *Atomic Energy Research Establishment*, Report AERE-R.6469.
Powell, M. J. D. (1980), in L. C. W. Dixon, E. Spedicato, and G. P. Szegö, Ed., *Nonlinear Optimization, Theory and Algorithms*, Birkhäuser, Boston, p. 279.
Prausnitz, J. M. (1969), *Molecular Thermodynamics of Fluid-Phase Equilibria*, Prentice-Hall, Englewood Cliffs, NJ.
Prausnitz, J. M., T. F. Anderson, E. A. Grens, C. A. Eckert, R. Hsieh, and J. P. O'Connell (1980), *Computer Calculations for Multicomponent Vapor-Liquid and Liquid-Liquid Equilibria*, Prentice-Hall, Englewood Cliffs, NJ.
Prigogine, I., and R. Defay (1954), *Chemical Thermodynamics*, translated by D. H. Everett, Longmans, London.
Putnam, W. E., and J. E. Kilpatrick (1953), *J. Chem. Phys.*, **21**, 951.
Raju, B. N., and C. S. Krishnaswami (1966), *Ind. J. Technol.*, **4**, 99.
Ralston, A., and P. Rabinowitz (1978), *A First Course in Numerical Analysis*, 2nd ed., McGraw-Hill, New York.
Ratajczykowa, I. (1972), *Bull. Acad. Polonaise Sci., Série Sci. Chim.*, **20**, 691.
Redlich, O., and A. T. Kister (1948), *Ind. Eng. Chem.*, **40**, 345.
Redlich, O., and J. N. S. Kwong (1949), *Chem. Rev.*, **44**, 233.
Reed, T. M., and K. E. Gubbins (1973), *Applied Statistical Mechanics*, McGraw-Hill, New York.
Ridler, G. M., P. F. Ridler, and J. G. Sheppard (1977), *J. Phys. Chem.*, **81**, 2435.
Robinson, R. A., and R. H. Stokes (1965), *Electrolyte Solutions*, 2nd ed. (revised), Butterworths, London.
Rossini, F. D., D. D. Wagman, W. H. Evans, S. Levine, and I. Jaffe (1952), *Selected Values of Chemical Thermodynamic Properties, Circular of the National Bureau of Standards 500*, Washington, DC, February.
Rossini, F. D., K. S. Pitzer, R. L. Arnett, R. M. Braun, and G. C. Pimentel (1953), *Selected Values of Physical and Thermodynamic Properties of Hydrocarbons and Related Compounds*, Carnegie Press, Pittsburg, PA.
Rowlinson, J. S. (1969), *Liquids and Liquid Mixtures*, 2nd ed., Butterworths, London.
Samuels, M. R. (1971), *Ind. Eng. Chem. Fundam.*, **10**, 643.
Samuels, M. R., and J. D. Eliassen (1972), *Ind. Eng. Chem. Proc. Des. Dev.*, **11**, 383.
Sanderson, R. V., and H. H. Y. Chien (1973), *Ind. Eng. Chem. Proc. Des. Dev.*, **12**, 81.
Sargent, R. W. H., and B. A. Murtagh (1973), *Math. Prog.*, **4**, 245.
Schneider, D. R., and G. V. Reklaitis (1975), *Chem. Eng. Sci.*, **30**, 243.
Schott, G. L. (1964), *J. Chem. Phys.*, **40**, 2065.
Schubert, E., and H. Hofmann (1975), *Chem.-Ing.-Tech.*, **47**, 191.
Schubert, E., and H. Hofmann (1976), *Int. Chem. Eng.*, **16**, 132.
Scott, R. L. (1956), *J. Chem. Phys.*, **25**, 193.
Scully, D. B., (1962), *Chem. Eng. Sci.*, **17**, 977.
Seider, W. D., R. Gautam, and C. W. White, III (1980), in R. G. Squires and G. V. Reklaitis, Eds., *ACS Symp. Ser. No. 124*, Am. Chem. Soc., Washington, DC, p. 115.

Shapiro, N. Z. (1964), *On the Behaviour of a Chemical Equilibrium System When Its Free Energy Parameters are Changed*, Report RM-4128-PR, The RAND Corporation, Santa Monica, California.
Shapiro, N. Z. (1969), *SIAM J. Appl. Math.*, **17**, 83.
Shapiro, N. Z., and L. S. Shapley (1965), *SIAM J. Appl. Math.*, **13**, 353.
Shapley, M., L. Cutler, J. DeHaven, and N. Z. Shapiro (1968), *Specifications for a New Jacobian Package for the RAND Chemical Equilibrium Program*, RAND Corporation Memo RM-5426-PR, Santa Monica, California.
Shreve, R. N., and J. A. Brink, Jr. (1977), *Chemical Process Industries*, 4th ed., McGraw-Hill, New York.
Shunk, F. A. (1969), *Constitution of Binary Alloys, Second Supplement*, McGraw-Hill, New York.
Silcock, H. L. (1979), Ed., *Solubilities of Inorganic and Organic Compounds*, Vol. 3, Parts 1–3, Pergamon, Oxford.
Skjold-Jørgensen, S., P. Rasmussen, and Aa. Fredenslund (1982), *Chem. Eng. Sci.*, **37**, 99.
Skorobogatov, G. A. (1961), translation from *Investiya Akademii Nauk, SSSR, Otdelenie Khimicheskikh Nauk*, No. 10, October, p. 1763 (p. 1644 in translation).
Smith, B. D. (1959), *A. I. Ch. E. J.*, **5**, 26.
Smith, J. M., and H. C. Van Ness (1975), *Introduction to Chemical Engineering Thermodynamics*, 3rd ed., McGraw-Hill, New York.
Smith, W. R. (1966), M.A.Sc. Thesis, University of Toronto.
Smith, W. R. (1969), *Can. J. Chem. Eng.*, **47**, 95.
Smith, W. R. (1976), *Ind. Eng. Chem. Fundam.*, **15**, 227.
Smith, W. R. (1978), *Ind. Eng. Chem. Fundam.*, **17**, 69.
Smith, W. R. (1980a), in H. Eyring and D. Henderson, Eds., *Theoretical Chemistry: Advances and Perspectives*, Vol. 5, Academic, p. 185.
Smith, W. R. (1980b), *Ind. Eng. Chem. Fundam.*, **19**, 1.
Smith, W. R., and R. W. Missen (1967), *Can. J. Chem. Eng.*, **45**, 346.
Smith, W. R., and R. W. Missen (1968), *Can. J. Chem. Eng.*, **46**, 269.
Smith, W. R., and R. W. Missen (1974), *Can. J. Chem. Eng.*, **52**, 280.
Smith, W. R., and R. W. Missen (1979), *Chem. Eng. Educ.*, **13**, 26. (This paper contains an extensive bibliography to literature up to 1976.)
Sommerfeld, J. T., and C. T. Lenk (1970), *Chem. Eng.*, **77** (9), 136.
Stadtherr, M. A., and L. E. Scriven (1974), *Chem. Eng. Sci.*, **29**, 1165.
Stephen, H., and T. Stephen, Eds. (1979), *Solubilities of Inorganic and Organic Compounds*, Vol. 1, Parts 1 and 2, Vol. 2, Parts 1 and 2, Pergamon, Oxford.
Stone, E. E. (1966), *J. Chem. Educ.*, **43**, 241.
Storey, S. H. (1965), *Can. J. Chem. Eng.*, **43**, 168.
Storey, S. H., and F. Van Zeggeren (1964), *Can. J. Chem. Eng.*, **42**, 54.
Storey, S. H., and F. Van Zeggeren (1967), *Can. J. Chem. Eng.*, **45**, 323.
Storey, S. H., and F. Van Zeggeren (1970), *Can. J. Chem. Eng.*, **48**, 591.
Strelzoff, S. (1970), in G. A. Danner, Ed., *Methanol Technology and Economics*, Chemical Engineering Progress Symposium Series, No. 98, Vol. 66, pp. 54–68.
Stull, D. R., and G. C. Sinke (1956), *Advances in Chemistry Series*, No. 18, *Thermodynamic Properties of the Elements*, American Chemical Society, Washington, DC.
Stull, D. R., E. F. Westrum, Jr., and G. C. Sinke (1969), *The Chemical Thermodynamics of Organic Compounds*, Wiley-Interscience, New York.
Suzuki, I., H. Komatsu, and M. Hirata (1970), *J. Chem. Eng. Jap.*, **3**, 152.
Swartz, C. J. (1969), *J. Chem. Educ.*, **46**, 308.
Swinnerton, J. W., and W. W. Miller (1959), *J. Chem. Educ.*, **36**, 485.
Tenn, F. G., and R. W. Missen (1963), *Can. J. Chem. Eng.*, **41**, 12.
Timmermans, J. (1950, 1965), *Physico-Chemical Constants of Pure Organic Compounds*, Vol. I (1950), Vol. II (1965), Elsevier, Amsterdam.
Timmermans, J. (1959, 1960), *The Physico-chemical Constants of Binary Systems in Concentrated Solutions*, Vol. 1–4, Interscience, New York.

References

Urey, H. C. (1947), *J. Chem. Soc.*, 562.
van Laar, J. J. (1910), *Z. Physik. Chem.*, **72**, 723.
Van Ness, H. C. (1959), *Chem. Eng. Sci.*, **11**, 118.
Van Ness, H. C. (1964), *Classical Thermodynamics of Non-Electrolyte Solutions*, Pergamon, New York.
Van Rysselberghe, P. (1967), *Chem. Eng. Sci.*, **22**, 706.
Van Zeggeren, F., and S. H. Storey (1970), *The Computation of Chemical Equilibria*, Cambridge University Press.
van't Hoff, J. H. (1884), *Etudes de Dynamique chimique*, Frederik Müller, Amsterdam.
van't Hoff, J. H. (1898), *Lectures on Theoretical and Physical Chemistry*, Part I, *Chemical Dynamics*, translated by R. A. Lehfeldt, Edward Arnold, London.
Verhoek, F. H. (1969), *J. Chem. Educ.*, **46**, 140.
Villars, D. S. (1959), *J. Phys. Chem.*, **63**, 521.
Villars, D. S. (1960), in G. S. Bahn and E. E. Zukoski, Ed., *Kinetics, Equilibria, and Performance of High Temperature Systems, Proceedings of First Conference*, Butterworths, London, p. 141.
Vonka, P., and R. Holub (1971), *Collect. Czech. Chem. Commun.*, **36**, 2446.
Vonka, P., and R. Holub (1975), *Collect. Czech. Chem. Commun.*, **40**, 931.
Wagman, D. D., et al. (1965 to 1973), *Selected Values of Chemical Thermodynamic Properties*, National Bureau of Standards, Technical Notes 270-1 (1965), 270-2 (1966), 270-3 (1968), 270-4 (1969), 270-5 (1971), 270-6 (1971), 270-7 (1973), Washington, DC.
Walden, P. (1954), translated by R. Oesper, *J. Chem. Educ.*, **31**, 27.
Walsh, G. R. (1975), *Methods of Optimization*, Wiley, New York.
Weast, R. C., Ed. (1979–1980), *CRC Handbook of Chemistry and Physics*, CRC Press, Boca Raton, Florida.
White, C. W., III, and W. D. Seider (1981), *A. I. Ch. E. J.*, **27**, 466.
White, W. B. (1967), *J. Chem. Phys.*, **46**, 4171.
White, W. B., S. M. Johnson, and G. B. Dantzig (1958), *J. Chem. Phys.*, **28**, 751.
Whitwell, J. C., and S. R. Dartt (1973), *A. I. Ch. E. J.*, **19**, 1114.
Wichterle, I., J. Linek, and E. Hala (1973), *Vapor-Liquid Equilibrium Data Bibliography*, supplement I (1976), supplement II (1979), Elsevier, Amsterdam.
Wicks, C. E., and F. E. Block (1963), *Thermodynamic Properties of 65 Elements—Their Oxides, Halides, Carbides, and Nitrides*, U.S. Bureau of Mines, Bulletin 605, Washington, DC.
Wilde, D. J., and C. S. Beightler (1967), *Foundations of Optimization*, Prentice-Hall, NJ.
Wilf, H. S. (1962), *Mathematics for the Physical Sciences*, Wiley, New York.
Wilson, G. M. (1964), *J. Am. Chem. Soc.*, **86**, 127.
Wohl, K. (1946), *Trans. Am. Inst. Chem. Engrs.*, **42**, 215.
Zeleznik, F. J., and S. Gordon (1960), NASA Technical Note D-473.
Zeleznik, F. J., and S. Gordon (1962), NASA Technical Note D-1454.
Zeleznik, F. J., and S. Gordon (1966), *Can. J. Phys.*, **44**, 877.
Zeleznik, F. J., and S. Gordon (1968), *Ind. Eng. Chem.*, **60** (6), 27.
Zwolinski, B. J., et al. (1963), *Selected Values of Properties of Chemical Compounds*, Manufacturing Chemists Association Research Project, Chemical Thermodynamic Properties Center, Texas A & M University, College Station, Texas.
Zwolinski, B. J., et al. (1974, tables extant to), *Selected Values of Properties of Hydrocarbons and Related Compounds*, American Petroleum Institute Research Project 44, Thermodynamics Research Center, Texas A & M University, College Station, Texas.

ANSWERS TO SELECTED PROBLEMS

CHAPTER 2

2.4 $x_{H_2} = 0.026$, $x_{CH_4} = 0.627$, $x_{C_2H_4} = 0.340$.

2.7 On basis of given feed and **b**, and $SO_2 + \frac{1}{2}O_2 = SO_3$, $\xi = 6.97$.

CHAPTER 3

3.1 129.65 kJ mole^{-1}.

3.3 -632.20 kJ mole^{-1}.

3.7 (a) -77.7 kJ mole^{-1}.
(b) The difference is slight; for example, the standard chemical potential on the molarity scale is only 0.0072 kJ mole^{-1} greater than the result in (a).

3.9 (a) 0.320; (b) 0.285.

CHAPTER 4

4.1 For $\{H_2, H, N_2, N, NH, NH_2, NH_3\}$, $\{x_i\} = \{0.0663, 0.5778, 0.3544, 0.0015, 6.42 \times 10^{-5}, 1.32 \times 10^{-6}, 2.18 \times 10^{-8}\}$; this corresponds to 81.34% fractional conversion of H_2; if N_2 is inert, the result is 81.33%.

4.3 For $\{CH_4, S_2, CS_2, H_2S, H_2\}$, $\{x_i\} = \{3.31 \times 10^{-4}, 0.0153, 0.3281, 0.6270, 0.0293\}$, and $n_t = 3.046$.

4.6 For $\{CH_4, Cl_2, CH_3Cl, CH_2Cl_2, CHCl_3, CCl_4, HCl\}$, $\{x_i\} = \{0.2955, 7.58 \times 10^{-11}, 0.2854, 0.0426, 0.00145, 7.04 \times 10^{-7}, 0.3750\}$.

4.7 Note that rank $(\mathbf{A}) = 3 \neq M$. Any row of $\mathbf{A}$ may be omitted. This type of situation is discussed in Section 9.3, following Section 2.3.3.

4.8 For $\{CO_2, H_2O, CO, H_2, CH_4, C_2H_4, C_2H_2, C_2H_6, O_2\}$, $\{x_i\} = \{0.0615, 0.1716, 0.1566, 0.6028, 0.00751, 1.076 \times 10^{-8}, 3.56 \times 10^{-11}, 1.88 \times 10^{-8}, 6.15 \times 10^{-22}\}$.

Answers

CHAPTER 6

6.4 For {C(gr), C(g), C_2(g), C_3(g), C_4(g), C_5(g), CH, CH_2, CH_3, CH_4, CHN, CN, C_2H_2, C_2H_4, C_2N_2, C_4N_2, H, H_2, HN, H_2N, H_3N, N, N_2}, $\{n_i\}$ = {0, 0.0783, 0.0415, 0.0969, 0.00102, 1.77×10^{-4}, 0.00545, 2.11×10^{-4}, 1.06×10^{-4}, 1.60×10^{-6}, 0.2178, 0.1749, 0.0700, 4.02×10^{-7}, 0.00226, 7.06×10^{-5}, 2.2609, 0.6874, 3.11×10^{-4}, 1.69×10^{-5}, 7.38×10^{-7}, 0.00272, 0.7998}.

6.7 For {H, HCl, HF, H_2, Cl, ClF, Cl_2, F, F_2, N, N_2}, $\{x_i\}$ = {0.00604, 0.1804, 0.6145, 0.0302, 0.0244, 6.60×10^{-6}, 1.35×10^{-4}, 7.73×10^{-4}, 1.87×10^{-9}, 1.16×10^{-6}, 0.1435}.

6.11 (a) Taking $\mathbf{b} = (1, 2000/18.016 + 2, 1000/18.016 + 2)^T$, $m[\text{Mg}^{2+}(\ell)] = 1.31 \times 10^{-4}$; $m[\text{MgOH}^+(\ell)] = 1.44 \times 10^{-5}$; $m[\text{H}^+(\ell)] = 3.75 \times 10^{-11}$; $m[\text{OH}^-(\ell)] = 2.76 \times 10^{-4}$; $x[\text{H}_2\text{O}(\ell)] = 0.99999$. The solubility of Mg(OH)_2 is $m[\text{Mg}^{2+}(\ell)] + m[\text{MgOH}^+(\ell)] = 1.45 \times 10^{-4}$

6.12 For ethyl alcohol, acetic acid, ethyl acetate, and water, the respective liquid-phase mole fractions are $\{x_i\}$ = {0.0584, 0.0656, 0.4189, 0.4572}, and the gaseous phase mole fractions are $\{x_i\}$ = {0.0834, 0.0238, 0.6048, 0.2880}.

6.13 Consider the system to be {($C_6H_6(\ell)$, $C_7H_8(\ell)$, $C_8H_8(\ell)$, $C_8H_{10}(\ell)$, C_6H_6(g), C_7H_8(g), C_8H_8(g), C_8C_{10}(g)), (C_6H_6, C_7H_8, C_8H_8, C_8H_{10})}, with formula matrix

$$\mathbf{A}^* = \begin{pmatrix} 1 & 0 & 0 & 0 & 1 & 0 & 0 & 0 \\ 0 & 1 & 0 & 0 & 0 & 1 & 0 & 0 \\ 0 & 0 & 1 & 0 & 0 & 0 & 1 & 0 \\ 0 & 0 & 0 & 1 & 0 & 0 & 0 & 1 \end{pmatrix}$$

Take the respective standard chemical potentials to be (see problem 3.1) $\boldsymbol{\mu}^\circ = (RT\ln(0.1570), RT\ln(0.0482), RT\ln(0.0166), RT\ln(0.0109), 0, 0, 0, 0)^T$. The resulting mole fractions are $\mathbf{x}$ = (0.00600, 0.0194, 0.4385, 0.5361, 0.0628, 0.0624, 0.4852, 0.3896)^T.

CHAPTER 7

7.5 The respective mole fractions are $\mathbf{x}$ = (0.00538, 0.0183, 0.4389, 0.5414, 0.0564, 0.0592, 0.4877, 0.3967)^T.

7.7 If the Davies equation is used for individual ionic species and if nonionic species are treated as ideal, the respective mole numbers for the species in the problem statement for Example 6.3 are $\{n_i\}$ = {0.9418, 0.0309, 0.0273, 0.0273, 0.0273, 9.05×10^{-10}, 55.479}.

7.8 If the Davies equation is used for individual ionic species and if nonionic species are treated as ideal, for part a, the molalities (in the same order as that given in the solution to Problem 6.11a) are $\{1.37 \times 10^{-4}, 1.45 \times 10^{-5}, 3.76 \times 10^{-11}, 2.89 \times 10^{-4}\}$, and the mole fraction of H_2O is the same as in Problem 6.11a.

CHAPTER 8

8.1 -76.45 kJ mole^{-1}.

8.11 The first five columns of the sensitivity matrix, $RT(\partial n_i/\partial \mu_j^\circ)$ ($i = 1, 2, \ldots, 10;\ j = 1, 2, \ldots, 5$) are given by

-0.0398	0.00870	0.0122	-1.05×10^{-5}	6.89×10^{-4}
0.00870	-0.0809	0.0829	-1.05×10^{-5}	0.00478
0.0122	0.0829	-0.1177	2.31×10^{-5}	-0.00429
-1.05×10^{-5}	-1.05×10^{-5}	2.31×10^{-5}	-0.00141	6.98×10^{-4}
6.89×10^{-4}	0.00478	-0.00429	6.98×10^{-4}	-0.00677
2.09×10^{-5}	1.60×10^{-4}	1.91×10^{-4}	4.74×10^{-6}	3.53×10^{-4}
-0.00139	-0.00971	0.00836	1.64×10^{-5}	0.0125
-0.00107	-0.00667	0.00593	-6.16×10^{-6}	-6.51×10^{-4}
-0.00383	-0.0268	0.0231	-9.06×10^{-6}	-0.00277
-0.00212	-0.0129	0.0572	-1.52×10^{-5}	-0.00202

CHAPTER 9

9.1 For $\{P(s), VCl_2(s), CrCl_3(s), SiCl_4(\ell), PCl_3(\ell), Fe_2Cl(g), SiCl_4(g), PCl(g)\}$, $\{n_i\} = \{26.28, 4.4, 4.1, 4.9998, 4.51994, 27.85, 1.96 \times 10^{-4}, 5.983 \times 10^{-5}\}$; all other solid species are zero; all other gaseous and liquid species are $< 10^{-24}$.

9.3 For Example 4.3, the Lagrange multipliers for $\{C, H, O, N\}$ are, respectively, $\{\lambda_i/RT\} = \{-17.232, -1.761, -1.480, -1.693\}$. Using these values and equation 9.5-1, we obtain $x_H = 2.41 \times 10^{-5}$, $x_O = 1.53 \times 10^{-5}$, and $x_{OH} = 6.68 \times 10^{-4}$. The correct solution of the full 10-species problem gives $x_H = 2.45 \times 10^{-5}$, $x_O = 1.47 \times 10^{-5}$, and $x_{OH} = 6.56 \times 10^{-4}$.

9.5 Note that the results are independent of pressure. The respective mole fractions are $\{x_i\} = \{0.1934, 0.3284, 0.1678, 0.2071, 0.1033\}$.

9.8 At $P = 1$ atm, the mole fractions of $\{C_5H_8, C_5H_{12}, H_2$, 2-methyl-1-butene, 3-methyl-1-butene, 2-methyl-2-butene$\}$ are $\{x_i\} = \{0.0202, 0.0458, 0.4872, 0.00239, 0.3677, 0.0768\}$. At $P = 10$ atm, the mole fractions are $\{0.00217, 0.2747, 0.3637, 0.00192, 0.2957, 0.0617\}$.

9.11 $T = 2110$ K.

Answers 345

9.15 Take $\mu^*(M) = \mu^*(H) = \mu^*(A) = \mu^*(OH) = 0$, $\mu^*(M_mH_hA_a)/RT = -\ln K_{mha}$, $\mu^*(H_2O) = -\ln K_w$. Assuming ideal solution behavior, using the Raoult convention for H_2O and the Henry convention for the remaining species and assuming $V = 1$ liter, we obtain molarities of $\{0.00424, 0.00236, 5.25 \times 10^{-5}, 0.00124, 0.00293, 2.23 \times 10^{-4}, 1.17 \times 10^{-5}, 5.26 \times 10^{-4}, 4.23 \times 10^{-12}, 55.5062\}$.

9.16 Take $\mu_f^* = 0$ for the first five species and $\mu_k^*/RT = -\ln K_k - \text{NH}(k)\ln(H^+)_m$ for the kth complex species, where (H^+) is determined by the specified pH. Ignore H from the species and element lists. Alter the chemical potential subroutine DFE in the VCS computer program of Appendix D to compute chemical potentials, using the molarity scale. This yields molarities corresponding to the given species list of $\{1.54 \times 10^{-4}, 3.90 \times 10^{-5}, 5.20 \times 10^{-4}, 0.1265, 0.0194, 6.31 \times 10^{-8}, 0.00157, 4.24 \times 10^{-4}, 8.86 \times 10^{-8}, 0.00166, 0.00178, 2.55 \times 10^{-4}, 2.92 \times 10^{-4}, 5.93 \times 10^{-4}, 4.20 \times 10^{-4}, 5.48 \times 10^{-5}, 0.00185, 8.07 \times 10^{-6}, 6.97 \times 10^{-5}, 1.15 \times 10^{-5}, 2.19 \times 10^{-5}, 1.18 \times 10^{-8}\}$.

Author Index

Adams, B., 183, 191, 199, 336
Alcock, C.B., 71, 337
Allison, D.O., 335
Anderko, K., 73, 336
Anderson, T.F., 339
Apse, J.I., 37, 123, 224, 225, 231, 334
Aris, R., 28, 334
Arnett, R.L., 339

Bahn, G.S., 334, 335, 336, 337, 338, 341
Baier, R.W., 338
Bakemeier, H., 7, 334
Balzhiser, R.E., 99, 334
Bard, Y., 197, 334
Barin, I., 71, 334
Barner, H.E., 71, 334
Barnhard, P., 127, 228, 334
Bauer, S.H., 219, 335
Bazaraa, M.S., 100, 334
Beightler, C.S., 137, 335, 341
Belcea, I.M., 338
Benedict, M., 160, 334
Bergman, T., 9
Berthelot, M., 10, 334
Berthollet, C.L., 9
Bigelow, J.H., 184, 208, 334
Bishnoi, P.R., 338
Björnbom, P.H., 30, 31, 35, 197, 219, 334
Block, F.E., 71, 341
Blum, E.H., 6, 334
Boll, R.H., 212, 334
Bos, M., 12, 82, 140, 141, 334
Boublik, T., 71, 334, 336
Boynton, F.P., 126, 168, 334
Braun, F., 10, 334
Braun, R.M., 339
Brewer, L., 337
Brink, J.A., Jr., 98, 99, 340
Brinkley, S.R., Jr., 11, 12, 17, 90, 122, 126, 127, 130, 131, 132, 164, 228, 334, 335, 337

Bromley, L.A., 163, 335
Brown, L.P., 336
Brown, T.H., 336
Browne, H.N., 144, 335
Buthod, P., 336

Callen, H.B., 10, 335
Caroli, S., 71, 338
Castillo, J., 136, 169, 335
Ceram, H.S., 61, 335
Chao, K.C., 338
Chappelear, P.S., 337
Charlot, G., 71, 335
Chase, M.W., 336
Chaston, S.H.H., 151, 335
Chien, H.H.Y., 25, 99, 140, 233, 339
Core, T.C., 126, 335
Cowperthwaite, M., 164, 335
Cruise, D.R., 12, 140, 143, 144, 208, 335
Cumme, G.A., 234, 335
Curl, R.F., Jr., 339
Curnutt, J.L., 336
Cutler, L., 340

Dalzell, W.H., 338
Damköhler, G., 11, 12, 83, 335
Danner, G.A., 334, 340
Dantzig, G.B., 219, 335, 341
Dartt, S.R., 38, 341
Davidon, W.C., 114
Davies, C.W., 163, 335
De Donder, Th., 10, 26, 46, 335
Defay, R., 10, 26, 40, 199, 339
De Haven, J.C., 219, 335, 340
de Heer, J., 10, 335
Dembo, R.S., 59, 335
Deming, H.D., 144, 335
Denbigh, K.G., 6, 40, 46, 49, 58, 63, 72, 147, 174, 199, 230, 335
Din, F., 71, 335
Dinkel, J.J., 137, 184, 335

347

Dixon, L.C.W., 338, 339
Dodge, B. F., 171, 177, 335
Duff, R.E., 219, 335
Duffin, R.J., 136, 335
Dunsmore, H.S., 233, 335

Eckert, C.A., 339
Edse, R., 11, 12, 83, 335
Eliassen, J.D., 184, 189, 334, 339
Elliott, R.P., 73, 335
Engelder, C.J., 24, 335
Erickson, W.D., 12, 335
Ericksson, G., 12, 126, 127, 169, 207, 208, 212, 335
Ermenc, E.D., 8, 335
Evans, W.H., 339
Everett, D.H., 339
Eyring, H., 337, 340

Farmer, C.H., 336
Feenstra, T.P., 166, 172, 229, 335
Fickett, W., 164, 335
Fletcher, R., 100, 103, 114, 335
Folkman, J., 165, 335
Fredenslund, Aa., 340
Fried, V., 334, 336

Gamson, B.W., 337
Gautam, R., 61, 136, 163, 168, 213, 233, 336, 339
George, B., 99, 136, 169, 233, 336
Gibbs, J.W., 10, 11, 336
Giggenbach, W.F., 12, 336
Gordon, S., 12, 13, 118, 126, 127, 133, 135, 136, 168, 184, 205, 206, 208, 336, 341
Gould, R.F., 337
Greenkorn, R.A., 338
Grens, E.A., 339
Grossman, I.E., 136, 169, 335
Gubbins, K.E., 11, 154, 155, 339
Guggenheim, E.A., 10, 336
Guldberg, C.M., 10, 336

Hadley, G., 60, 336
Hala, E., 71, 334, 336, 341
Hamer, W.J., 71, 336
Hancock, J.H., 61, 207, 336
Hansen, M., 73, 336
Harned, H.S., 71, 163, 336
Harvey, K.B., 6, 336
Hawkins, A.W., 127, 228, 334
Heidemann, R.A., 61, 336
Helgeson, H.C., 71, 336
Henderson, D., 337, 340

Hildebrand, J.H., 157, 336
Hirata, M., 340
Hochstim, A.R., 183, 191, 199, 336
Hofmann, H., 24, 339
Holman, K.L., 229, 336
Holub, R., 13, 90, 118, 126, 150, 151, 160, 164, 169, 170, 184, 232, 336, 341
Hougen, O.A., 230, 336
Hsieh, R., 339
Hu, A.T., 336
Huff, V.N., 122, 133, 228, 336
Huggins, C.M., 339
Hutchison, H.P., 82, 140, 141, 336

Jaffe, I., 339
Jenne, E.A., 338
Johansen, E.S., 118, 336
Johnson, S.M., 341
Jones, L.H., 224, 225, 226, 231, 337

Kandiner, H.J., 127, 337
Karapet'yants, M. Kh., 71, 337
Karapet'yants, M.L., 71, 337
Kaskan, W.E., 337
Kay, W.B., 161, 337
Kelley, K.K., 71, 337
Kemper, J.T., 335
Kilpatrick, J.E., 160, 339
Kim, J.J., 163, 339
King, E.G., 71, 337
Kister, A.T., 156, 339
Klein, M., 13, 118, 337
Knacke, O., 71, 334
Kobe, K.A., 11, 13, 151, 171, 232, 337
Komatsu, H., 340
Kramer, J.R., 8, 337
Krishnaswami, C.S., 126, 339
Kubaschewski, O., 71, 337
Kubert, B.R., 126, 337
Kusik, C.L., 163, 338
Kwong, J.N.S., 160, 339

Lahiri, A.K., 206, 337
Laitenen, H.A., 7, 337
Lakshmanan, R., 137, 184, 335
Larsen, A.H., 184, 191, 337
Latimer, W.M., 71, 337
Laurer, P.R., 334
Le Chatelier, H., 9, 10, 184, 337
Leeper, R.H., 336
Lehfeldt, R.A., 341
Leland, T.W., 11, 13, 162, 337
Lenk, C.J., 8, 340
Levenspiel, O., 15, 337

AUTHOR INDEX

Levin, E.M., 73, 337
Levine, H.B., 126, 133, 135, 337
Levine, S., 339
Lewis, G.N., 9, 11, 40, 71, 73, 161, 229, 337
Lindauer, M.W., 9, 337
Linek, J., 341
Linke, W.F., 71, 337
Lippman, R.F., 339
Lu, C.S., 184, 337
Lund, E.W., 10, 337
Luus, R., 6, 334
Lynn, R.E., Jr., 151, 171, 232, 337

Ma, Y.H., 141, 208, 337
McBride, B.J., 127, 205, 206, 208, 336
McDonald, R.A., 336
McDowell, R.S., 224, 225, 226, 231, 337
McKay, H.A.C., 223, 338
McKeown, J.J., 173, 338
McKittrick, P.S., 335
McMurdie, H.F., 337
Madeley, W.D., 37, 127, 136, 205, 208, 337
Mah, R.H.S., 28, 334
Mahan, B.H., 23, 337
Manning, F.S., 336
Margrave, J.L., 224, 337
Margules, H., 156, 339
Marquardt, D.W., 140, 338
Martell, A.E., 71, 338
Martin, F.J., 150, 338
Mash, C.J., 71, 338
Masi, J.F., 336, 338
Mason, E.A., 159, 160, 338
Mathewson, C.H., 230, 338
Meershoek, H.Q.J., 12, 82, 140, 141, 334
Mehrotra, A.K., 12, 233, 338
Meissner, H.P., 140, 144, 163, 338
Meites, L., 71, 338
Mellon, E.K., 10, 338
Mentzer, R.A., 160, 338
Michels, H.H., 168, 338
Midgley, D., 233, 335
Mihail, R., 230, 231, 338
Milazzo, G., 71, 338
Miller, W.W., 37, 340
Missen, R.W., 6, 7, 12, 14, 19, 24, 25, 30, 37, 56, 57, 115, 140, 143, 144, 154, 156, 182, 202, 203, 219, 221, 222, 224, 225, 231, 334, 338, 340
Moffatt, W.G., 73, 338
Moore, W.G., 28, 338
Morrell, V.E., 336
Motzkin, T.S., 61, 207, 336

Murtagh, B.A., 136, 339
Mysels, K.J., 10, 338

Naphtali, L.M., 12, 140, 167, 338
Nernst, W., 9, 11, 338
Neumann, K-K., 183, 191, 199, 338
Noble, B., 16, 24, 30, 338
Nordstrom, D.K., 12, 162, 169, 214, 338

O'Connell, J.P., 339
Oesper, R., 341
Oliver, P., 28, 39, 338
Oliver, R.C., 126, 127, 206, 207, 208, 338
Othmer, H.G., 61, 338
Otterstedt, J-E.A., 156, 338
Owen, B.B., 71, 336

Partington, J.R., 9, 338
Passey, U., 137, 338
Pedley, J.B., 71, 339
Pemberton, R.C., 71, 338
Petersen, D.E., 339
Peterson, E.L., 335
Phillips, D.J., 335
Pick, J., 336
Pimentel, G.C., 339
Pings, C.J., 184, 228, 337, 339
Pitzer, K.S., 161, 163, 337, 339
Polak, J., 336
Polansky, R.B., 224, 337
Porter, G.B., 6, 336
Powell, M.J.D., 136, 167, 339
Prausnitz, J.M., 7, 11, 52, 55, 56, 74, 154, 156, 158, 159, 171, 336, 339
Prigogine, I., 10, 26, 40, 199, 339
Prophet, H., 336
Putnam, W.E., 160, 339

Rabinowtiz, P., 76, 100, 110, 111, 226, 339
Radu, C.V., 338
Ragatz, R.A., 336
Raju, B.N., 126, 339
Ralston, A., 76, 100, 110, 111, 226, 339
Randall, M., 9, 40, 71, 73, 161, 229, 337
Rasmussen, P., 340
Ratajczykowa, I., 200, 339
Redlich, O., 156, 160, 339
Reed, T.M., 11, 154, 155, 339
Reklaitis, G.V., 24, 31, 339
Ridler, G.M., 27, 339
Ridler, P.F., 339
Robbins, C.R., 337
Robinson, R.A., 71, 74, 163, 336, 339
Rosen, E., 126, 169, 335

Rossini, F.D., 71, 230, 339
Rowlinson, J.S., 154, 337, 339
Rubin, L.C., 160, 334
Rylance, J., 71, 339

St. Gilles, P. de, 10, 334
Samuels, M.R., 12, 127, 184, 189, 334, 339
Sanderson, R.V., 25, 99, 140, 233, 339
Sargent, R.W.H., 136, 339
Sather, G.A., 337
Saunders, S.G., 335
Scheuerman, R.V., 71, 334
Schmorak, J., 337
Schneider, D.R., 24, 31, 339
Schneiderman, S.B., 168, 338
Schott, G.L., 337, 339
Schroder, W., 152
Schubert, E., 24, 339
Scott, R.L., 161, 336, 339
Scriven, L.E., 61, 136, 335, 340
Scully, D.B., 136, 339
Seider, W.D., 13, 61, 118, 136, 163, 168, 213, 218, 233, 336, 339, 341
Shapiro, N.Z., 61, 165, 182, 184, 334, 335, 340
Shapley, L.S., 61, 340
Shapley, M., 184, 340
Sheppard, J.G., 339
Shetty, C.M., 100, 334
Shipman, C.W., 141, 208, 337
Shreve, R.N., 98, 99, 340
Shunk, F.A., 73, 340
Silcock, H.L., 71, 340
Sinke, G.C., 151, 340
Skjold-Jørgensen, S., 156, 340
Skorobogatov, G.A., 223, 224, 225, 340
Smith, B.D., 219, 340
Smith, J.M., 56, 199, 232, 340
Smith, R.M., 71, 338
Smith, W.R., 6, 12, 13, 14, 19, 24, 25, 30, 36, 45, 60, 100, 115, 118, 140, 143, 144, 182, 184, 192, 194, 197, 202, 203, 204, 205, 208, 218, 219, 221, 340
Sommerfeld, G.T., 8, 340
Spedicato, E., 338, 339
Spurling, T.H., 159, 160, 338
Squires, R.G., 339
Stadtherr, M.A., 136, 340
Standart, G., 336
Stephanou, S.E., 126, 337, 338
Stephen, H., 71, 340
Stephen, T., 71, 340
Stokes, R.H., 71, 74, 339
Stone, E.E., 82, 140, 141, 340

Storey, S.H., 13, 118, 120, 121, 122, 136, 167, 183, 184, 340, 341
Strelzoff, S., 7, 340
Stull, D.R., 71, 98, 99, 151, 230, 232, 336, 340
Suzuki, I., 99, 171, 340
Svrcek, W.Y., 338
Swartz, C.J., 6, 340
Swinnerton, J.W., 37, 340
Syverud, A.N., 336
Szegö, G.P., 338, 339

Tenn, F.G., 154, 340
Timmermans, J., 71, 230, 340
Tizard, H.T., 338
Toguri, J.M., 37, 127, 136, 205, 208, 337
Tsai, D.H., 336, 338

Urey, H.C., 224, 341

van Laar, J.J., 156, 341
Van Ness, H.C., 56, 156, 158, 199, 232, 340, 341
Van Rysselberghe, P., 6, 335, 341
van't Hoff, J.H., 9, 10, 11, 184, 341
Van Zeggeren, F., 13, 118, 120, 121, 122, 136, 167, 184, 340, 341
Verhoek, F.H., 7, 341
Vilim, O., 336
Villars, D.S., 12, 140, 143, 144, 341
Vonka, P., 13, 90, 118, 126, 150, 151, 160, 164, 169, 170, 184, 232, 336, 341

Waage, P., 10, 336
Wagman, D.D., 71, 151, 152, 166, 339, 341
Walden, P., 9, 341
Walker, L.C., 336
Walsh, G.R., 47, 58, 100, 105, 107, 108, 114, 341
Watson, K.M., 336
Weast, R.C., 151, 224, 341
Webb, G.B., 160, 334
Westrum, E.F., Jr., 340
White, C.W., III, 218, 339, 341
White, W.B., 12, 90, 122, 125, 126, 127, 133, 137, 138, 146, 194, 341
Whitwell, J.C., 38, 341
Wichterle, I., 71, 73, 336, 341
Wicks, C.E., 71, 341
Wilde, D.J., 137, 338, 341
Wilf, H.S., 88, 341
Williams, M.M., 335
Wilson, G.M., 157, 341
Wohl, K., 156, 341

AUTHOR INDEX

Yachter, M., 150, 338
Yost, W., 337

Zeleznik, F.J., 12, 13, 118, 126, 127, 133, 135, 136, 168, 184, 336, 341

Zener, C., 335
Zukoski, E.E., 335, 336, 337, 338, 341
Zuman, P., 71, 338
Zwisler, W.H., 164, 335
Zwolinski, B.J., 71, 77, 150, 200, 230, 341

Subject Index

Acentric factor, 161, 162
Activity, 11, 56
  definitions, 54
  and equilibrium constant, 62, 64, 225
  mean-ion, 55
  of solid, 64
Activity coefficient, 54, 56, 154, 156
  in algorithm for nonideal system, 164–166
  compositional dependence, 155
  definitions, 54–55
  of electrolytes, 162–163
  and excess chemical potential, 155
  from excess free energy, 155
  and fugacity, 158
  mean-ion, 55, 74, 163
  pressure dependence, 56, 156
  from Redlich-Kwong equation, 170
  in regular solution, 171
  temperature dependence, 56, 155–156
Adiabatic change, 5, 6, 231
Adiabatic reaction temperature, 231, 232
Adiabatic system, 41
Affinity, 9–10, 46
Algorithm(s), 3, 119
  for aqueous systems, 169
  classification, 118–119, 127, 133, 136, 140
  first-order, 119, 120–122, 136, 140, 141, 167
  flow charts, 83, 92, 128, 145, 165
  general-purpose, 11, 75, 76, 100, 117, 119, 140, 153
  gradient-projection, 119, 120–122, 167
  for ideal systems, 117–148, 228
  linear programming, 202
  line search, 103
  Marquardt's, 140
  for nonideal systems, 153, 164–169
  and nonnegativity constraint, 204–213
  for relatively simple systems, 81–83, 88–92
  reviews of, 118
  second-order, 119, 122–136, 141, 167–169
  simple optimization, 115
  special-purpose, 100, 108, 117
  stoichiometric-coefficient or stoichiometry, 14, 23, 214, 216
  see also BNR algorithm; Brinkley algorithm; NASA algorithm; Nonstoichiometric algorithm; RAND algorithm; Stoichiometric algorithm; VCS algorithm
Analytical chemistry, 2, 7, 23
Aqueous system, 169, 214

Balancing equations, 23, 37
BASIC, 76
  computer program, 23, 82, 92
  program in, initial estimate for, 203–204
  see also Computer input data; Computer output; Computer program listings
Basis amount of system 16, 17
Batch system, 15
Benedict-Webb-Rubin equation, 160
BNR algorithm, 122–136, 184
  computer program, 127, 137
  flow chart (RAND variation), 128
  FORTRAN computer program listing, 278–291
  illustrative example, 137–139
  initial estimate for, 201, 202
  input data, 137
  and minor species, 218
  modification for composition variable other than mole fraction (RAND variation), 147, 151
  modification when $C \neq M$, 213–214
  for nonideal system, 165, 168
  and nonnegativity constraint, 144
  number of linear equations, 144
  output for, 138
  and singularities, 204, 205, 206, 208, 209, 211, 212

353

and testing for presence of nonideal phase, 213
user's guide, 137, 292–294
Brinkley algorithm, 12, 119, 122, 123, 127–133, 136
  classification, 133, 140
  as equilibrium-constant method, 133
  extension to multiphase system 131, 149
  and NASA algorithm, 135–136
  number of unknowns, 133
  and RAND algorithm, 130, 132, 133, 135–136
  working equations, 132
Bromley equation, 163

C:
  case when not equal to M, 166, 201, 213–214
  as number of components, 23, 25
  as number of linearly independent element-abundance constraints, 29, 213
  as rank of formula matrix, 19, 22, 24, 46, 75
Calculators, programmable, 3, 75, 76
  see also HP-41C calculator
Canonical form of stoichiometric matrix, 24, 25, 58, 142, 202
Charge balance, 4, 69, 148, 166
Charged species, 25
Chemical cell, see Electrochemical cell
Chemical economics, 8–9
Chemical element, 14, 15
  see also Elements
Chemical equations, 14, 18
  balancing, 23, 37
  complete set, 22, 25, 27
  concept of, 20, 27
  linear independence of, 27, 28, 29
  number of, 25, 29
  procedure for determining, 23–25
  see also Stoichiometric equation(s)
Chemical equilibrium:
  concept of, 9, 11
  conditions for, see Equilibrium conditions
  historical sketch, 9–11
  as including phase equilibrium, 2, 154
  nature of, 1–2
  problem, 28, 101, 136, 139, 181, 182, 226
    geometric-programming formulation of, 136–137
    mathematical dual, 59
  relation to kinetics (rates), 2
Chemical kinetics, see Kinetics

Chemical potential, 10, 11, 40, 48–57
  definitions, 43
  excess, 155
  expression and equilibrium constant, 62
  expressions for, 48–55
  of ideal gas, 49
  of ionic species, 73–74
  of liquid, 50
  nondimensional form, 57
  of nonideal gas, 49–50
  numerical assignment, 56–57, 65, 72, 73
  pressure dependence of, 43, 49
  of solid, 50
  of species:
    in ideal-gas solution, 51
    in ideal solution, 51–53, 82, 117
    in nonideal solution, 54–55, 57, 154, 155–163
  standard:
    consistent set from equilibrium constants, 214–217
    and conventional absolute entropy, 67
    effect on equilibrium composition, 182, 194, 196–198, 200
    effect on extent of reaction, 188, 191–192
    as free energy data, 214
    and free energy of formation, 65, 72
    and free energy function, 65, 66–67
    function of T, 49
    function of T and P, 52
    as parameter, 4, 173, 184, 228
    relations among forms, 53
  temperature dependence of, 43
Chemical processes, 7
  ammonia synthesis, 7, 171, 177
  chlorination of methane, 98
  esterification of ethyl alcohol, 10, 25–26, 99, 152, 171, 232–233
  Fischer-Tropsch, 38
  lead-acid cell, 63–64, 69, 72
  methanol synthesis, 7, 38, 170
  partial oxidation of natural gas, 34, 98, 149
  production of acrylonitrile, 98
    carbon disulfide, 98
    cyclohexane, 190
    ethyl amines, 232
    ethylene, 37, 38, 149–150, 175, 176
    hydrogen, 98, 99, 200
    refrigerants CCl_3F and CCl_2F_2, 99
    synthesis gas, 38, 98, 149, 200
    zinc, 37, 229–230
  steam-cracking of ethane, 99, 200

steam-reforming of natural gas, 98
sulfur dioxide oxidation, 2, 7, 38, 93–95, 175, 176, 215, 231
see also Chemical reactions; System
Chemical (reaction) equilibrium analysis:
  applications, 5–9
  concerns, 1
  importance, 2
  mathematical aspects, 2–4
  reviews, 13
Chemical reactions:
  chlorination:
    of coal ash, 233
    of ferrophosphorus, 229
  combustion:
    of hydrazine, 137–139, 146–147, 194–195
    of propane, 83–84, 95–96, 97, 229, 232
  dehydrogenation:
    of n-butane, 221, 222, 230
    of n-butene, 178, 221–222, 230
    of ethane, 37, 38, 149–150, 175, 176
  dissociation of hydrogen, 77–78, 79, 86, 87, 96–97
  formation:
    of ethanethiol and diethylsulfide, 150–151, 169–170, 262–264
    of hydrogen bromide, 28–29
    of isoprene, 230–231
  hydrodealkylation of toluene, 31–32, 197–198, 219
  hydrogen isotope exchanges, 225–226
  isomerization of hexane, 230
  methane and nitrogen, 67–69, 149
  nitrogen isotope exchanges, 226
  oxidation of ammonia, 16, 21–22, 23, 46, 47
  polymerization of carbon, 79–81
  see also Chemical processes; System
Chemical species, 17, 44
  see also Species
Chemical stoichiometry, 14, 18–27, 28, 108
  elementary treatment, 19
  general treatment, 19–23
  genesis of, 19
  introductory concepts, 18–19
  see also Compositional restrictions; Stoichiometry
Chemical substance, 17, 221
  see also Substances
Chemical system:
  closed, *see* Closed system
  designation of, 17

heterogeneous, 44
homogeneous, 42, 44
stoichiometric description, 40
thermodynamic description, 42–44
Chemical thermodynamics, 40–48
Closed system, 1, 14, 16, 18, 42, 44
  constraint, 4, 15–18, 40, 44, 45
  definition, 15, 17
Coefficient matrix, 17, 130, 140, 191, 194
  β and singularities, 205, 206, 207, 208
  nonsingular, 180, 181
  singular, 127, 145, 204, 213
Common gases, sources of data, 70, 71, 72
Component(s):
  definition, 23
  number of, 23, 25, 37
    as rank of formula matrix, 19, 25
    in relation to number of elements, 46, 75, 166, 213–214
  set of, 23, 25, 26, 37
Component species, 11, 131, 133, 135, 142, 143
  in initial estimate of mole numbers, 202, 203
  set of, 24, 217
Compositional restrictions, 30, 201, 218
  in standard form, 27–36
Composition variables other than mole fraction, 52, 55, 117, 147–148
Compressibility factor, 56, 159, 161, 162
Computation of equilibrium:
  historical sketch, 11–13
  problem of, 2–4
  reviews of, 13
Computer, 3, 11, 12, 13, 76, 101
  rounding errors, 122, 126
  small, 3, 75, 76, 100, 117
Computer input data:
  BASIC programs, 95–96, 150, 242–243, 262–264
  BNR program, 137
  FORTRAN program for generating stoichiometric equations, 246
  HP-41C, 84
  VCS program, 146, 148, 210
Computer output, 96, 137
  BASIC programs, 97, 150, 243, 264
  BNR program, 138
  FORTRAN program for generating stoichiometric equations, 246
  VCS program, 146, 148, 210
Computer program listings, 4, 82, 92
  BASIC, 240–241, 256–261, 270–276

BNR (FORTRAN), 278–291
FORTRAN for generating stoichiometric
  equations, 244–245
HP-41C, 236–237, 248–250, 265–267
VCS (FORTRAN), 296–323
Computer programs, 4, 14, 28–29, 85, 147
  BASIC, 23, 82, 92
  for calculating equilibrium for relatively
    simple systems, 247–276
  for constraints other than (T,P), 228
  convergence criteria used, 116
  FORTRAN, 23, 127, 145
  general-purpose, 115, 117
  for generating stoichiometric equations,
    235–246
  HP-41C, 23, 82, 84, 92
  modifications for molality, 147–148
  nonstoichiometric, 94
  for RAND/BNR algorithm, 127, 137
  stoichiometry, 216
  VCS, 27, 145
  see also Computer program listings
Conservation, 14, 15, 18, 19
Constrained minimization methods, 105–109
Constraints, 4–5, 105
  additional, 218
  charge-balance, 4
  compositional, 32, 218
  inequality, 58
  linear, 103, 107, 108, 121
  mandatory, 4
  mass-balance, 3, 4, 11, 12
  nonlinear, 121, 122
  number of, 29, 179
  overriding, 4
  satisfaction of, 106, 107
  see also Closed system, constraint; Element-
    abundance constraints; Nonnegativity
    constraint; Thermodynamic constraints
Constraint set, 101, 103, 107, 178, 218
Convergence, 82, 103, 104, 124, 140, 144
  criteria, 115–116
  difficulties, 166, 228
Conversion, 2, 7, 175, 184
  equilibrium, 6, 175, 176, 177, 178, 221,
    222
Corresponding states theory, 155, 160–162

Data, sources of, 70, 71, 72–73
Davies equation, 163, 166, 229
Degrees of freedom, stoichiometric, 18, 29–31
Descarte's rule of signs, 88
Descent method, 103, 111, 114, 115, 144
  method of steepest descent, 104

Distinguishability, 17, 219, 222
Dummy index, 89

Effects excluded, 4
Electrochemical cell, 7, 40, 42, 64–65, 69, 72
  cell process, 64, 65, 69
  electrode potentials, 7, 65, 69, 72, 74
  electrode processes, 7, 69
  electromotive force (emf), 2, 7, 56, 64
  fuel cell, 7
  lead-acid cell, 63–64, 69, 72
  Nernst equation, 7, 64
  standard emf, 65, 69, 176
Electrolyte(s), 53, 56, 63, 72, 147, 155
  activity coefficient, 162–163
  mean-ion activity, 55
  mean-ion activity coefficient, 55, 163
  mean-ion molality, 53, 55
  strong, 54
  system, relation of C and M, 166
Element-abundance constraints, 29, 36, 123,
    124, 130, 139
  case of linearly dependent, 213
  and classification of algorithms, 118
  and initial estimate, 122, 201
  number of, 46, 213
  satisfaction of, 118, 121, 126, 127, 128,
    133, 136
Element-abundance equations, 15, 17, 20, 27,
    34
  maximum number of linearly independent,
    16
Element-abundance vector, 16, 205
  definition, 17
  effect on equilibrium solution, 194
  effect on extent of reaction, 189–190
  hypothetical, 129
  modified, 31, 33, 213, 218
  as parameter, 4, 173, 184, 228
Elemental species, 134, 135, 202
Elements, 65, 135
  conservation of, 14, 15
  in designation of chemical system, 17
  number of, 11, 15
    in relation to number of components, 46,
      75, 166, 213–214
    in relation to rank of formula matrix, 46,
      75, 106
  sources of data, 70, 71, 72, 73
  standard state of, caution, 73
Endothermic reaction, 8, 10, 175, 176, 186
Energy balance, 5
Energy conversion, 7
Enthalpy, 42, 327

SUBJECT INDEX

change, standard, 63, 66
of combustion, standard, 72
excess, 72
of formation, 56, 65, 66
function, 66
in ideal solution, 52
of isotopic exchange, 225, 231
of mixing, 56
partial molar, 43, 56
of reaction, 174–186
of solution, 56, 72
as thermodynamic constraint, 5, 6, 133, 226, 228
Entropy, 5, 41, 65, 67, 186, 225
Environmental chemistry, 8
Equation of state, 3, 5, 56, 155, 158–160
Benedict-Webb-Rubin, 160
ideal gas, 49
ideal-gas solution, 51
Redlich-Kwong, 150, 160, 169–170, 232
virial, 159
Equilibrium conditions, 15, 40, 62–63, 118, 123
classical forms, 46, 110, 139
equivalence of formulations, 45, 48
formulations of, 44–48
and nonnegativity constraint, 57–60
nonstoichiometric formulation, 45, 46–47, 48, 58, 59, 73, 192
satisfaction of, 128, 130, 133, 135, 136
stoichiometric formulation, 45–46, 48, 58–59, 62, 63, 64, 73, 184, 190
Equilibrium constant, 10, 40, 65, 136
and Brinkley algorithm, 130, 132, 133
and chemical potential expression, 62
forms, 62–64
free energy data from, 201, 214–217
for isotopic exchange reactions, 222, 224, 225, 231
pressure dependence of, 63, 177
temperature dependence of, 63
Equilibrium-constant expressions, 10, 11, 12, 62, 63
number of, 183
Equilibrium-constant method, 12, 118, 132, 133
Equilibrium model, 1, 2, 8
and kinetic model, 2
Equilibrium state, 1, 2
Errors, rounding, 122, 126
Errors in standard chemical potential vector, **u***, effect on equilibrium composition, 183, 188, 196–198, 200
Excess function, 7, 56–57, 72

chemical potential, 155
enthalpy, 56
free energy, 72, 155–158, 162
molar enthalpy, 72
molar volume, 57
partial molar volume, 57
volume, 56
Existence of solution, 40, 60, 61, 117
Exothermic reaction, 175, 176, 186
Extensive quantity, 27, 44, 60
Extent of reaction, 12, 26, 45, 64, 77, 79
definition, 27
effect of **b** on, 189–190
effect of inert species on, 178, 188
effect of parameter changes on, 184–192
effect of pressure on, 177
effect of temperature on, 174–175
effect of **u*** on, 188, 191
introduction as ξ parameter, 19, 27
optimal value, 189–190
variables, number of, 139

Feasibility, 18
First-order method, 102, 104, 106, 109, 119, 120, 140, 141, 167
First-variation method, 104
Flow charts for algorithms, 83, 92, 128, 145, 165
Formula matrix, 16, 22, 27, 29, 30, 35, 139
compatible, 31, 34, 35, 39, 218
definition, 17
finding linearly dependent rows of, 213
modified, 31, 33, 39, 213, 218
and phase equilibrium, 152
rank, 19, 25
and sensitivity analysis, 184
unit matrix form, 24
Formula vector, 20, 21, 22, 23, 34, 88, 137
definition, 17
and isomers, 219
FORTRAN computer programs, 23, 127, 145
modifications for molality, 147–148
see also Computer input data; Computer output; Computer program listings
Fractional conversion, *see* Conversion
Free energy:
change, standard, 62, 65, 66
data:
from equilibrium constants, 214–217
errors in, 173
excess, 72
of formation, standard, 65, 72, 74, 152, 217
function, 65, 66–67, 73, 224
Gibbs, 127

information, availability of, 65–69
of reaction, 10, 174
standard, 66
see also Gibbs function; Helmholtz function
Free-energy-minimization method, 12, 118, 127, 144
Free enthalpy, 41
Fugacity, 11, 50, 52, 56, 150, 158
Fugacity coefficient, 55, 56, 150, 169–170

Gaseous system, 147, 164, 177, 178, 183, 184, 187
Gauss-Jordan reduction, 24, 37
Geochemistry, 169
Geometric programming, 136–137, 184
Gibbs-Duhem equation, 44, 48, 156
Gibbs function, 3, 4, 41, 43, 44, 61, 120, 139, 181
  definition, 327
Gradient method, 104, 106
Gradient-projection method, 108, 119, 120–122, 167
Gradient vector, 104, 109
Graphical solution, 77, 78–81, 84–86, 87, 88, 94

Hand Calculation, 3, 11, 13, 29, 75, 76, 224
  for initial estimate, 202
  for stoichiometric equations, 23
Heat capacity, 56, 72, 198, 227
Helmholtz function, 5, 41
  definition, 326
Henry convention, 52–53, 65, 72, 74, 153
  and activity, 54
  and activity coefficient, 55
  and equilibrium constant, 64
  and excess function, 57
  and standard state, 158
Hessian matrix, 101, 105, 109, 111
  as approximately diagonal matrix, 143
  definition, 327
  of G, 60, 141
  inversion of, 142
Heterogeneous system, 44
  see also Multiphase system
Homogeneous function, 42, 44
Homogeneous system, 42, 44
  see also Single-phase system
HP-41C calculator, 76
  computer programs for, 23, 82, 84, 92
  initial estimate for programs, 203
  see also Computer input data; Computer program listings; User's guide

Ideal behavior, 154
Ideal gas, 49
Ideal-gas solution, 51, 63, 153, 183, 199
Ideal solution, 88, 124, 133, 154, 182
  characteristics, 52
  chemical potential expressions, 51–53, 82, 117
  and equilibrium constant, 63
  Henry convention, 52–53, 153
  and nonnegativity constraint, 58, 59
  Raoult convention, 52, 153
  and sensitivity analysis, 186, 187, 188, 189, 191, 192, 193, 194, 196, 199
  and singularities, 209, 211–212
  for test for presence as phase, 59, 211–212
  and uniqueness of equilibrium composition, 60, 61
Ideal system, 154, 182, 184
  algorithms for, 117–148
Inert species, 17, 25, 89, 90, 123
  amount as parameter, 173, 184, 228
  distinct from reacting species, 89
  effect on equilibrium, 178, 188, 191–192, 194
  effect on total moles, 199, 200
Infeasibility, 18
Infinitely dilute solution, 52, 152
Initial estimate, 84, 103, 201–204, 213
  hand calculation, 202
  of Lagrange multipliers, 96, 203–204
  linear programming procedure, 202–203
  of mole numbers, 201–203
Inorganic chemistry, 6, 23
Inorganic compounds, 214
  sources of data, 70, 71, 72, 73
Input data, *see* Computer input data
Intensive quantity, 57, 60
Intensive state, 43
Internal energy, 42
Ionic species, 2, 53, 166
  in designation of chemical system, 17
  standard chemical potential, 73–74
Ionic strength, 163, 166
Irreversible change, 42
Isomeric degeneracy, 220, 221, 222
Isomerization, effect on equilibrium, 182, 219, 221, 222
Isomers, 6, 17, 201, 219–222
  accessible forms, 220
  chemical potential, 220, 221
  equilibrium distribution, 219, 221, 230
Isotope effect, equilibrium (EIE), 223, 225–226, 231

SUBJECT INDEX

Isotope-exchange reactions, 17, 222, 225
   equilibrium constants for, 222, 224–225, 231
Isotopes, 17, 201, 222–226
   atomic (molar) abundance, 224, 225
   in distinguishable molecular species:
      calculation of equilibrium distribution, 222–226
      temperature dependence of equilibrium distribution, 222, 225–226
   natural distribution, 222
   number of, 223
   random distribution, 222
Iteration, 101
   equation, 81, 103, 111, 124, 226
   index, 103
   loop, inner and outer, 227, 228
   scheme, 228

Jacobian matrix, 11, 113, 173, 180
   definition, 328
JANAF Thermochemical Tables, 71

Kay's rule, 162
Kinetics, 1, 6, 9, 10
   kinetic model and equilibrium model, 2
   mechanisms, 22, 27, 28
   rates, 2, 18
Kirchoff equation, 174
Kronecker delta, definition, 330
Kuhn-Tucker conditions, 58, 60

Lagrange multiplier method, 86, 88, 102, 105–107
   first-order, 102, 106, 107
   second-order, 102, 107, 109
Lagrange multipliers, 3, 12, 45, 47, 106, 121, 129
   initial estimate, 96, 201, 203–204
   number of, 85, 127
Lagrangian, 47, 106
Le Chatelier-Braun principle, 10
Le Chatelier's principle, 10, 174
Lewis-Randall fugacity rule, 52
Linear algebra, 3, 14
   row and column operations, 24, 25, 26, 28
Linear equations:
   number of, 124–125, 126, 141, 168, 184, 193, 194
   solution of, 3, 19, 24, 101, 109, 180, 204
   underdetermined sets of, 216
Linear independence:
   of columns of matrix, 35, 205, 206, 207, 209
   of element-abundance equations, 16
Linearly dependent columns and singularities, 205, 207, 208, 209
Linear programming:
   problem, 202, 205, 208, 209
   procedure for initial estimate, 202–203
   value, 202
Liquid, 50, 54, 154

Margules equation, 56, 156, 157
Mass number, 223
Mass transfer, 2, 15, 25, 27, 36
Matrix:
   augmented, 18
   β, 205, 206, 207, 208
   canonical form, 24
   compatible, 31, 34, 35, 39, 218
   determining rank, 24, 25
   diagonal, 121
   identity, 24, 35
   inverse, 105
   inversion, 120, 184
   null, 205
   unit, 24, 25, 213, 214, 216
   see also Coefficient matrix; Formula matrix; Hessian matrix; Jacobian matrix; N matrix; Projection matrix; Rank of matrix; Sensitivity matrix; Stoichiometric matrix
Metallurgical applications, 12, 117, 144, 205
Minimization:
   of function, 101, 110, 116
   of Gibbs function, 3, 120, 127, 139, 218
   point of view, 139
   see also Optimization
Minimization method(s), 139
   and classification of algorithms, 118
   constrained, 105–109, 119
   RAND algorithm as, 127
   unconstrained, 104–105, 119–120
   VCS algorithm as, 140
Minimization problem, 103–109
   as class of numerical problem, 102
   constrained, 102, 103, 110, 120, 218
   formulations, 45–46
   unconstrained, 45, 102, 103, 139
   see also Optimization problem
Minimum, 44, 61
   necessary conditions for, 44, 45, 47, 101, 104, 105, 139, 179
   sufficient conditions for, 44, 101
Minor species, 194, 197, 201, 217–218
Molality, 52, 53–54, 55, 63

and RAND(BNR) algorithm, 147, 151
and VCS algorithm, 147–148
Molarity, 53, 55, 63
and RAND(BNR) algorithm, 147
and VCS algorithm, 147
Molecular formula, 3, 15, 17, 20, 53
involving isotopes, 223
as parameter, 4
Mole fraction, 51, 52, 117, 147
Moles, total number of, 51, 89, 123, 127, 187
Monte Carlo technique, 197
Multiphase system, 12, 44, 123, 143, 144, 192, 200
and Brinkley algorithm, 149
and RAND(BNR) algorithm, 149
and uniqueness, 60, 61
Multispecies phase, 60, 123, 126, 143, 200
and numerical singularities, 204, 205, 206, 208, 209, 211–213

NASA algorithm, 119, 122, 123, 133–136
and Brinkley algorithm, 135–136
and RAND algorithm, 135–136
Necessary condition, 100
for feasibility, 18
for minimum, 101, 104, 105, 179
for minimum in G, 44, 45, 47, 139
for uniqueness, 60
Nernst equation, 7, 64
Newton-Raphson method, 3, 82, 91, 102, 110–111
iteration equations, 129, 131, 226, 228
N matrix, 22, 27, 28, 29, 108–109, 139
see also Rank of matrix; Stoichiometric matrix
Noncomponent species, 23, 58, 142, 217
Nonelectrolytes, 7–8, 56, 147, 155, 162
Nonideal behavior, 153, 154
Nonideal gas, 49–50, 154
Nonideal solution, 154
chemical potential expressions, 54–55, 57
excess functions, 56–57
and fugacity, 55
coefficient, 169–170
Henry convention, 54, 55
and nonnegativity constraint, 57, 58
and numerical singularities, 212–213
Raoult convention, 54, 55
test for presence as phase, 59–60, 212–213
and uniqueness, 61
Nonideal system, 122, 124, 127, 140, 154, 181
algorithms for, 153, 164–169
flow chart, 165

Nonlinear-equation method, 136, 139
Nonlinear-equation problems, 102, 109–113, 116
Nonlinear equations:
sensitivity analysis for, 173, 180
solution of, 3, 11, 100, 109–113, 226
Nonlinear problems, 101
Nonnegativity constraint, 4, 117, 144
implications of, 57–60
non-binding, 44
and numerical singularities, 204–213
satisfaction of, 60, 201, 217
Nonstoichiometric algorithm, 118, 119, 167–168
comparison with stoichiometric algorithm, 140
convergence criteria, 116
first-order, 119, 120–122
flow charts, 92, 128
initial estimate, 201–204
number of variables, 140
relatively simple system, 88–92
second-order, 119, 122–139, 167, 169
see also BNR algorithm; Brinkley algorithm; NASA algorithm; RAND algorithm
Nonstoichiometic formulation of equilibrium conditions, 45, 46–47, 48, 58, 59, 73, 192
Nonstoichiometric formulation of equilibrium problem, 76, 173, 183
and sensitivity analysis, 192–195, 196
see also Nonstoichiometric algorithm; Relatively simple system
Nonuniqueness, 60, 61
Numerical analysis, 3, 111
Numerical methods, 100–116, 119
Numerical problem, 101–103, 127
Numerical singularities, 201, 204–213
classification, 208, 209
of coefficient matrix β, 205, 206, 207, 208
multispecies phases, 211–213
obviation in BNR algorithm, 208
single-species phases, 204–211

Objective function, 111, 136
Open system, 42
Optimal feed composition, 184, 189–190, 199, 200
Optimization, 3
see also Minimization
Optimization problem, 3, 12, 47, 116
sensitivity analysis for, 173, 178–181
see also Minimization problem
Optimized stoichiometry, 119, 120, 141

SUBJECT INDEX

Organic chemistry, 6
Organic compounds, sources of data, 70, 71, 72
Output, *see* Computer output

Parameter changes:
  effects on equilibrium composition, 190–198, 200
  effects on mole fraction at equilibrium, 200
  effects for simple system, 184–190
  overall effects for simple system, 173, 174–178
  qualitative effects of, 173, 179, 181–183
  quantitative effects of, 173, 180–181
Parameter(s), 3, 4–5
  auxiliary, 112, 113
  set of, 4, 173, 228
  see also Step-size parameter
Parameter-variation method, 102, 111–113, 166
Partial molar quantities, 44
  enthalpy, 43, 56
  entropy, 186
  heat capacity, 227
  volume, 43, 56, 158
Partial pressure, definition, 51
Path, 1, 41
Phase:
  incipient formation, 59, 60
  test for presence, 57–60, 211–213
Phase designation, 17, 63, 137
Phase diagrams, 8, 73
Phase equilibrium, 56
  bibliographies, 72
  as chemical equilibrium, 2, 154
  formula matrix in, 152
  free-energy data for, 152
  liquid-liquid, 156
  solid-vapor, 50
  specified by stoichiometric matrix, 218
  vapor-liquid, 36, 50, 152, 156, 160, 171–172, 230
Plug flow, 15
Potential functions, 40–41
Practical considerations, 201
Precision of computed composition, 196
Pressure:
  critical, 161
  dependence of free-energy quantities on, 72, 156
  derivative for chemical potential, 43, 49
  derivative for Gibbs function, 43
  effect on equilibrium, 177, 186–187, 191–192, 194–195

effect on equilibrium constant, 63, 177
effect on total moles, 187, 199
as parameter, 4–5, 173, 184
reduced, 161
reference, 49
sensitivity coefficient, 198, 227
standard state, 49
unit of, 49
Program listings, *see* Computer program listings
Programs, *see* Computer programs
Projection matrix, 107, 120, 121
  as stoichiometric matrix, 122
Projection method, 102, 107–108
  as stoichiometric technique, 122
  variable-metric, 136
Prompting statement, 96
Protonic charge, in designation of chemical system, 17

Quadratic form, 60, 61, 74
Quadratic function, 104
Quadratic programming, 136
Quasi-Newton method, 167, 169

R, 18–22
  definition, 30
  as maximum number of linearly independent stoichiometric equations, 25
  as number of degrees of freedom, 30
  as number of noncomponents, 23
  relation to F_s, 30
  relation to rank (**N**), 27, 29, 215, 218
RAND algorithm, 119, 122, 123–127, 133, 136, 137
  and Brinkley algorithm, 130, 132, 133, 135–136
  composition variable other than mole fraction, 147, 151
  computer program, *see* BNR algorithm
  extension to multiphase system, 126, 149
  flow chart, 128
  as method of solving nonlinear equations, 125, 127
  as minimization method, 125–126, 127
  modifications of, 125–126
  and NASA algorithm, 135–136
  for nonideal system, 127, 168
  number of equations, 126, 127
  number of variables, 127
Rank of matrix, 3, 16
  determining, 23, 25
  matrix **A**, 19, 25
  matrix **N**, 27, 29, 30, 33

Raoult convention, 52, 65, 72, 153
  and activity, 54
  and activity coefficient, 55
  and equilibrium constant, 64
  and excess function, 57
  and standard state, 158
Raoult's law, 52
Reaction equilibrium, 2, 154
Reaction isotherm, 63, 65–66
Reaction mechanism, 22, 28
Redlich-Kister equation, 156
Redlich-Kwong equation, 150, 160, 169–170, 232
Regular solution, 157, 171–172
Relatively simple system, 75–99, 100, 141, 213
  algorithms for, 75, 81–83, 88–92
  characteristics, 76
  choice of formulation, 76
  definition, 75
  nonstoichiometric formulation, 84–96
  stoichiometric formulation, 77–84
Restricted equilibrium problems, 218–219
Restrictions, see Compositional restrictions
Reversible change, 42
Rockets, 2, 7, 11, 12, 113, 150

Second law of thermodynamics, 3, 40, 41
Second-order method, 102, 104, 107, 109, 119, 120, 141, 167–169
Second-variation method, 104, 105, 107, 111, 125
Sensitivity analysis, 173, 179, 183–184, 198, 228
Sensitivity coefficient, 191, 193, 194, 197, 228
  in calculating effect of parameter changes, 194
  first-order, 196
  pressure, 198, 227
  second-order, 195–196, 200
  temperature, 198, 227
Sensitivity matrix, 173, 180, 181, 197, 228
  first-order, 180, 183
  literature review, 183–184
  second-order, 183
  for (T,P) problem, 183–196
Simple system ($R = 1$), 19, 138, 173, 183, 214–215
  effect of pressure and inerts on total moles, 199
  effects of parameter changes on extent of reaction, 184–190
  overall effects of parameter changes, 174–178

Single-phase system, 42, 44, 60, 75, 120
Single-species phase, 12, 126, 140, 144
  and nonnegativity constraint, 58
  and numerical singularities, 204–211
  test for presence, 58
Singularities, see Numerical singularities
Solid, 50, 54, 63, 64, 154
Solid-gas systems, 183
Solubility parameter, 158, 171–172
Solute, 52
Solvent, 52, 53
Special topics, 201
Species:
  list, 17
  number of, 11, 15, 89
  reacting, distinct from inert, 89
  see also Charged species; Chemical species; Component species; Elemental species; Inert species; Ionic species; Minor species; Noncomponent species
Species present in small amounts, see Minor species
Spontaneous process, 41
Standard deviation, 196, 197
Standard form:
  of compositional restrictions, 27–36
  of equilibrium problem, 218, 219
  of stoichiometry, 33
Standard problem, 4, 226
Standard state, 51, 73, 177
  fugacity, 158
  Henry convention, 52, 54
  infinitely dilute solution, 152
  Raoult convention, 52, 54
State function, 40, 42
Statistical mechanics, 11, 67, 154, 155, 160, 162, 222
Statistical thermodynamics, 154, 222
Statistical weight, 204, 217
Steady state, 2
Step-size parameter, 82, 91, 103, 122, 144
  choosing, 104, 111
  computation of, 113–115
  in initial estimate, 203
Stoichiometric algorithm, 12, 108, 118, 119, 139–145, 167
  BASIC, 150
  comparison with nonstoichiometric algorithm, 140
  convergence criteria, 116
  equilibrium-constant, 214
  first-order, 119, 120
  flow charts, 83, 145
  initial estimate, 201–203
  number of variables, 140

SUBJECT INDEX 363

relatively simple system, 81–83, 84
second-order, 119, 120, 167, 168, 169
and singularities, 208
see also VCS algorithm
Stoichiometric coefficient, 20, 130, 131, 140, 142
convention regarding sign, 21–22
Stoichiometric-coefficient algorithm, 14, 23
see also Stoichiometry, algorithm
Stoichiometric degrees of freedom, 18, 29–31
Stoichiometric elimination, method of, 108–109, 110
Stoichiometric equation(s), 11, 12, 45
number of, 25, 27, 215
representation, 77
single, 76, 77–78, 144, 288
system involving two, 78–81
and VCS algorithm, 143
see also Chemical equations
Stoichiometric formulation of equilibrium conditions, 45–46, 48, 58, 62–63, 73, 184, 190
Stoichiometric formulation of equilibrium problem, 76, 140, 173, 183
and sensitivity analysis, 184–192, 199
see also Relatively simple system; Stoichiometric algorithm
Stoichiometric matrix, 28, 30, 33, 35, 39
canonical form, 24, 25, 58, 142, 202
complete, 22, 24, 25, 34, 35, 38, 39, 69, 73
modified complete, 31, 32, 33
nonuniqueness of, 35, 142
optimized choice of, 142–143
projection matrix as, 122
rank, 27, 29, 30, 33
reduction to standard form, 28–29
situations, incorrect rank of, 27
see also **N** matrix
Stoichiometric methods, 12, 140
Stoichiometric procedure, 23–26
Stoichiometric restrictions, *see* Compositional restrictions
Stoichiometric vector, 19, 21, 22, 122, 142
Stoichiometry, 14, 30
algorithm, 214, 216
see also Stoichiometric-coefficient algorithm
optimized, 119, 120, 141
standard form, 33
see also Chemical stoichiometry
Substances, common, sources of data, 70, 71
see also Chemical substance
Sufficient condition, 18, 44, 60, 101, 165

System:
acetaldehyde, 198–199
aqueous aluminum sulfate, 53–54
aqueous cadmum ion, 73–74
aqueous chloride ion, 72
aqueous chlorine, 72, 147–148, 151, 172
aqueous nitrogen, 74
aqueous phosphoric acid, 37, 213–214, 216–217, 229
aqueous phosphoric acid + $CaHPO_4$, 166–167, 172
aqueous sulfuric acid, 74
benzene, 73
benzene-toluene-xylene, 37, 230
blast-furnace problem, 37, 208, 210, 211
chlorine dioxide, 73
crude-styrene-unit product, 152, 171–172
deuterized ammonia, 226
deuteromethanes, 37, 225, 226, 231
hydrogen-carbon oxygen, 11
ligand-metal complexing (biological application), 234
metal-liquid, complexing involving stability constants, 233–234
one-element, 84–88
solubility calculations, 151–152, 172
sulfur dioxide, 73
see also Chemical processes; Chemical reactions

Taylor series, 82, 104, 121
Tektronix 4052 computer, 96
Temperature:
critical, 161
dependence of free-energy quantities on, 72, 155–156
derivative for chemical potential, 43
derivative for equilibrium constant, 63
integrated form, 174
derivative for Gibbs Function, 43
derivative for standard emf, 176
derivative for standard enthalpy of reaction, 174
effect on equilibrium, 174–176, 185–186, 191, 194, 222, 225–226
effect on standard free energy of reaction, 174
as parameter, 4–5, 173, 184
reduced, 161
sensitivity coefficient, 198, 227
Terminology, 16–18, 19, 22, 23
Test for presence of phase, 57–60, 211–213
Thermodynamic constraints, 4, 6, 41, 226, 228
other than (T, P), 74, 133, 201, 226–228

Thermodynamic data files, 169
Thermodynamic derivatives, calculation of, 173, 198
Thermodynamic isotope effect, 223
  see also Isotope effect
Third law of thermodynamics, 11, 67
Transformation, 20, 108, 120

Unconstrained minimization methods, 45, 104–105, 108, 110, 139
Uniqueness of solution, 60–61, 88, 117, 191, 207, 221
Unsteady state, 1
User's guide, 4
  BNR computer program, 137, 292–294
  HP-41C computer programs, 84, 203, 238–239, 251–255, 268–269
  VCS computer program, 146, 324–325

van Laar equation, 56, 156, 157
van't Hoff equation, 63, 174, 175, 176, 225
Vapor pressure, 50
Variable-metric method, 167
Variables:
  independent, 40, 42, 45, 74
  mole numbers as, 127, 136
  number of, 118–119, 127, 140
  thermodynamic, 40
Variance, 196, 197, 217
VCS algorithm, 119, 141–145, 195, 208, 211
  computational advantage, 144–145
  computer program, 27, 145
  flow chart, 145
  FORTRAN computer program listing, 296–323
  illustrative example, 146–147
  initial estimate for, 201, 202
  input data, 137, 146, 148, 210
  as method for solving nonlinear equations, 144
  as minimization algorithm, 144
  and minor species, 217–218
  modification for composition variable other than mole fraction, 147–148
  and multiphase problem, 144
  no modification when $C \ne M$, 213
  for nonideal system, 165
  output for, 146, 148, 210
  user's guide, 146, 324–325
Vector(s):
  column, 16, 101
  composition, 29
  gradient, 104, 109
  Lagrange multiplier, 47, 179
  linear dependence and independence of, 3
  -matrix form, manipulation, 3, 14, 15, 107
  parameter, 179, 185
  solution, 179
  species-abundance, 16, 17
  species-abundance-change, 18
  transpose of, 16
  see also Element-abundance vector; Formula vector; Stoichiometric vector
Virial coefficients, 159–160
Virial equation, 159–160
Volume:
  change, standard, 63
  critical, 162
  excess, 56, 57
  fraction, 157–158, 172
  molar, 49, 56, 57, 158
  as parameter, 5, 74
  partial molar, 43, 56, 57, 158

Weierstrass theorem, 60
Wilson equation, 56, 157, 171
Wohl equation, 156–157
Work, 41, 64